SLOW VIRUS DISEASES OF ANIMALS AND MAN

NORTH-HOLLAND RESEARCH MONOGRAPHS

FRONTIERS OF BIOLOGY

VOLUME 44

Under the General Editorship of

A. NEUBERGER

London

and

E. L. TATUM

New York

NORTH-HOLLAND PUBLISHING COMPANY
AMSTERDAM · OXFORD

SLOW VIRUS DISEASES
OF ANIMALS AND MAN

Edited by

R. H. KIMBERLIN

A.R.C. Institute for Research on Animal Diseases, Compton, Newbury, Berkshire, U.K.

1976

NORTH-HOLLAND PUBLISHING COMPANY, AMSTERDAM · OXFORD

AMERICAN ELSEVIER PUBLISHING COMPANY, INC. – NEW YORK

© *North-Holland Publishing Company* – 1976

North-Holland ISBN for this series: 0 7204 7100 1
North-Holland ISBN for this volume: 0 7204 0418 5
American Elsevier ISBN: 0444 11066 6

PUBLISHERS:
NORTH-HOLLAND PUBLISHING COMPANY – AMSTERDAM
NORTH-HOLLAND PUBLISHING COMPANY LTD. – OXFORD

SOLE DISTRIBUTORS FOR THE U.S.A. AND CANADA:
AMERICAN ELSEVIER PUBLISHING COMPANY, INC.
52 VANDERBILT AVENUE, NEW YORK, N.Y. 10017

PRINTED IN THE NETHERLANDS

General preface

The aim of the publication of this series of monographs, known under the collective title of '*Frontiers of Biology*', is to present coherent and up-to-date views of the fundamental concepts which dominate modern biology.

Biology in its widest sense has made very great advances during the past decade, and the rate of progress has been steadily accelerating. Undoubtedly important factors in this acceleration have been the effective use by biologists of new techniques, including electron microscopy, isotopic labels, and a great variety of physical and chemical techniques, especially those with varying degrees of automation. In addition, scientists with partly physical or chemical backgrounds have become interested in the great variety of problems presented by living organisms. Most significant, however, increasing interest in and understanding of the biology of the cell, especially in regard to the molecular events involved in genetic phenomena and in metabolism and its control, have led to the recognition of patterns common to all forms of life from bacteria to man. These factors and unifying concepts have led to a situation in which the sharp boundaries between the various classical biological disciplines are rapidly disappearing.

Thus, while scientists are becoming increasingly specialized in their techniques, to an increasing extent they need an intellectual and conceptual approach on a wide and non-specialized basis. It is with these considerations and needs in mind that this series of monographs, '*Frontiers of Biology*' has been conceived.

The advances in various areas of biology, including microbiology, biochemistry, genetics, cytology, and cell structure and function in general will be presented by authors who have themselves contributed significantly to these developments. They will have, in this series, the opportunity of bringing together, from diverse sources, theories and experimental data, and of integrating these into a more general conceptual framework. It is unavoidable, and probably even desirable, that the special bias of the individual authors will become evident in their contributions. Scope will also be given for presentation of new and challenging ideas and hypotheses for which complete evidence is at present lacking. However, the main emphasis will be on fairly complete and objective presentation of the more important and more rapidly advancing aspects of biology. The level will be advanced, directed primarily to the needs of the graduate student and research worker.

Most monographs in this series will be in the range of 200–300 pages, but on occasion a collective work of major importance may be included somewhat exceeding this figure. The intent of the publishers is to bring out these books promptly and in fairly quick succession.

It is on the basis of all these various considerations that we welcome the opportunity of supporting the publication of the series '*Frontiers of Biology*' by North-Holland Publishing Company.

E. L. TATUM
A. NEUBERGER, *Editors*

List of contributors

Cho, H. J. – Animal Pathology Division, Animal Diseases Research Institute (Western), P.O. Box 640, Lethbridge, Alberta T1J 3Z4, Canada

Crawford, T. B. – Pioneering Research Laboratory, Agricultural Research Service, U.S. Department of Agriculture and Institute of Comparative Medicine, Washington State University, Pullman, Washington 99163, U.S.A.

Dickinson, A. G. – A.R.C. Animal Breeding Research Organisation, Edinburgh, and Moredun Research Institute, Edinburgh, U.K.

Fraser, H. – A.R.C. Animal Breeding Research Organisation, Edinburgh, and Moredun Research Institute, Edinburgh, U.K.

Georgsson, G. – Institute for Experimental Pathology, University of Iceland, Keldur, Reykjavík, Iceland

Gorham, J. R. – Pioneering Research Laboratory, Agricultural Research Service, U.S. Department of Agriculture and Institute of Comparative Medicine, Washington State University, Pullman, Washington 99163, U.S.A.

Harter, D. H. – Department of Neurology, The Medical School, Northwestern University, Chicago, Illinois 60611, U.S.A.

Henson, J. B. – International Laboratory for Research on Animal Diseases, P.O. Box 47543, Nairobi, Kenya

Hunter, G. D. – A.R.C. Institute for Research on Animal Diseases, Compton, Newbury, Berkshire, U.K.

Kimberlin, R. H. – A.R.C. Institute for Research on Animal Diseases, Compton, Newbury, Berkshire, U.K.

Marsh, R. F. – Department of Veterinary Science, University of Wisconsin, Madison, Wisconsin 53706, U.S.A.

McGuire, T. C. – Pioneering Research Laboratory, Agricultural Research Service, U.S. Department of Agriculture and Institute of Comparative Medicine, Washington State University, Pullman, Washington 99163, U.S.A.

Millson, G. C. – A.R.C. Institute for Research on Animal Diseases, Compton, Newbury, Berkshire, U.K.

Nathanson, N. – Department of Epidemiology, School of Hygiene and Public Health, Johns Hopkins University, Baltimore, Maryland, U.S.A.

Outram, G. W. – A.R.C. Animal Breeding Research Organisation, Edinburgh, and Moredun Research Institute, Edinburgh, U.K.

Padgett, G. A. – Pioneering Research Laboratory, Agricultural Research Service, U.S. Department of Agriculture and Institute of Comparative Medicine, Washington State University, Pullman, Washington 99163, U.S.A.

Pálsson, P. A. – Institute for Experimental Pathology, University of Iceland, Keldur, Reykjavík, Iceland.

Panitch, H. – Department of Epidemiology, School of Hygiene and Public Health, Johns Hopkins University, Baltimore, Maryland, U.S.A.

Pétursson, G. – Institute for Experimental Pathology, University of Iceland, Keldur, Reykjavík, Iceland.

Thormar, H. – New York State Institute for Basic Research in Mental Retardation, 1050 Forest Hill Road, Staten Island, New York 10314, U.S.A.

Contents

Part I   Slow virus diseases

Part II   Visna-maedi

The disease

The virus

Virus–host interactions

Part III   Aleutian disease of mink

The disease

The virus

Virus–host interactions

Part IV   Scrapie

The disease

The virus

Relationship of scrapie with other diseases

Part V   Conclusions

Part I

Slow virus diseases

Dr. Bjorn Sigurdsson

CHAPTER 1

General introduction to some slow virus diseases

Richard H. KIMBERLIN

1.1 Introduction

A convenient way of introducing the subject of 'slow virus diseases' would be to give a concise and unambiguous definition. Unfortunately no such definition exists and any attempt to formulate one in precise terms would be futile because this is not an area with clearly discernable boundaries. Consequently there are many different opinions of how the term should be used. This difference is reflected in the very wide variety of conditions which have been classified as slow virus diseases in three recent reviews of the subject (Fuccillo et al., 1974; Hotchin, 1974; Zeman and Lennette, 1974). The situation is made worse by the phrase itself which can be interpreted literally to mean a *virus disease* which is *slow* to develop, but clearly any definition in these terms is meaningless. However the basic concept behind the term slow virus disease was more precise than this: the difficulty is, it is not precise enough to enable the term to be used unambiguously.

The purpose of this chapter is to briefly examine the concept of slow virus disease in its historical context since this is the only satisfactory way to approach the subject. No attempt will be made to give an exhaustive account or to describe all possible examples of slow virus disease. Instead, attention will be focussed on a few general points and on one particular interpretation of the concept which provides the reasons for selecting the small number of diseases to be described in great detail in the subsequent chapters.

The word 'virus' is used in this discussion in its original sense of a filterable, transmissible, replicating agent so as to include scrapie and the scrapie-like

Slow virus diseases of animals and man, edited by R. H. Kimberlin
© *North-Holland Publishing Company 1976*

agents, which are undoubtedly virus-like, but whose chemical nature is still undefined. The term 'disease' denotes a clinically recognisable syndrome which can be described in terms of clinical signs and histopathological lesions.

1.2 Acute and chronic viral disease

Historically, the origin of slow virus diseases is set against the background of 'acute disease' and 'chronic disease' and it is appropriate to consider briefly the meaning of these terms as applied to viruses.

The course of an acute virus infection usually runs as follows. Once the virus has entered a susceptible host there follows an almost immediate and rapid phase of virus multiplication. At some stage during this phase the host defence mechanisms, in the form of the immune system and perhaps also the production of interferon, are stimulated. If the initial infection is severe or if the host defences are at a low level, the virus is likely to produce a clearly recognisable disease before the host either dies or finally clears the infection. Alternatively, the host defences may clear the infection before any clinical disease develops. The essential features of this type of virus–host interaction are (1) the relatively predictable outcome of the infection which either leads to death of the host or to the elimination of infection with a much enhanced resistance to a subsequent infection by the same type of virus, and (2) the short time scale of these events which may sometimes occupy only a few days.

In marked contrast, a chronic infection produces a far less rapid development of disease and the outcome of the infection is often indecisive. After a long and unpredictable incubation period an acute clinical episode may develop, sometimes followed by complete recovery; however, the virus may not always be cleared from the host even though the host is producing virus-specific antibodies. The virus may, therefore, persist and produce recurrent disease. Alternatively, a rather ill-defined and often protracted period of illness can result or there may even be no obvious disease at all, despite the continued presence of virus. This situation indicates a greater complexity in virus–host interactions than that found in acute diseases and the time scale often extends over a period of months or years.

It is emphasised that in applying the terms 'acute' and 'chronic' to disease (terms which simply mean 'coming rapidly to a crisis' and 'lingering') one is not trying to describe totally distinct biological phenomena but rather to indicate different ends of a broad spectrum of viral (and other microbial) diseases. It should also be made clear that acute and chronic viral diseases are not necessarily caused by different viruses. On the contrary some viruses which usually cause an acute disease may persist in the host almost indefinitely and may later be the cause of a totally different clinical disease; for example it is now believed that a childhood infection with measles may appear years later as subacute sclerosing panencephalitis (SSPE).

This is an important consideration because it stresses that 'slowness' and 'fastness' are not intrinsic attributes of the virus alone nor indeed of the host. It is the precise way in which virus and host interact that determines how rapidly disease develops. It is for this reason that the term 'slow virus' may be inappropriate since it has quite the wrong connotations. Similarly in using the term 'slow virus diseases' it is not the viruses which are necessarily slow but the development of the diseases they cause.

1.3 Slow infections – Sigurdsson's concept

The origin of this concept dates back to the mid-1930s when four chronic diseases appeared in Icelandic sheep, namely pulmonary adenomatosis, maedi (progressive pneumonia), visna and paratuberculosis (Johne's disease). These diseases occurred in sheep which were kept in close contact with some Karakul sheep that had been imported two years earlier from an experimental farm in Halle, Germany (2.1). By the early 1940s these diseases reached epidemic proportions in Iceland and they were only brought under control by applying an extensive slaughter policy in the affected areas.

Dr. Bjorn Sigurdsson and his colleagues in Iceland made a detailed study of these diseases and from their observations he evolved the concept of slow diseases which could be distinguished from chronic diseases. His studies of visna and maedi in particular revealed one major distinguishing feature of slow diseases which was this; once clinical signs of disease have appeared the disease then follows a regular progressive course which always ends in serious illness and usually death. This situation is in marked contrast to the unpredictable progress of chronic diseases and Sigurdsson himself stressed that the slow diseases 'follow a (clinical) course which is just as regular as the course of the acute infections only the time factor is different'. However, he went on to point out that he did not regard slow diseases 'simply as slow-motion pictures of the chain of events occurring in the acute infections' and he considered it likely 'that at least some phases of these slow infections differ essentially from the corresponding phases in the acute diseases'.

Sigurdsson's concept was introduced in the last of a series of 3 lectures delivered in London in March 1954. Towards the end of the lecture he suggested the following criteria for slow diseases (Sigurdsson, 1954).

(1) A very long initial period of latency lasting from several months to several years.

(2) A rather regular protracted course after clinical signs have appeared usually ending in serious disease or death.

(3) Limitation of the infection to a single host species and anatomical lesions in only a single organ or tissue system.

Twenty-two years later these criteria are still widely quoted in this, the original, form. However, today two qualifications need to be made. First, the word 'latency' has acquired a rather different meaning and in the context of the first criterion should refer to the disease, not the virus. Secondly, Sigurdsson himself anticipated that the last criterion may have to be modified as knowledge increased. Certainly in terms of experimentally induced disease it is no longer true that infection is limited to a single host species and diseases such as visna–maedi and scrapie have been found to occur naturally in goats as well as sheep (see 2.5 and 10.4.2). Today one can also make the case that the close similarity between members of the so called 'subacute spongiform encephalopathies' justifies the suggestion that, for example, Creutzfeldt–Jakob disease and transmissible mink encephalopathy may represent scrapie in man and mink, respectively (see 15.5).

1.4 Slow virus diseases today

When Sigurdsson formulated his criteria he was concerned with slow infections that produce overt clinical disease and he made no particular distinction between viral, bacterial or any other kind of microbial infection. The present discussion is concerned with viral (or at least ultrafilterable) agents and in 1954 Sigurdsson suggested the following as examples of slow virus disease:

Rida (Icelandic form of scrapie) in sheep
Visna in sheep
Maedi (progressive pneumonia) in sheep
Infectious adenomatosis (jaagsiekte) in sheep
'Bittner' mammary carcinoma in mice
'Gross' leukemia in mice

In addition he pointed out that certain traits of the Rous sarcoma, Lucké's renal carcinoma of the Leopard frog and Shope's rabbit papilloma were reminiscent of the slow virus diseases and should perhaps be included.

It is interesting that a large proportion of these diseases are caused by oncogenic viruses and are rarely listed as slow virus diseases today. A modern list of naturally occurring slow virus diseases would vary considerably with different authors but the following diseases are commonly included:

Scrapie in sheep

Visna in sheep

Maedi in sheep

Kuru in man

Creutzfeldt–Jakob disease in man

Transmissible mink encephalopathy (TME)

Aleutian disease of mink (AD)

Progressive multifocal leucoencephalopathy (PML) in man

Subacute sclerosing panencephalitis in man

Multiple sclerosis in man

Taking the two lists together we have an extremely heterogenous collection of conditions with little in common as a group except that all more or less fit Sigurdsson's criteria for slow virus diseases. Out of this apparent diversity a number of important points can be made.

Firstly, comparing the two lists one can see that five human diseases appear in the second whereas none appeared in the first. Herein lies the historical significance of Sigurdsson's concept. By drawing attention to a group of chronic animal diseases in which incubation periods could be several years, Sigurdsson emphasised the possibility that many chronic progressive diseases in humans may also have a viral aetiology and the search has been going on ever since. The most dubious member in the second list is multiple sclerosis (MS) because it has not been shown to be caused by a virus (even in terms of the broad definition given earlier) and the disease is not usually progressive; the common pattern is one of repeated relapses and remissions with an overall deterioration in condition that may only be apparent after many years. But the fact that MS is sometimes included among the slow virus diseases illustrates the point made above.

Secondly, with such a heterogenous group of diseases, (caused by a wide range of viruses and involving a variety of pathogenic mechanisms) it is obviously not possible to use Sigurdsson's criteria to define slow virus diseases as a group quite distinct from the general area of chronic diseases. Present day virology makes much use of terms like persistent infections, latent infections, slow infections but these terms are not necessarily mutually exclusive. Slow infections are persistent in the sense that recoverable virus is present in the host for most of the time between infection and disease, and latent in the sense that there are no overt signs of disease during the incubation period. Therefore, the term 'slow virus diseases' is of very limited value if used in a taxonomic sense; it is better employed to describe one part of the chronic disease spectrum.

Finally, the difference between the two lists illustrates that the use of the term 'slow virus disease' has undergone some change. For example, the enormous increase in our knowledge of tumour viruses and their oncogenic effects has led to a separate grouping of the neoplastic diseases since they have much in common. In particular, the ability of many of the tumour viruses to integrate their genomes directly, or through DNA transcripts, into the host genome has meant that some members of this group are frequently classified as latent infections, which in many respects can be considered apart from slow infections (see below). However, it would be unwise to emphasise such differences unduly since it is known that although visna is not a neoplastic disease it is, nevertheless, caused by an RNA virus which may have oncogenic potential (see 5.3.2).

1.5 Additional criteria for slow virus diseases?

It has already been suggested that the term 'slow virus disease' is of limited use in a taxonomic sense because it cannot be used to classify any one group of viral agents or any one type of disease process. There is, however, one important feature of slow virus diseases which was discussed by Sigurdsson in his original paper but which has been rather overlooked.

In the second of his criteria, Sigurdsson emphasised the regular course of the disease after clinical signs first appeared. However, he also noted that with visna and maedi the 'incubation period, although extremely long, apparently does not vary within very wide limits'. It is quite clear from the context of this remark that Sigurdsson was thinking of an overall regularity of slow virus diseases from infection to death, not just a predictable clinical course.

In general the time relationships between infection and the development of disease are difficult to follow in natural slow virus diseases. Many factors probably influence the incubation period such as the strain and dose of virus, the age and genetic constitution of the host. In addition, one often has no knowledge of the incidence of infection compared to the incidence of disease and in many cases the disease may occur but rarely in any one host population. Only in an epidemic situation does one have a good opportunity to study the incubation period of a natural slow virus disease.

However, experimental transmission of the disease to the same or a different host species overcomes many of these problems particularly in those cases where the relevant genetic variation of the host can be controlled by selection or inbreeding. Under these conditions the degree of uniformity of incubation period can be studied and in some cases this is very striking despite the long incubation periods involved. The idea of overall predictability in the course of slow virus diseases is very important because the corollary is that the aetiology and pathogenesis is relatively simple. In other words, slow virus disease in a genetically susceptible host is caused by the introduction of a virus and the host is limited in its ability to do much about it. Hence there is a clear difference between this kind of simple situation and the more complex one in which a persistent or latent infection may cause a late onset disease only after some other triggering event has taken place; a secondary infection or an altered physiological state in the host.

This extension of Sigurdsson's original concept emphasises the great importance of animal hosts in the study of slow virus diseases since it is only in animals that the aetiology of these diseases can be completely investigated. For example, there is evidence that PML is caused by the presence of a papovavirus in adult brain but the disease is usually found in persons who have some other chronic condition (e.g., Hodgkin's disease or chronic lymphocytic leukemia) that impairs immune responses. This suggests a more complex aetiology involving perhaps a latent infection with the papovavirus which is activated during times of immune suppression. The point is an animal model of PML is not yet available to explore the aetiology in more detail and the inoculation of the virus into hamsters produces tumours, not PML.

The precise aetiology of SSPE is also in doubt. Again there is good evidence that the disease is caused by a virus, in this case measles virus, but it is not known if SSPE is simply due to a rare type of measles virus, or to an unusual host response or whether a second virus is involved. The lack of a suitable experimental model hampers the study of some aspects of this disease.

To investigate a complex aetiology of this type it is necessary to control any relevant genetic variation in the host. For example, twenty years ago, scrapie in sheep appeared to be an unpredictable disease since the inoculation of scrapie-infected tissue produced disease in only some sheep and in those that did succumb, the incubation period was very variable (see 10.4.1). The significant step was transmitting the disease to mice, which were much more uniform in their response to infection, particularly inbred strains. With the discovery of the gene *sinc* in mice it was realised that a different genotype within a host species could have a considerable effect on incubation period (see 14.5). Lines of sheep have been selectively bred to control these pre-determined (genetic) host factors with the result that a far greater uniformity of response to infection can be obtained in sheep than was previously apparent (see 10.7). Hence we can justifiably discuss scrapie (at least in its experimental form) as a slow disease of simple aetiology in the sense that the presence of the virus (or agent) in a susceptible host is basically all that is required.

The same 'simple aetiology' has been demonstrated in Aleutian disease of mink, at least in mink of the Aleutian genotype (*aa*) (see 7.3.1). Ex-perimental visna and maedi appear to be less predictable diseases than either Aleutian disease of mink or scrapie. However it should be noted that no attempt has been made to produce a line of sheep, uniformly susceptible to these diseases and no other experimental animal host is yet available to examine this point further.

1.6 Scope of the book

To summarise briefly, the phrase 'slow virus diseases' is a useful descriptive term applicable to a number of chronic diseases which are characterised by a long incubation period and a predictable clinical course. It has been suggested that the concept of predictability can be extended (in some cases at least) to include the entire period from infection to death. Moreover it is clear that in these cases, the virus is the major aetological factor (given that the host is susceptible) and that the slowness of the disease must be considered in terms of the precise interaction between virus and host with secondary factors playing a minor role. Slow virus diseases have little in common as a group in terms of either the viruses or the diseases they cause. However, they constitute an important group of diseases, not least because their inherent predictability makes them relatively accessible to investigations of the variety of pathogenic mechanisms which underly the slowness of disease development.

The importance of animal models of slow virus diseases has already been emphasised and is further stressed in the choice of diseases discussed in this book, namely, visna–maedi in sheep, Aleutian disease of mink and scrapie in sheep. Not only are these diseases transmissible to their natural or to experimental hosts but they are natural diseases of animals. Consequently, much is known of their epidemiology in addition to the nature of the causal viruses and the pathogenesis of disease. They are in fact the best understood of the slow virus diseases from this general stand point.

1.6.1 Visna–maedi

Visna and maedi are two distinct clinical syndromes involving the brain and lung, respectively, but they are discussed together because they are caused by what appears to be the same virus. The majority of studies have been concentrated on visna since it is a demyelinating disease with some implications in human demyelinating conditions such as multiple sclerosis. Visna is readily transmitted to sheep but for many years only Icelandic sheep were believed to be susceptible to experimental infection and it is very recently that random-bred American lambs have been used to study the disease. Because of the problems of working with sheep and the lack of other susceptible host species a particular emphasis in visna studies has been on the virus, which multiplies readily in cell cultures in which it can easily be assayed. It is of great interest that visna virus has a structure similar to oncogenic RNA viruses and, moreover, exhibits some oncogenic properties in vitro, and yet there is no hint that visna and maedi are neoplastic diseases.

1.6.2 Aleutian disease of mink

Aleutian disease appeared quite dramatically in the 1940s in blue mink which were originally derived from Aleutian mink. Although it is now known that most types of mink (if not all) are susceptible to the experimental disease, mink homozygous for the Aleutian gene are highly susceptible and give the shortest incubation periods. Therefore in this case the disease appeared in a host that was particularly susceptible genetically and it often occurred in epidemic proportions. Much, therefore, is known of the epidemiology of the disease. However, no laboratory animal model of the disease has been found and it was many years before in vitro assays of the virus were developed. Detailed characterisation of the virus has therefore only recently become possible with the advent of better purification techniques and methods of propagating the virus in cell culture. However, the dramatic nature of the host response to infection has facilitated detailed studies of the disease in mink. As a consequence Aleutian disease of mink has become important as a model immune-complex disease.

1.6.3 Scrapie

Scrapie has the longest history of the diseases discussed in this book and in many ways it is the most unusual. Some people might object to its inclusion in a work on slow virus diseases on the grounds that it does not appear to be caused by a conventional virus. However, despite the long held view that it is a genetic disease there can be no doubt that scrapie is caused by a transmissible, filterable, replicating agent so in these terms it is a virus. The preference of many workers (present author included) to use the term agent rather than virus is simply to avoid prejudging its molecular nature which is not known but which has been clearly shown to be somewhat beyond the established spectrum of viruses.

Apart from sheep, scrapie can be transmitted to a wide variety of host species including many laboratory rodents. The mouse has been extensively used since it is the most convenient host for bio-assay of the agent; this being the only available method. Much is now known of the genetic factors which control incubation period (in mice, and to a lesser extent in sheep) of the wide diversity of scrapie agents in nature and of the general features of pathogenesis. One remarkable feature of pathogenesis is the apparent complete lack of any specific immunological response to infection. However, quite apart from its intrinsic interest, scrapie has assumed considerable importance as the best known example of a group of diseases which have become known as the 'subacute spongiform encephalopathies'. This group includes transmissible mink encephalopathy, kuru and Creutzfeldt–Jakob disease. It is of special interest that the last two diseases occur in man.

1.7 Final comments

It has already been emphasised that the three diseases which form the main subject of this book have little in common; they are caused by quite different viruses and involve fundamentally different types of virus–host interactions. It is therefore obvious that even in the rather strict interpretation of slow virus diseases described in this chapter we are not seeking to understand just one pathogenic mechanism underlying slowness, but a diversity of slow, virus–host interactions.

The purpose of describing three such diseases in one volume is to examine some of these interactions at all levels of study from the natural history of clinical diseases to the molecular events taking place between virus and host. It is apparent that much remains to be learned of slow virus diseases but, in view of the considerable progress that has been made, it is perhaps an appropriate time to present and analyse what is already known, to outline those areas where knowledge is lacking, and to illustrate the need to understand these diseases as representatives of one part of a much wider field of chronic progressive disease which so much concerns modern medicine.

References

FUCCILLO, D. A., KURENT, J. E. AND SEVER, J. L. (1974) Annu. Rev. Microbiol., 28, 231.
HOTCHIN, J., ed. (1974) Progr. Med. Virol., 18, pp. 1–371.
SIGURDSSON, B. (1954) Br. Vet. J., 110, 341.
ZEMAN, W. AND LENNETTE, E. H., eds. (1974) Slow Virus Diseases. (Williams and Wilkins, Baltimore) pp. 1–145.

Part II

Visna-maedi

The disease

The virus

Virus–host interactions

CHAPTER 2

Maedi and visna in sheep

Páll A. PÁLSSON

2.1 Historical introduction

A chronic pneumonia of sheep formerly unknown in Iceland became prevalent on many farms in the 1940s.

This disease was referred to in Icelandic by the name 'maedi' which means dyspnoea. When first recognized in 1939 by Gíslason (1947) the disease had already spread from farm to farm in two different districts wide apart, one in the north-eastern part, the other in the south-western part of the country (see Fig. 2.1).

Maedi was without doubt introduced to Iceland following the importation of 20 apparently healthy Karakul sheep bought by the Icelandic Government from Halle, Germany in the year 1933 (Gíslason, 1947, 1966). The animals were kept in quarantine for two months, and then sent to 14 farms in various parts of the country. Apparently, at least 2 of the rams of the imported flock carried the infection and gave rise to epizootics in two widely separated districts of the country, starting from two foci where the rams were kept.

The flock of Karakul sheep in Halle from which the imported animals were purchased had apparently been self-contained for a long time, and in this flock no losses from a disease resembling maedi has ever been observed (Straub, personal communication, 1971).

By the time maedi was first recognized in Iceland the disease had already spread to many flocks. The long preclinical period of the disease and its insidious onset facilitated this unnoticed spread.

Another reason for the delayed recognition of maedi in Iceland was that one of the two rams apparently also carried the agent of another lung disease, jaagsiekte (epizootic adenomatosis), and this infection had become

Slow virus diseases of animals and man, edited by R. H. Kimberlin
© *North-Holland Publishing Company 1976*

18   *P. A. Pálsson*

PRIMARY FOCI OF MAEDI

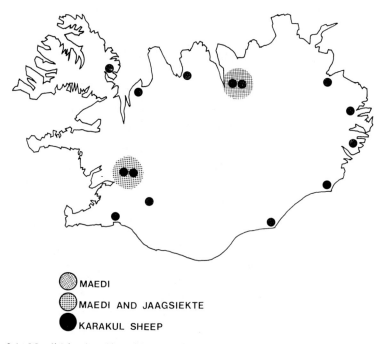

○ MAEDI

◉ MAEDI AND JAAGSIEKTE

● KARAKUL SHEEP

Fig. 2.1. Maedi (visna) and jaagsiekte were introduced into Iceland in 1933 with 20 Karakul sheep imported from Germany. They were distributed to 14 farms. Maedi spread over wide areas from two of these farms.

widespread and caused extremely heavy losses in many of the same flocks where maedi later became prevalent. Although many of these flocks had probably become infected with both diseases at the same time, jaagsiekte having a shorter incubation period and a shorter clinical course, over-shadowed maedi altogether during the first years and therefore attracted most attention. After a few years the incidence of jaagsiekte declined rapidly and maedi which develops more slowly came into prominence and caused annual losses of 15–30%. On many farms these two distinct diseases were found simultaneously in the same flock, and sometimes even in the same animal.

Parasitic pneumonia was also common in many of these flocks which further complicated diagnosis.

A similar insidious and unnoticed spread of maedi has recently been experienced in other countries (Hoff-Jørgensen, 1974a, b; Krogsrud, 1974).

Sheep farming practices in Iceland were conducive to the rapid and wide spread of maedi. During the winter months sheep flocks are housed, in spring they graze pastures close to the farms, but during the summer the sheep used to roam freely on common unfenced pastures on the hills and in the mountains. Sheep from different flocks and different districts could therefore mingle freely on the pastures and have close contact during the autumn roundups, after which they were kept by the thousands in big collecting folds for 1–3 days while being sorted out into their original flocks belonging to individual owners. This traditional method of sheep farming also created great difficulties when introducing control measures to prevent further spread of maedi within the country.

It should be emphasized that sheep farming has always been the main agricultural industry in Iceland. Therefore this epizootic soon caused great concern.

The Icelandic sheep is a primitive breed of hill sheep, originally brought to the country by the first settlers more than thousand years ago. Since then the sheep population, now approximately 800,000 winterfed sheep, has lived in almost complete isolation as importation of other breeds has been extremely rare and consisted each time of only a few animals. It is therefore probable that the sheep population in Iceland differs genetically from other breeds of sheep. When a new contagious disease was introduced into this virgin population, the heavy losses encountered were not surprising.

'Visna', which in Icelandic means wasting, is a name given to a slow, progressive viral encephalomyelitis of sheep in Iceland. This disease was first observed on several farms in the southwestern districts of Iceland in the early 1940s. Visna was only found in sheep flocks where maedi had already been causing losses for some years. Usually only a few sheep in the flock were affected but sometimes the losses from visna exceeded those of maedi in the same flock (Snorrason, personal communication). Cases of visna were always observed in the field in animals that were also found to be affected with maedi at various stages. A certain relationship between these two diseases was therefore suspected from the beginning, although it was only proved many years later in animal experiments.

It is, however, noteworthy that in the two main epizootics of maedi which occurred simultaneously in two different parts of Iceland in the years 1933–1951, maedi was only accompanied by visna on farms in the south-western part of the country; it was never observed in the north-eastern part of Iceland, where maedi was also prevalent for many years. Since then, visna has never been observed in the restricted epizootics of maedi, that occurred in the years 1954–1965 in the western part of Iceland.

Later studies and experience have shown that maedi and visna are probably two clinical and pathological entities caused by the same viral infection (Pálsson, 1972).

2.2 Clinical features

2.2.1 Maedi

Clinical signs of maedi appear only in adult sheep, usually more than 3–4 years old. The first sign is a slowly advancing listlessness and loss of condition. These signs are usually observed in pregnant or nursing ewes, and often become apparent when the sheep have been exposed to stress such as inclement weather or excessive physical strain. An early sign is that the sheep become dyspnoeic; after exertion, the respiration becomes very rapid and shallow. The respiratory rate is sometimes excessively high, 80–120/min.

As the disease progresses the respiration even at rest becomes gradually more and more difficult or laboured; the nostrils are dilated and flank breathing or pumping aided by the accessory respiratory muscles, accompanied by characteristic rhythmic jerks of the head, are commonly observed. Mouth breathing has occasionally been described and in some cases dry coughing is observed.

In contrast to jaagsiekte (epizootic adenomatosis), maedi never produces any appreciable amount of fluid in the lung, which in jaagsiekte results in nasal discharge and frequent spasmodic coughing. In uncomplicated cases of maedi the body temperature and the pulse rate remain within the normal range.

Ewes affected with maedi often give birth to small and weak lambs, and their milk production is apparently decreased. Abortion is frequently observed in advanced cases.

As the disease advances and the sheep lose condition, the haemoglobin content of the blood decreases and in later stages values as low as 7–8 gm/100 ml blood are found, compared with 12–14 gm/100 ml blood for normal sheep of the same age. The anaemia is of the hypochromic type. The number of white cells in the blood of sheep affected with maedi varies but often prolonged leucocytosis of a lymphocytic type is observed. However, when the disease advances the lymphocyte count is often only slightly above normal figures.

Chaugham and Singh (1970) reported a significant increase in erythrocyte sedimentation rate in sheep affected with maedi. They also found that sera from affected sheep had a shift to the right in the Wiltman reaction. Both these observations were thought to be of value in the diagnosis of maedi.

Non-protein nitrogen of blood plasma is normal. Total protein in plasma is decreased to about 6 gm protein/100 ml of plasma against about 6.5 gm/100 ml in normal Icelandic sheep. It has been reported that the albumin fraction is low but that the globulin content is normal or somewhat increased (Sigurdsson et al., 1952).

The low values for total protein are probably an indication of the general wasting of the organism encountered in advanced cases of maedi. If affected sheep are well fed and attended to, the progress of the disease can in some cases be retarded for a while, after which the disease progresses with loss of condition and increased respiratory distress to a fatal conclusion.

Under natural conditions sheep affected with maedi very often succumb to a terminal acute bacterial pneumonia.

The general practice among the farmers was to destroy all cases encountered in the autumn and early winter. In order to reveal the disease in its early stages in the autumn they exposed the flock to physical strain in various ways. Those animals that then showed laboured respiration for an abnormal length of time were disposed of. Cases that occurred in the flock after March were left alive in the hope that some would be able to rear their lambs next spring.

It is doubtful whether recoveries ever occur after the disease has become clinically evident, and there is no experimental evidence that a sheep can overcome the disease, regardless of the clinical course.

If affected sheep meet with no particular hardship, they may be expected to survive for 3–8 months after the first symptoms are noticed. In certain cases and under experimental conditions, they may, however, survive much longer, even for years (Sigurdsson et al., 1952).

2.2.2 Visna

Visna is very insidious in its onset and, in the field, is never found in sheep under two years of age. It occurs without any recognizable antecedent fever or other acute signs.

One of the first signs noticed is that the sheep lags behind when the flock is driven, especially when going uphill or over uneven ground, and sometimes the animal falls for no evident reason. An early sign of visna is a slight aberration in gait, especially of the hind-quarters, in sheep made to trot. Gradually, stumbling and weakness of the hind limbs becomes more apparent. At the same time there is some loss of weight. Later the power to extend the fetlocks becomes impaired, usually in one hind limb more than the other, and often the sheep is seen resting the distal end of the metatarsus of the affected limb on the ground. In some cases the head is kept in an unnatural position, tilted a little over to one side. A fine trembling of the lips and facial muscles is sometimes observed as an early sign. Tremor of the head, grinding of teeth or itching as seen in scrapie (rida) are not observed (Pálsson and Sigurdsson, 1958). The paresis of the limbs progresses slowly, the sheep tends to lie down even when grazing and rising becomes difficult. Sometimes the disease progresses in waves with slight intervening remissions. Gradually the animal becomes paralytic so that it cannot rise unaided. If hand fed it may survive for some time in this condition. Even at this stage the sheep remains alert, and difficulties in feeding, defecation or micturition are not seen. Fever does not occur. Although appetite seems to remain unaffected the sheep gradually loses weight.

The course of the clinical disease is protracted, usually several months to one year may elapse before the stage of paraplegia or even total paralysis is reached. Under experimental conditions the clinical course can even last for years.

Clinical laboratory tests on natural cases of visna have shown an elevated gammaglobulin level in the cerebrospinal fluid (Sigurdsson et al., 1961) and an increased number of mononuclear cells varying from 40 up to 2000 per mm^3 of cerebrospinal fluid. Pleocytosis of the cerebrospinal fluid varies but remains throughout the course of the disease in most cases. In sheep infected experimentally, increased numbers of mononuclear cells are found in the cerebrospinal fluid at different times during the pre-clinical phase (Sigurdsson et al., 1957; see 4.1.3).

2.3 Pathological features

2.3.1 Maedi

The pathology of maedi and visna has been described in detail (Sigurdsson and Pálsson, 1958; Sigurdsson et al., 1962; Georgsson and Pálsson, 1971).

In maedi the anatomical changes are confined to the thorax, except for lesions secondary to emaciation. When the thoracic cavity is opened, the lungs collapse less than normally, especially the diaphragmatic lobe. In a

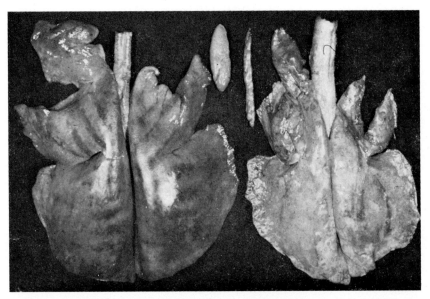

Fig. 2.2. Maedi lungs (1370 g) and corresponding mediastinal lymph node on the left and normal lungs (450 g) and mediastinal lymph node on the right.

minority of cases fibrous adhesions are found covering a portion of the surface of the lungs. The lungs are somewhat increased in size and heavy, weighing 2–3 times more than normal depending on the stage of the lesions (Fig. 2.2). Often distinct lesions are found in lungs of sheep that never showed any clinical signs of maedi. The actual increase in size of the lungs is, however, not so noticeable as is the increase in their weight. The shape of the lungs remains normal. The colour of the maedi lung differs considerably from the normal pinkish-red colour. It varies from dull grayish-blue to grayish-brown (beige).

To the touch the affected lungs are uniformly more compact than normal and feel like a rubber sponge, but less elastic. The whole lungs are usually remarkably uniform in consistency, although often the lesions are most conspicuous in the diaphragmatic lobes. Thus, the pathological changes seem to progress evenly in the whole organ, in contrast to most other lung lesions where diseased and normal parts may be easily distinguished. On the cut surface the tissue is rather dry and is homogenous in colour and consistency. Maedi lungs have always a much softer consistency than that found in ordinary pneumonic hepatization. The trachea and the larger bronchi are normal except for some thick mucous which, however, is not found in all cases. The tracheobronchial and mediastinal lymph nodes are always greatly enlarged, often weighing 3–5 times more than normal (Sigurdsson et al., 1952, 1953; Fig. 2.2).

The main histological changes found in maedi are chronic interstitial inflammation with thickening and infiltration of the interalveolar septa, hyperplasia of smooth muscle fibres, and slight fibrosis (see Ch. 4). Proliferation of lymphatic tissue is always found throughout the lung and is usually most marked close to the bronchi or bronchioles. There is epithelial proliferation of the small bronchi accompanied by epithelialization of some alveoli in far advanced cases.

In smears from diseased lungs, large mononuclear cells are often seen to contain one or more protoplasmic round inclusion bodies, 1–3 microns in size. They stain with a soft grayish-blue colour by Giemsa, and are rather characteristic (Sigurdsson et al., 1952).

2.3.2 Visna

In sheep affected with visna macroscopic lesions are seldom found on post-mortem examination. In old cases there is a marked atrophy of the skeletal muscle of the one or both hind limbs. In a few cases a slight hyperaemia of the meninges is observed and small yellowish spots can sometimes be seen on the cut surface of the brain or spinal cord.

Microscopically visna is characterized by extensive, microglial infiltration of a destructive character (see Ch. 4). The primary lesions of the central nervous system are meningeal and subependymal infiltrates and proliferation of round cells. In advanced cases extensive lesions can be found throughout the brain, brainstem, pons, medulla and spinal cord. The lesions, on becoming confluent, give rise to large lesions which tend to necrose and form cavitations. Around these lesions extensive perivascular cuffs of lymphocytes and plasma cells are found. Peripheral nerve-roots sometimes show diffuse lymphocytic infiltration.

There is not necessarily any relationship between the duration of the clinical disease and the severity of the lesions (Sigurdsson et al., 1962; Pálsson, 1966; Ressang et al., 1966; see 4.1.3).

2.4 Problems of diagnosis

2.4.1 Maedi

The clinical signs found in sheep affected with maedi are not pathognomonic, as in the late stages they resemble several other chronic lung diseases. In some cases it may also be difficult to differentiate diffuse verminous pneumonia in older sheep from maedi in its earlier stages.

Maedi is only found in adult sheep. An unusually protracted course and progressive loss in weight of several animals in the flock give an indication of maedi, but in the single animal, the disease can only be diagnosed with certainty on autopsy. The heavily increased weight of the lungs and the lymph nodes, the characteristic diffuse, compact, dry and homogenous lesions throughout the lungs, with the peculiar grayish-brown colour, will distinguish maedi from most other chronic lung diseases. Finally the diagnosis can be confirmed by histological examination.

Various serological tests and cell-culture methods for isolating the virus from tissues such as lung, spleen and mediastinal lymph nodes are also available to confirm the diagnosis of suspected cases (see 2.6.2 and Ch. 5).

2.4.2 Visna

The clinical signs observed in sheep suffering from visna, especially in the first stages, can easily be confused with other afflictions of the central nervous system (CNS) such as abscesses, traumatic lesions, lesions caused by parasites, etc. In more advanced cases the protracted course, the rather characteristic gradually progressive paresis of the hind limbs in the absence of a rise in body temperature, accompanied by pleocytosis of the cerebrospinal fluid, will support a proper diagnosis. Histological examination will reveal the characteristic lesions of the CNS.

The use of serological tests, and the isolation of virus from the CNS (particularly choroid plexus) will, if necessary, confirm the diagnosis (see 2.6.2 and Ch. 5).

It should be emphasized that in slowly progressive diseases such as maedi and visna, where the course of the disease often lasts for several months, even years, there is the risk that various concurrent and faster developing infections may be present as well. Such concurrent, unrelated infections seem to be common, and have often caused considerable confusion in the past, sometimes with the result that maedi and visna have been overlooked.

2.5 Geographical occurrence

2.5.1 Maedi

A number of progressive pneumonias of sheep and also of goats (Rajya and Singh, 1964) with clinical manifestations and with pathological features resembling maedi have been described under various names by a number of authors in several countries. The results obtained in neutralization and complement fixation tests with a small number of sera from maedi-like diseases in various countries (Kenya, Holland, U.S.A., Denmark, Norway and Germany) suggest that the agents causing these lung diseases are closely related to maedi virus (Thormar, 1966; de Boer, 1970a; Wandeira, 1971; Pétursson, personal communication).

Although there is apparently some divergence between the descriptions of these various lung diseases, they can probably be attributed to some extent to differences in sheep breeds and in sheep farming methods practised in the different countries.

However, more work is needed to clarify definitely the relationship between the agents causing slow progressive interstitial pneumonia of sheep found in different parts of the world.

South Africa

From the literature it appears that the first description of progressive pneumonia of sheep resembling maedi was given by D. T. Mitchell in 1915. Later de Kock (1929) reported similar findings in a number of sheep at the Graaf–Reinet Experimental Station. He stresses the histological difference between jaagsiekte on one hand which he described as multiple papilliform cyst–adenomata of the lung, and on the other hand the chronic pneumonia with lymphoid infiltration and secondary hyperplasia of bronchiolar and alveolar epithelium, which later has been referred to as Graaf–Reinet disease.

Had this distinction made by de Kock been more widely known and accepted, a confusion still existing with regard to progressive pneumonia and jaagsiekte might have been avoided.

Jaagsiekte appears to have been widespread in South Africa appearing sometimes side by side with progressive pneumonia. The losses seem to have been low, 1.6–3%, but what part of these losses can be attributed to progressive interstitial pneumonia is difficult to tell. No cases of Graaf–Reinet disease have been encountered for a long time in South Africa (Tustin, 1969).

U.S.A.

In the state of Montana and other Northwestern states, progressive pneumonia of range sheep has been known since 1915 according to Marsh (1923a, b) who described the disease as chronic interstitial pneumonia, with peribronchial and perivascular accumulation of mononuclear cells. Clinically the disease was characterized by a slowly progressive dyspnoea with gradual weakness and emaciation. Progressive pneumonia was mainly seen in sheep four years of age or older, with an incidence of 2–5%.

When Cowdry and Marsh (1927) compared the pathological lesions found in Montana progressive pneumonia ('Lungers') with certain forms of jaagsiekte in South Africa they came to the conclusion that they were similar. Later Marsh (1966), however, made a clear distinction based on histological studies between progressive pneumonia and jaagsiekte. The reason for the authors' earlier statement was that lesions from the Graaf–Reinet form of jaagsiekte had been compared with the Montana disease. Marsh (1966) stated that the histologic picture of progressive pneumonia corresponds quite closely to that described for maedi, la bouhite, the Graaf–Reinet form of jaagsiekte and dampigheid or zwoegerziekte.

Recently, progressive pneumonia has been confirmed in other parts of the U.S.A. A viral agent very similar to maedi–visna virus has been isolated from lungs affected with Montana disease (Kennedy et al., 1968; Takemoto et al., 1971).

Holland

According to de Boer (1970a) a progressive pneumonia of sheep in the northern part of Holland called 'zwoegers' was described in 1918. Koens (1943) gave a detailed description of the clinical and pathological features of zwoegerziekte and pointed out its close relationship with Montana sheep disease and the lung disease in South Africa described by Mitchell (1915).

Bos (1951) drew attention to the close similarities between lesions found in zwoegerziekte and maedi. Ressang et al. (1968) studied the histopathological lesions found in lungs affected with zwoegerziekte and grouped this progressive interstitial pneumonia together with maedi, Montana progressive pneumonia, la bouhite and Graaf–Reinet disease.

De Boer (1970a) isolated virus from affected lungs, and demonstrated that various morphological, biological and biophysical properties of viruses isolated from sheep with zwoegerziekte were similar to those of maedi. He succeeded in transmitting zwoegerziekte by inoculating this virus which had been propagated in tissue culture, into healthy sheep.

Zwoegerziekte appears to be fairly widespread in Holland especially along the coast. On some farms in these areas annual losses of up to 15% have been ascribed to zwoegerziekte. A serological relationship of maedi and zwoegerziekte has been reported (Thormar, 1966).

France

Lucam (1942, 1944) described a lung disease of sheep in France, which he named 'lymphomatose pulmonaire maligne', but which was known by the farmers as 'la bouhite'. The clinical course was characterized by a progressively increasing dyspnoea and emaciation lasting 6–8 months. Lucam came to the conclusion that this disease probably was similar to Montana progressive pneumonia and Graaf–Reinet disease. He described the disease as an interstitial pneumonia with nodular or diffuse accumulation of cells of the lymphocytic type. La bouhite is apparently known in several parts of France and in the district of Mont de Marsan 5 to 20 % of single flocks can be affected.

India

Rajya and Singh (1964) reported several cases of maedi found in a survey of a great number of lungs of sheep and goats collected on farms and slaughterhouses. Some of these lungs were affected with pulmonary adenomatosis as well.

   Later several authors have reported the presence of maedi from various parts of India. Sharma (1972) in a survey of a great number of lungs found 4 % of sheep lungs and 5 % of lungs from goats affected with maedi. Bhagwan et al. (1973) studied maedi in goats and sheep and found that the overall incidence was 0.5 and 0.8 %, respectively.

   These findings were based on pathological lesions and not confirmed by serological tests.

Germany

Seffner and Lippmann (1967) reported some cases of slow progressive pneumonia in a flock of Merino sheep near Leipzig.

   The affected lungs showed a uniform increase in consistency, and the histological pattern was characterized by cellular infiltration of the inter-alveolar septa, hypertrophy of inter-alveolar muscle fibres and proliferation of the lymphoid follicles.

   Weiland and Behrens (1970) reported similar findings in some flocks of the Texel breed, and so did Straub (1970) and Schaltenbrand and Straub (1972) in a flock of Merino sheep, where they were able to confirm the diagnosis and its relationship with maedi by serological tests. All these authors noted that the features of their cases were common to those of Montana disease and maedi. In the years 1970–1972, twenty sheep flocks in Germany were slaughtered because of maedi.

Kenya

Wandera (1970) studied pneumonia as a cause of serious losses of sheep in Kenya. He described a progressive interstitial pneumonia manifested by dyspnoea, increased respiration rate, fits of coughing and emaciation. The characteristic lesions were progressive interstitial pneumonia, often with pulmonary lymphoid hyperplasia. Sera from some of the clinical cases contained significant titres of neutralizing antibody against the Icelandic maedi virus strain.

Denmark and Norway

Hoff-Jørgensen (1971, 1974a) and Krogsrud (1974) reported lung disease of sheep closely resembling maedi in several flocks of sheep in Denmark and Norway, respectively. Pathological changes characteristic of visna–maedi infection were seen on post-mortem examination. Agents isolated from organs of affected sheep, when propagated in cell cultures, produced a cyto-pathogenic effect, similar to that described for visna–maedi virus, and they could be neutralized by anti-visna and anti-maedi sera.

In both countries sheep of the Texel breed were mainly affected, and it is believed that the disease was introduced into Norway by imported Texel sheep in 1967. The incidence of the disease within these two countries is still unknown, but in single flocks a high percentage of the adult sheep have shown positive titres of complement-fixing antibodies. Eradiction of the infection by killing sheep showing positive serum reactions is being attempted.

Hungary

Süveges and Széky (1973) observed a chronic progressive interstitial pneu-monia resembling maedi in three Merino-type sheep flocks in south and west Hungary. The main clinical signs were progressive dyspnoea, and emaciation. The disease did not respond to therapy and always ended fatally after a course of 2–6 months. The lungs were enlarged and their weight increased 2 or 3 times, and the microscopic picture of the lung lesions corresponded to maedi or chronic progressive interstitial pneumonia. On one farm, 35 ewes from a flock of 420 became ill in the period from September 1971 to February 1972. It was thought probable that the disease had been introduced into the country by imported breeder rams.

Other countries

Nobel et al. (1973) reported a few cases of maedi found when studying lung diseases of imported sheep of Merino breed from Rumania, slaughtered

at the Jerusalem abattoir. Histological examination revealed lesions resembling those found in maedi and maedi-like diseases. The relatively low
incidence of the disease found in the material examined is thought to be due
to the fact that 90% of the animals were only one year old.

Papadopulos et al. (1971) reported a lung disease of sheep in Greece
resembling progressive interstitial pneumonia, and in Bulgaria an apparently
similar lung disease has been reported by Pavloff (1963).

Mitrofanov and Yartsev (1973) reported progressive interstitial pneumonia
(maedi) in a sheep flock in Kirgizia, U.S.S.R. The pathological lesions
resembled those of maedi and inclusion bodies were reported in the cytoplasm of histocytes and reticuloblasts. This appears to be the first report of
maedi in Krigizia.

Bellavance et al. (1974) reported lesions in sheep lungs from Quebec,
Canada, very similar to the lesions found in maedi–visna. Neutralizing
antibodies against maedi–visna virus were demonstrated.

Iceland
Maedi was introduced into Iceland by a Government importation of 20
sheep of the Karakul breed in the summer of 1933 as described at the

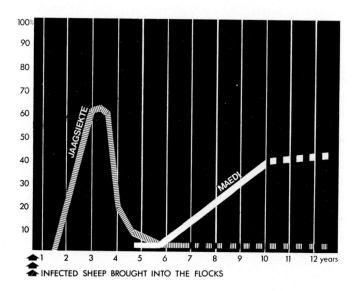

Fig. 2.3. Annual losses of sheep in two different flocks, one affected with jaagsiekte, the
other affected with maedi.

beginning of this chapter. The epidemiological studies by Gíslason (1947) showed that two of the imported rams were carriers of the infection of maedi, and one of them was also carrier of jaagsiekte (see Fig. 2.1).

This particular ram showed distinct symptoms of a lung disease in spring 1934, and never returned from the common summer pastures. It was therefore considered that he had possibly died from the disease. Next winter losses from jaagsiekte began on the farm where the ram had been kept and also on some of the neighbouring farms in the south-western part. The following years jaagsiekte spread over vast areas, causing heavy losses, on some farms exceeding 50% annually.

DISTRICTS AFFECTED WITH MAEDI (VISNA) AND JAAGSIEKTE

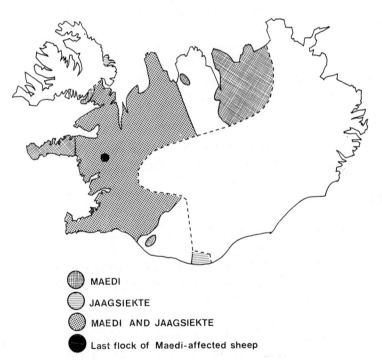

⊕ MAEDI

⊖ JAAGSIEKTE

⊕ MAEDI AND JAAGSIEKTE

● Last flock of Maedi-affected sheep

Fig. 2.4. The epizootics reached maximal spread in 1944, when the lung diseases were prevalent on most farms in areas raising approximately 60% of the sheep in Iceland. Eradication of these sheep diseases was attempted by systematic slaughtering of all sheep in affected areas, followed by restocking with young sheep from unaffected parts. This was carried out in the period 1937–1965. In 1965 the last flock affected with maedi was slaughtered.

No wonder, therefore, that jaagsiekte overshadowed maedi completely to begin with in these districts. Pasteurellosis and pneumonia due to lung worms were often seen which added further complications. The disastrous losses from jaagsiekte in the flocks often declined after 3–4 years and the sheep farmers became somewhat hopeful (see Fig. 2.3). But then, approximately 6 years after the Karakul sheep had been imported, another lung disease behaving differently from jaagsiekte appeared in these same flocks, i.e., maedi.

This disease was found to be much more insidious in its onset both with regard to individual sheep and the flock as a whole.

The imported Karakul ram in the north-eastern part of the country acted as a carrier for maedi only (see Fig. 2.1). In this district the disease spread from farm to farm in an insidious way, and losses in individual flocks increased more gradually than experienced in the south-western part.

During the following years maedi spread to new areas in spite of control measures. Jaagsiekte on the other hand declined at least in its recognized clinical form and as time passed it became of less importance in the districts where it had previously caused such heavy losses.

When the epizootic of maedi was at its peak the disease was more or less prevalent on most farms in approximately 60% of the sheep-raising districts of the country (Fig. 2.4).

During these years (1933–1944), the total number of winterfed sheep declined from approximately 730,000 to 450,000 sheep.

2.5.2 *Visna*

Visna or visna-like diseases have been observed in sheep and sometimes in goats from various countries, i.e. Iceland (Sigurdsson et al., 1957), India (Rajya and Singh, 1964; Sharma, 1972), Holland (Ressang et al., 1966), Kenya (Wandera, 1970), Germany (Schaltenbrand and Straub, 1972; Weinhold, 1974; Dahme et al., 1973), Hungary (Süveges and Széky, 1973), Norway (Krogsrud, 1974), Denmark (Hoff-Jørgensen, 1974b), and in U.S.A. (Cork et al., 1974). In many of these reports the clinical and pathological findings have been supported either by serological tests or confirmation of a viral agent in the central nervous system.

It is noteworthy that in all the above-mentioned countries natural cases of visna in sheep have been found along with maedi or maedi-like diseases in the same flock or even in the same animal. Apparently the incidence of clinical cases of visna is low, and the disease is rare even in countries like

Holland where zwoegerziekte has been quite common in sheep flocks in some parts of the country for a long time.

In Iceland, visna was eradicated in the year 1951. But in the period from approximately 1940 to 1951 the disease was quite well known on many farms over wide areas in the south-western and southern parts of the country. Often only few sheep were affected on each farm, but sometimes the losses from visna exceeded those of maedi in the same flock.

When clinical signs of visna became evident the farmers used to dispose of the animal as soon as possible as no cure was known, and affected sheep deteriorated within a few months to a fatal end under usual farm conditions.

It is worth mentioning that when a flock of young sheep naturally affected with maedi were sacrificed, virus could be demonstrated in the choroid plexus in approximately 25% of the cases, although none of these sheep showed clinical signs of nervous disorder. This viral agent was similar to maedi virus isolated from the lungs of the same sheep (Gudnadóttir et al., 1968).

2.6 Evidence for a viral aetiology of visna–maedi

2.6.1 Transmission experiments: cultivation of virus

Epidemiological studies by Gíslason (1947) from the very first years of the epizootic indicated that maedi was a contagious disease. It was often possible to trace the appearance of the disease into flock after flock following the addition of new animals. On the basis of such records the incubation period of maedi was estimated to be at least 2–3 years.

Later maedi was successfully transmitted to healthy sheep by direct contact between healthy and diseased animals, by contaminating their drinking water with faeces from diseased animals and by injecting material from typically affected lungs and lymph nodes intranasally, intrapulmonarily and intravenously (Sigurdsson et al., 1953). Repeated attempts to transmit the disease to various species of laboratory animals have failed.

Visna was first transmitted by intracerebral inoculation of brain and spinal cord material from natural cases of visna in the year 1951. The transmitted disease appeared to be similar clinically and pathologically to the natural cases. Later the disease was transmitted serially by intracerebral inoculation from sheep to sheep, and in this way the infective agent was maintained in the laboratory after the disease had been eradicated in the field in 1951 (Sigurdsson et al., 1957).

In 1957 a previously unknown viral agent was isolated in tissue culture from five brains of transmitted cases of visna, and the methods of cultivation were described.

Tissue culture passage of this virus was found to produce the disease when inoculated into healthy sheep and the virus was consistently recovered from inoculated sheep (Sigurdsson et al., 1960; Thormar and Pálsson, 1967).

In 1958 a viral agent was isolated from lungs affected with maedi brought to the laboratory from two flocks affected with maedi on the west coast of Iceland (Sigurdsson, personal communication). The following year apparently the same agent was isolated from a large number of lungs of natural cases of maedi. The fact that the virus could consistently be isolated from lungs with maedi lesions and that serum from maedi cases neutralized the virus, suggested an aetiological relationship with the disease (Sigurdardóttir and Thormar, 1964).

Maedi and visna viruses showed a characteristic cytopathic effect in tissue culture, and the chemical and physical properties of these two viruses have been found to be almost identical (Thormar and Helgadóttir, 1965; see also Chs 3 and 5).

Gudnadóttir and Pálsson (1967) reported the successful transmission of maedi to healthy sheep by inoculating intrapulmonarily three strains of virus isolated in tissue culture from maedi-affected lungs. Visna lesions were also produced in some of the animals, indicating their close relationship. Viral agents closely related to maedi–visna have been isolated in Holland from sheep with zwoegerziekte, and transmission experiments in sheep have shown a very close relationship to maedi (de Boer, 1970a, b). In flocks with zwoegerziekte, visna-like meningoencephalitis has occasionally been reported (Ressang et al., 1966). From these cases viruses with properties similar to zwoegerziekte virus have been isolated.

Recently viral agents with properties very similar to maedi and visna viruses have been isolated from cases of progressive pneumonia in Montana (Kennedy et al., 1968; Lopes et al., 1970; Takemoto et al., 1971) and from sheep of the Texel breed, showing a lung disease resembling maedi, in West Germany, Denmark and Norway (Straub, 1970; Hoff-Jørgensen, 1971; Krogsrud, 1974). These viruses isolated in various laboratories seem in fact to be strains of the same virus, and it is likely that eventually they can all be classified as a single virus (Thormar et al., 1974).

It is also noteworthy that by inoculating healthy sheep intracerebrally by visna virus propagated in tissue culture, lesions indistinguishable from maedi lesions were demonstrated (Gudnadóttir and Pálsson, 1965).

These findings support previous statements of the relationship between the visna and maedi virus, as an infectious agent.

2.6.2 Serological tests

It has been shown repeatedly in sheep inoculated experimentally with maedi–visna or zwoegerziekte virus that antibodies are formed at various times after inoculation. These antibodies seem to remain detectable for years and probably throughout the lifespan of the animal. It is, however, still unknown how early these antibodies can be detected in natural cases of maedi–visna or zwoegerziekte (Thormar et al., 1966).

In experimental cases with maedi–visna, virus neutralizing antibodies emerge in different sheep with a remarkable variation in time, but usually they are detectable 2–3 months after inoculation of high-titer virus (Gudnadóttir and Pálsson 1965, 1966, 1967). A neutralization test in visna was first described by Sigurdsson et al. (1960).

Complement-fixing antibodies are found both in natural cases of maedi–visna and in sheep experimentally infected with maedi or zwoegerziekte virus (Gudnadóttir and Kristinsdóttir, 1967; de Boer, 1970a). These complement-fixing antibodies are usually first detected 3–4 weeks after inoculation of virus, and seem to persist often at comparatively high levels for the rest of the animal's life.

The complement fixation test appears therefore to be a fairly easy and rather sensitive method to detect the disease even in its earlier stages, and has been used quite extensively in surveys of sheep sera in districts where maedi cases could be expected or to examine sheep flocks for the presence of positive reactors (Hoff-Jørgensen, 1974a; Krogsrud, 1974).

In sheep experimentally infected with visna and maedi viruses (Thormar, 1969) and with zwoegerziekte virus (de Boer, 1970b) antibodies that could be detected by immunofluorescent staining were found at an early stage of the infection. High titers of antibody of this type were developed in a relatively short time and the serum titers apparently remained elevated throughout. The immunofluorescent staining procedure has been described in detail (de Boer, 1970b) and it was found that the indirect immunofluorescence test was a convenient method for detecting antibodies against zwoegerziekte virus. This technique has been used successfully to detect the incidence of maedi virus infection in sheep flocks in Holland (de Boer, personal communication).

Terpstra and de Boer (1973) have demonstrated precipitating antibodies by the Ouchterlony technique in sera from sheep infected intrapulmonarily or intracerebrally with maedi–visna virus. Precipitating activity could be detected within 2 to 8 weeks after infection and persisted for years. This test is thought to be a valuable tool for the early diagnosis of maedi and visna in sheep as the test is rapid and simpler to apply than the serum tests referred to above. So far it appears that this technique has not been applied as a field test.

Occasionally in all these different serological tests, some sera do not react positively, even though definite histological and pathological lesions are found and the presence of viral agent established. So far no satisfactory explanation of this discrepancy has been offered.

2.7 Natural transmission

Maedi disease has an extremely insidious onset and in its early or preclinical stages it is rarely transmitted except from the ewe to its lamb.

On common pastures communicability of maedi appears to be low even in its clinical stage. However, when an affected sheep was housed with healthy ones during the winter, even for a short period, this usually resulted in spread of the infection and that was how the disease most often spread in Iceland. Often several years elapse after the infection is brought into the flock until the first losses from maedi are observed. During the next three to four years the mortality increases rapidly, so that after 9–10 years the annual mortality was often found to be 20–30% (Fig. 2.3).

Although it is believed that the disease has been transmitted experimentally by feeding faeces from diseased sheep to healthy ones via the drinking water, an indirect spread of maedi was exceptionally rare in Iceland. While the restocking of affected farms was being carried out (see 2.10), healthy lambs 5–6 months old were often brought to these farms just 1 or 2 weeks after the old diseased flock had been disposed of. Usually the premises were not disinfected. In spite of this there is no record indicating that the newly bought healthy animals became infected through the sheep sheds, permanent pastures, etc. on these farms.

Lambs born and reared by ewes suffering from maedi often became affected at a comparatively young age. This also appears to be the case in other maedi-like diseases. In zwoegerziekte the presence of virus in milk from affected ewes has repeatedly been confirmed and virus has occasionally been isolated from lungs of lambs only 4–5 months old. It is therefore assumed that lambs can occasionally become infected orally shortly after birth (de Boer, 1970a).

Attempts to demonstrate a transplacental transmission of the disease experimentally have so far given negative results (Gudnadóttir, 1974; de Boer, 1970a).

In advanced cases of maedi and zwoegerziekte the presence of the viral agent can be demonstrated regularly in various organs. Occasionally maedi virus has also been demonstrated in nasal swabs from such sheep. During the clinical course fits of dry coughing are occasionally seen, and thick mucous is often seen in the larger bronchi. Transmission of maedi by the respiratory route as a droplet spread of the infectious agent while animals are in close contact is considered from field experience to be the most likely way the disease is spread naturally. However, experimental evidence to support this statement is still inconclusive.

2.8 Susceptibility of different breeds

Maedi–visna have been reported to affect various breeds of sheep, i.e., Texel sheep, various Merino-type sheep, Karakul sheep, Icelandic sheep and goats of different breeds. From the data available it is very difficult indeed to form any opinion of the susceptibility of various breeds to maedi as sheep farming practices and other conditions which might influence the spread and the severity of the infection vary so greatly.

Epidemiological studies in Iceland indicate that certain strains within the Icelandic breed were more resistant than others (Pálsson, 1947; Gíslason, 1947) and crosses between Icelandic sheep and Border Leicester rams appeared to be particularly resistant. Therefore farmers were encouraged to acquire rams belonging to such breeds. Consequently for some years a number of rams belonging to these breeds were used for artificial insemination, and some ram semen from Great Britain was also imported for the same purpose. Increased resistance in this context implies delayed progression of the lesions in the lungs, resulting in more productive years for each ewe, rather than total resistance to the disease.

The decision to eradicate maedi by drastic depopulation of all sheep in affected parts of Iceland brought this work to an end (Pálsson, 1947; see 2.10).

2.9 Methods of disease control

When the disease had gained a foothold within a flock, most of the ewes succumbed to maedi when 4 to 5 years of age, or even younger, whereas normal Icelandic ewes often maintained satisfactory productivity until 9 to 10 years old. Although the clinical course of maedi can apparently, in some cases, be retarded for a while by liberal feeding and careful nursing, there is no indication that special methods of feeding or shepherding can influence the disastrous losses in an affected flock. Males and females seem to be equally susceptible.

Various therapeutic methods have been tried in order to cure or delay the disease, and some encouraging attempts were also made to prevent it by vaccination (Gíslason, 1947). All these efforts have, however, given inconclusive or negative results.

Animals infected with visna and maedi virus form antibodies that are detectable often in relatively high titers throughout the course of the disease. In spite of high titers of antibodies, pathological lesions progress continuously in affected organs and sometimes viraemia is observed (see Ch. 6).

There appears to be a certain response of the lymphoid system both in the preclinical and clinical stages of the disease, but little is known about the role of cellular immunity in these infections.

As long as the immune response to these diseases is so poorly understood, immunological methods of prevention – if at all applicable – are a distant goal. However, complement fixation tests and the indirect immunofluorescence technique have been used for detecting infected animals at an early stage of the infection, often with satisfactory results.

It is, however, still an open question whether repeated tests of animals in a particular flock and disposal of positive reactors will in the long run rid the flock of the infection. So far, the slaughter of all animals in an affected flock is the only certain method of eradicating the disease.

2.10 Eradication of maedi in Iceland

When it became apparent in the year 1935 that a previously unknown contagious, highly fatal lung disease was rampant, various control measures were instituted. However, the nature of the disease was little known at the time and not all of these attempts were equally effective.

In order to prevent free movement of sheep from affected areas, hundreds of miles of barbed wire fences were erected. Guards were continuously at work to prevent sheep from breaking through the fences and mingling with sheep from unaffected parts. Sheep that crossed these quarantine fences either way were slaughtered ruthlessly.

Later it was realized that some of the first fences built did not serve a

ERADICATION OF LUNG DISEASES
QUARANTINE AREAS

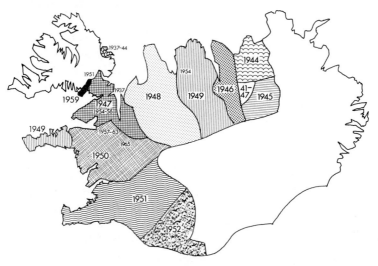

Fig. 2.5. Experimental slaughtering and restocking took place in small areas in the years 1937–1943. The general slaughtering and restocking program was carried out in the years 1944–1954. (Large letters). Recurrence of maedi in few places necessitated limited slaughtering on several farms in restricted areas in the years 1951–1965.

useful purpose as the disease had already been incubating in flocks behind the fences when they were erected. Although some attempts to arrest the spread of the disease failed, others were more successful and further spread could be prevented. This was of vital importance, when lambs from unaffected areas were needed for restocking.

In 1937 attempts were made to eradicate the lung diseases by slaughtering all sheep in two small areas and restocking them with healthy lambs (Fig. 2.5). Although these early attempts were not altogether successful these experiments proved to be valuable when plans were later formed for the general eradication of maedi.

The losses caused by maedi in affected areas continued year after year, and as more and more flocks became affected sheep farming became a hopeless task in spite of various government subsidies.

Faced with these problems the Government decided to combat the disease by slaughtering all sheep on every farm within the affected areas, and restocking the farms with healthy young sheep from non-affected parts. Two special Acts to this effect were adopted in the year 1941.

As the number of healthy young sheep that could be provided each year was limited and because of financial restrictions it was evident that the program of eradication would take several years to accomplish.

By the year 1943–1944 the spread of the lung diseases had been fairly well established and the affected parts of the country were already fenced. When the eradication began the fences were improved, repaired and extended where necessary, to give a number of quarantine areas (Fig. 2.5). Systematic slaughtering was begun in the first quarantine area in the autumn of the year 1944. After that slaughtering and restocking in additional areas was carried out each autumn, and by the end of the year 1952 all sheep in the last quarantine area had been slaughtered. The final restocking of these areas was accomplished by the end of the year 1954.

Owing to various local reasons, the slaughtering and restocking had to be carried out in the months of September and October each year. The young sheep from healthy areas therefore had to be brought to their new farms only one or two weeks after all the old diseased stock had been disposed of. Usually no disinfection of the premises was carried out, and would often have been a difficult task owing to the traditional stone and turf construction of many of the old sheep sheds. Two of the quarantine areas were, however, kept free of sheep for one year after the old stock had been slaughtered.

In certain districts, mainly in the western part of the country, maedi reappeared several times in the replacement flocks in the years between 1952 and 1965 as indicated in Fig. 2.5.

This recurrence of maedi in the new stock could in all but one case be traced back to contact with affected animals. The reason for this unexpected recrudescence of the infection was associated with the long and silent incubation period of maedi. It was later found that in these few instances some apparently healthy young sheep had already become infected by contact by the time they had been bought and transported to these areas for restocking purposes. Some years later these animals developed maedi and spread the disease among the new stock.

However, only a few farms were affected at a time, and the disease was again eradicated by slaughtering all affected and in-contact sheep in these districts. Sometimes, however, several thousand sheep had to be slaughtered.

The last appearance of maedi, which was restricted to only one flock occurred in 1965. The flock was immediately disposed of. Since then Iceland has been free of maedi.

It has been estimated that more than 105,000 sheep succumbed to maedi and more than 650,000 sheep had to be killed in order to eradicate the disease. The campaign lasted almost 30 years.

While eradicating maedi, visna was also eradicated in the year 1951 and jaagsiekte in the year 1952.

The decision to combat maedi by systematic slaughtering of all sheep in the affected parts was not unanimously supported at the beginning. It was a desperate attempt to save sheep farming, based more on optimism than knowledge of the nature of maedi and of the agent causing it (Fridriksson, 1970). This knowledge was only established several years later.

The eradication scheme and control measures were supervised and directed by a board of five members elected by the Parliament. All sheep owners within a quarantine area founded, as stipulated in the law, an association to deal with the slaughtering of their sheep and the restocking. In this way the sheep owners were made responsible for the task of gathering and slaughtering every single sheep of the old stock within the quarantine area. This task had to be accomplished before a given date in the autumn when the restocking began. It was also the duty of the sheep owners' association to select, purchase, transport and distribute the new young stock to the numerous persons entitled to acquire these sheep.

In this way sufficient manpower familiar with sheep farming and with great loyalty to this important task was secured. The final success of the eradication program is to a great part due to the cooperation and loyalty of the Icelandic sheep farmers towards the drastic control measures instituted.

For the first 10–12 years following restocking a close examination for maedi lesions was carried out on all sheep lungs of the new stock when brought to slaughter. Thousands of blood samples were also collected every year from adult sheep at the abattoires, and later tested for the presence of antibodies against maedi virus. These control methods were both convenient and valuable for detecting possible recurrence of maedi in the restocked districts.

Acknowledgement

The author is grateful to G. Pétursson for critical appraisal of the manuscript and for permission to reproduce maps and graph originally prepared for exhibition of maedi–visna at the Nordic Veterinary Congress, Reykjavík, 1974.

P. A. Pálsson

References

BELLAVANCE, R., TURGEON, D., PHANEUF, J. B. AND SAUVAGEAU, R. (1974) Can. Vet. J., 15, 293.

BHAGWAN, P. S. K., SING, N. P. AND SINGH, V. B. (1973) Ind. J. Anim. Health, 12, 45.

DE BOER, G. F. (1970a) Thesis, Elinkwijk, Utrecht.

DE BOER, G. F. (1970b) J. Immunol. 102, 414.

BOS, S. E. (1951) Thesis, Amsterdam.

CHAUHAM, H. V. S. and SINGH, C. M. (1970) Br. Vet. J., 126, 364.

CORK, L. C., HADLOW, W. J., CRAWFORD, T. B., GORHAM, J. R. and PIPER, R. C. (1974) J. Infect. Dis., 129, 134.

COWDRY, E. V. and MARSH, H. (1927) J. Exp. Med., 45, 571.

DAHME, E., STAVROU, D., DEUTSCHLÄNDER, N., ARNOLD, W. and KAISER, E. (1973) Acta Neuropathol. (Berl.), 23, 59.

FRIDRIKSSON, S. (1970) Árbók landbúnadarins, 21, 101.

GEORGSSON, G. and PÁLSSON, P. A. (1971) Vet. Pathol., 8, 63.

GÍSLASON, G. (1947) Iceland Ministry of Agriculture Publ., Reykjavík, p. 235.

GÍSLASON, G. (1966) Int. Encycl. Vet. Med. Vol. 3. (Green, Edinburgh) p. 1780.

GUDNADÓTTIR, M. (1974) Progr. Med. Virol., 18, 336.

GUDNADÓTTIR, M., GÍSLASON, G. and PÁLSSON, P. A. (1968) Res. Vet. Sci., 9, 65.

GUDNADÓTTIR, M. and KRISTINSDÓTTIR, K. (1967) J. Immunol., 98 663.

GUDNADÓTTIR, M. and PÁLSSON, P. A. (1965) J. Infect. Dis., 115, 217.

GUDNADÓTTIR, M. and PÁLSSON, P. A. (1966) J. Immunol., 95, 1116.

GUDNADÓTTIR, M. and PÁLSSON, P. A. (1967) J. Infect. Dis., 117, 1.

HOFF-JØRGENSEN, R. (1971) Slow virus (visna–maedi) in Danish sheep. Preliminary report Scottish–Scandinavian Conference on Infectious Diseases, Copenhagen, June 1971.

HOFF-JØRGENSEN, R. (1974a) Proc. 12th Nordic Vet. Congr., Reykjavík, p. 251.

HOFF-JØRGENSEN, R. (1974b) Medlemsbl. Dan. Dyrlægeforen., 57, 142.

KENNEDY, R. C., EDLUND, C. M., LOPEZ, C. and HADLOW, W. J. (1968) Virology, 35, 483.

DE KOCH, G. (1929) Union S. Africa, Dir. Vet. Serv. 15th Annu. Rep. 611.

KOENS, H. (1943) Thesis, Utrecht.

KROGSRUD, J. (1974) Proc. 12th Nordic Vet. Congr., Reykjavík, p. 250.

LOPEZ, C., EKLUND, C. M. and HADLOW, W. J. (1970) Bact. Proc., 70, 198.

LUCAM, F. (1942) Réc. Méd. Vét., 118, 273.

LUCAM, F. (1944) Bull. Acad. 120, 240.

MARSH, H. (1923a) J. Am. Vet. Med. Assoc., 64, 304.

MARSH, H. (1923b) J. Am. Vet. Med. Assoc. 62, 458.

MARSH, H. (1966) In: (L. Severi, Ed.) Proc. Int. Conf. on Lung Tumors in Animals. (Perugia, Italy) p. 285.

MITCHELL, D. T. (1915) 3rd and 4th Resp. Vet. Rec. S. Afr. 585.

MITROFANOV, V. M. and YARTSEV, N. M. (1973) In: Proc. 5th All Union Conference on Pathological Anatomy of Farm Animals. Moscow Vet. Acad., 199. Vet. Bull 44, 570 (1974).

NOBEL, T. A., NEUMANN, F. and KLOPFER, U. (1973) Q. Israel, Vet. Med. Assoc. 30, 19.

PÁLSSON, H. (1947) Icelandic Ministry of Agriculture, Reykjavík, p. 256.

PÁLSSON, P. A. (1966) Int. Encyc. Vet. Med. Vol. 5, (Green, Edinburgh) p. 3041.

PÁLSSON, P. A. (1972) J. Clin. Pathol., 25, suppl. (R. Coll. Pathol.) 6, 115.

PÁLSSON, P. A. and SIGURDSSON, B. (1958) 8th Scandinav. Vet. Congr. Sect. A. Rep. 8, Helsinfors, 1.

PAPADOPOULOS, C., SEIMENIS, A., FRANGOPOULOS, A. and MENACÉ, I. (1971) Vet. News– Greece, 3, 11.

PAVLOFF, N. (1963) Mh. Vet. Med., 18, 398.

RAJYA, B. S. and SINGH, C. M. (1964) Am. J. Vet. Res., 25, 61.

RESSANG, A. A., DE BOER, G. F. and DE WIJN, G. C. (1968) Pathol. Vet., 5, 353.

RESSANG, A. A., STAM, F. C. and DE BOER, G. F. (1966) Pathol. Vet., 3, 401.

SCHALTENBRAND, G. and STRAUB, O. C. (1972) Dtsch. Tieraerztl. Wochenschr. 79, 10.

SEFFNER, W. and LIPPMANN, R. (1967) Mh. Vet. Med., 22, 901.

SHARMA, D. N. (1972) Agra Univ. Res., 21, 85.

SIGURDARDÓTTIR, B. and THORMAR, H. (1964) J. Infect. Dis., 114, 55.

SIGURDSSON, B., GRÍMSSON, H. and PÁLSSON, P. A. (1952) J. Infect. Dis., 90, 233.

SIGURDSSON, B., KARCHER, D., VAN SANDE, M. and LOWENTHAL, A. (1961) In: Protides of the Biological Fluids. Proc. 8th Colloquium, Bruges (Elsevier Publ. Co., Amsterdam) 110.

SIGURDSSON, B. and PÁLSSON, P. A. (1958) Br. J. Exp. Pathol., 39, 519.

SIGURDSSON, B., PÁLSSON, P. A. and VAN BOGAERT, L. (1962) Acta Neuropathol., 1, 343.

SIGURDSSON, B., PÁLSSON, P. A. and GRÍMSSON, H. (1957) J. Neuropathol., 16, 389.

SIGURDSSON, B., PÁLSSON, P. A. and TRYGGVADÓTTIR, A. (1953) J. Infect. Dis., 93, 166.

SIGURDSSON, B., THORMAR, H. and PÁLSSON, P. A. (1960) Arch. Ges. Virusforsch., 10, 368.

STRAUB, O. C. (1970) Berl. Münch. Tierärztl. Wochensch., 83, 357.

SÜVEGES, T. and SZÉKY, A. (1973) Acta Vet. Acad. Sci. Hung., 23, 205.

TAKEMOTO, K. K., MATTERN, C. F. T., STONE, L. B., COE, J. E., HADLOW, W. J. and LAVELLE, G. (1971) J. Virol., 7, 301.

TERPSTRA, C. and DE BOER, G. F. (1973) Arch. Ges. Virusforsch., 43, 53.

THORMAR, H. (1966) In: (L. Severi, Ed.) Proc. Int. Conf. on Lung Tumors in Animals. (Perugia, Italy) p. 393.

THORMAR, H. (1969) Acta Pathol. Microbiol. Scand., 75, 296.

THORMAR, G., GÍSLASON, G. and HELGADÓTTIR, H. (1966) J. Infect. Dis., 116, 41.

THORMAR, H. and HELGADÓTTIR, H. (1965) Res. Vet. Sci., 6, 456.

THORMAR, H., LIN, F. H. and TROWBRIDGE, R. S. (1974) Progr. Med. Virol., 18, 323.

THORMAR, H. and PÁLSSON, P. A. (1967) In: Perspectives in Virology. (Academic Press, New York) p. 291.

TUSTIN, R. C. (1969) J. S. Afr. Vet. Med. Assoc., 40, 3.

WANDERA, J. G. (1970) Vet. Rec., 86, 434.

WANDERA, J. G. (1971) Adv. Vet. Sci. Comp. Med., 15, 251.

WEILAND, F. and BEHRENS, H. (1970) Dtsch. Tierärztl. Wochensch., 77, 373.

WEINHOLD, E. (1974) Sbl. Vet.-Med. B, 21, 32.

CHAPTER 3

The detailed structure of visna-maedi virus

Donald H. HARTER

3.1 Introduction

Visna and maedi viruses cause fatal slow diseases which affect the lungs and nervous system of sheep. They are distinguished from other slow infections, such as subacute sclerosing encephalitis, which may be considered to represent an aberrant response to a conventional virus which usually causes a rapidly evolving illness (Fucillo et al., 1974), and the transmissible spongiform encephalopathies where the viral nature of the causative agent has yet to be conclusively demonstrated (see Ch. 11). Visna and maedi agents are well-characterized viruses which share many features usually associated with oncogenic RNA viruses. The diseases they cause evolve slowly as a rule, rather than as an exception.

3.2 Experimental source of virus

Visna virus was first isolated in tissue culture by Sigurdsson and co-workers from the brain of a sheep afflicted with visna disease (Sigurdsson et al., 1960). Initial isolates were made by inoculating infected brain homogenate into cell cultures obtained from ependyma and white matter of normal sheep brain. Later, the virus was also isolated from cell cultures prepared by explanting the choroid plexus removed from the lateral ventricles of brains of visna sheep. Subsequently, virus was passaged numerous times in tissue cultures prepared from explanted or trypsin-dispersed choroid plexus from normal sheep. Tissue culture passaged visna virus has been found to cause

Slow virus diseases of animals and man, edited by R. H. Kimberlin
© *North-Holland Publishing Company 1976*

characteristic visna disease after inoculation into unaffected animals (Sigurdsson et al., 1960).

Shortly after the isolation of visna virus, a similar agent called maedi virus was recovered from the lungs of Icelandic sheep afflicted with a form of progressive interstitial pneumonia using similar sheep choroid plexus cell cultures (Sigurdardóttir and Thormar, 1964). Maedi was then transmitted to healthy sheep by injection of 3 serologically related strains of virus isolated in tissue culture from maedi-affected lungs (Gudnadóttir and Pálsson, 1967). Lesions in the central nervous system, indistinguishable from visna, were also observed.

Most studies of visna and maedi viruses have employed cell cultures derived from sheep choroid plexus (SCP). Two strains of visna virus (K485 and K796) and one strain of maedi virus (M88) have been used in studies of viral properties and virus–cell interactions.

3.3 *Methods of virus assay*

Visna and maedi viruses were initially quantitated by end-point assays which determined the amount of virus capable of producing cytopathic changes in SCP cells within 14–21 days after infection (Sigurdsson et al., 1960; Thormar, 1963a). The end point obtained appears to be affected by the passage level of the cells used as well as the strain of the viral inoculum (Thormar and Sigurdardóttir, 1962; Thormar, 1963b; Lopez et al., 1971).

A variety of plaque assays for visna virus have been reported. The first to be described involved maintenance of infected SCP cell monolayers under a carboxymethylcellulose overlay for 10 days followed by the addition of a sufficient number of BHK21-F cells to form a second confluent monolayer over the SCP cells (Harter and Choppin, 1967). Since BHK21-F cells are susceptible to visna virus-induced cell fusion but do not support visna virus replication, they were used as an indicator to detect foci of SCP cells capable of producing sufficient quantities of virus to promote cell fusion (see 5.2.1). Plaque formation was monitored by staining the BHK21-F monolayer with neutral red. Utilizing this method a PFU:TCIU ratio of 1.8 was obtained and the time required for assay was reduced. However, the method was not suitable for cloning virus preparations and did not detect foci of cells producing small quantities of progeny virus or virus incapable of causing cell fusion.

Subsequently, assays involving maintenance of infected SCP cell monolayers under semi-solid overlays of carboxymethylcellulose, agar or agarose were described (Harter, 1969). After 12–14 days, the overlay was removed and the cell sheet fixed and stained. Carboxymethylcellulose proved most satisfactory with a PFU : TCIU ratio of 2.5, but again the necessity of fixing and staining the cells precluded cloning by this method.

Recently a plaque assay has been described for visna and maedi viruses which does not require fixation of the infected SCP cell monolayer or disturbance of the gel overlay (Trowbridge, 1974). This assay utilized a soft agarose overlay supplemented with DEAE–dextran and secondary staining of the monolayer with neutral red. Maximal plaque formation occurred between 10 and 13 days and appeared to depend on the strain of virus employed. The PFU : TCIU ratio employing this method was 5.4. Heterogeneity in the size of plaques could be recognized; both large and small plaques were noted in cultures infected with visna and maedi viruses.

3.4 Purification of virus

Visna and maedi viruses have been concentrated and purified from medium harvested from infected tissue culture cells by standard methods. These have involved concentration by precipitation with zinc acetate or ammonium sulfate after preliminary clarification (Lin and Thormar, 1970a; Harter et al., 1971; Haase and Baringer, 1974). Purification is accomplished by cycles of isopycnic centrifugation in potassium tartrate or sucrose density gradients. Sonication and treatment with RNAase or DNAase have also been used in purification procedures (Lin and Thormar, 1970a). In gradients formed from sucrose solutions, the main visna virus band is located at a density of 1.15–1.16 g/ml. (Lin and Thormar, 1970a; Haase and Baringer, 1974). Using potassium tartrate solutions prepared in 0.1 M phosphate buffer, the major portion of virus infectivity was found at a density of 1.19 g/ml. (Harter and Choppin, 1967). A lower value (1.15–1.16 g/ml) has been obtained using potassium tartrate prepared in TNE buffer (0.01 M Tris, 0.1 M NaCl, 0.001 M EDTA) (Mountcastle et al., 1972). Electron microscopic studies of virus purified in this manner failed to disclose recognizable contaminants. Additional purification steps, such as rate zonal sedimentation in a sucrose density gradient or isoelectric focussing, did not change the polypeptide pattern from that obtained after isopycnic sucrose gradient centrifugation of the original virus preparation (Haase and Baringer, 1974).

3.5 Physico-chemical properties of the virus

Visna and maedi viruses are inactivated by ethyl ether, chloroform, metaperiodate, ethanol and phenol, as well as formaldehyde, oxidized spermine and trypsin (Thormar, 1965a; Kremzner and Harter, 1970).

The thermal stability of visna and maedi viruses are similar. If the virus is maintained in medium containing 1% sheep serum, 90% of infectivity is lost after 4 months at 4 °C, 9 days at 20 °C, 24–30 h at 37 °C and 10–15 min at 50 °C (Thormar, 1960, 1965a). Visna virus is rapidly inactivated at pH less than 4.2. Its infectivity is relatively stable at pH values between 5.1 and 10; it is most stable in a slightly alkaline pH (Thormar, 1960). Maedi virus appears more sensitive than visna virus to a low pH environment (Thormar, 1965a). Visna virus can be stored for months at −50 °C and can withstand several cycles of rapid freezing and thawing without appreciable loss of infectivity. The virus can withstand sonication (Thormar, 1966a), and is unaffected by exposure to RNAase or DNAase (Sigurdsson et al., 1960).

When exposed to ultraviolet (UV) irradiation, visna virus exhibits a degree of resistance similar to that found with avian oncornaviruses (Thormar, 1965a). Visible light is able to inactivate visna virus in the presence of toluidine blue (Thormar and Petersen, 1964); photo-inactivation of visna virus is relatively rapid and resembles that of vaccinia virus under the conditions employed.

Although erythrocytes from many different species have been tested, hemagglutination by visna and maedi virus has not been found, nor has hemadsorption been observed in virus–cell monolayers (Thormar, 1965a). Purified visna virions do appear capable, however, of inhibiting influenza virus hemagglutination in the same manner as has been observed with other viruses possessing neuraminic acid at their external surface (Compans, 1974; Compans and Harter, unpublished observations).

3.6 *Ultrastructure of visna–maedi virus*

Electron microscopic studies of visna–maedi virus infected sheep-derived tissue culture cells indicate the presence of two types of extracellular particles (Thormar, 1961; Coward et al., 1970) (Fig. 3.1). One is 65–100 nm in diameter and contains a 20–30 nm electron-dense core often surrounded by a membrane. At times, particles containing 2 or 3 cores may be seen. The other is larger (100–140 nm), lacks a central dense region and contains material which is similar in appearance to the cell cytoplasm. This form appears to originate from crescent-shaped budding structures at the cell membrane. It has not yet proven possible to separate or purify this structure.

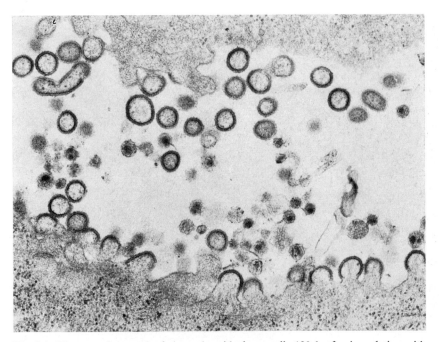

Fig. 3.1. Electron micrograph of sheep choroid plexus cells 120 h after inoculation with visna virus. Budding structures are seen arising from cell surface. Particles with electron-dense cores and core-less particles are present in the extra-cellular space. Preparation fixed in glutaraldehyde, post-fixed in osmium tetroxide, embedded in epoxy resin and stained with uranyl acetate and lead citrate. × 50,000 (Courtesy of Dr. J. E. Coward).

All of the structures observed in visna virus-infected cells (crescent forms, particles with an electron-dense core, and core-less particles) are antigenically related to visna virus as shown by indirect immunoferritin staining (Coward et al., 1972) using antiserum from a visna-afflicted sheep.

The smaller particle with the electron-dense central core appears to represent the infective virion because it is the only form recognized when material of high infectivity obtained after purification by density gradient centrifugation is examined by the thin-section technique (Coward et al., 1970; Fig. 3.2). Furthermore, it is these particles which attach to and penetrate newly inoculated tissue culture cells (Chippaux-Hyppolite et al., 1972). Although there is no direct evidence, it is tempting to believe that the larger, hollow particle may condense to form the smaller particle with the electron-dense core shortly after its release by budding from the cell surface. The outer membrane of both forms of particles appears to be 6 nm

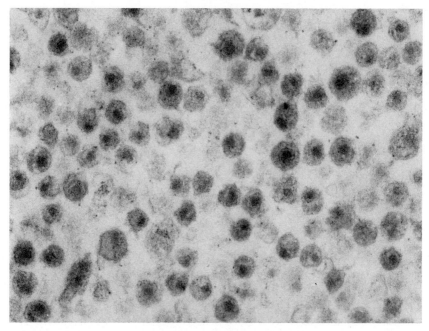

Fig. 3.2. Electron micrograph of visna virus particles purified by density gradient centrifugation. Preparation fixed in glutaraldehyde, post-fixed in osmium tetroxide, embedded in epoxy resin and stained with uranyl acetate and lead citrate. × 69,000 (Courtesy of Dr. J. E. Coward).

thick and to be composed of 2 layers (Chippaux-Hyppolyte, 1972) suggesting that it is derived from cell membrane.

When preparations of purified visna virus are negatively stained with phosphotungstate and examined with the electron microscope, pleomorphic spherical structures are seen varying in diameter from 900 to 1200 Å (Thormar and Cruickshank, 1965; Pautrat et al., 1971; Takemoto et al., 1973; Fig. 3.3). Projections 50–100 Å long appear to surround the outer membrane of the virus. These projections are not always readily resolved, possibly due to fragility during viral purification procedures. An internal membrane and a coiled structure have also been observed within negatively-stained particles.

Negative staining of visna virus particles disrupted under controlled conditions have indicated a spherical nucleoid 800 Å in diameter within the viral envelope. The nucleoid was found to contain filamentous structures

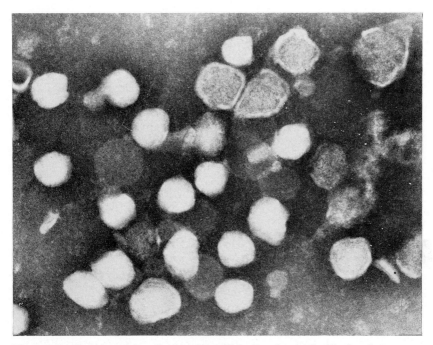

Fig. 3.3. Purified visna virions fixed in glutaraldehyde and stained with phosphotungstate. × 130,000 (Courtesy of Dr. R. W. Compans).

25 Å in diameter which appeared coiled into a nucleocapsid helix of 70–80 Å diameter (Pautrat et al., 1971).

3.7 Structural viral components

3.7.1 Nucleic acids (see also 5.2.3 and 4)

The predominant nucleic acid species present in visna virions appears to be a rapidly sedimenting (60–70 S), single-stranded RNA with a molecular weight of 10–12 × 10⁶ (Brahic et al., 1971; Lin and Thormar, 1971; Harter et al., 1971; Haase et al., 1974b). This high molecular weight RNA contains long stretches of polyadenylic acid (poly(A)) (Gillespie et al., 1973). The length of the poly(A) regions in 70 S visna virus RNA resembles that detected in the high molecular weight RNA of oncogenic viruses.

The high molecular weight RNA extracted from visna virus particles co-sediments with the high molecular weight RNA extracted from Rous sarcoma virus (RSV) (Haase et al., 1974b). The secondary structure of 70 S visna virus RNA is less stable than that of 70 S RSV RNA under conditions of reduced ionic strength. Complexity analysis of the 70 S RNA compared to that of poliovirus RNA reveals unique sequences of 7–10 $\times$ 10^6 daltons.

Heating 70 S visna virus RNA to 80 °C for 2 min releases subunits which co-migrate in polyacrylamide gel electrophoresis with the subunits of RSV RNA (Haase et al., 1974b). The molecular weight of these subunits is 2.8 $\times$ 10^6 indicating that there may be three or four subunits in the 70 S genome. Visna virus 70 S RNA can be completely dissociated into 35 S subunits by brief incubation at 37 °C in 2.5 $\times$ 10^{-4}M EDTA. Denaturation of 66 S visna RNA, prepared from virus harvested every 2 h, by heat, urea or dimethylsulfoxide (DMSO) produces a 36 S subunit (Brahic et al., 1973). When virus harvested after longer periods is used, denaturation results in heterogenous material sedimenting between 4 and 20 S. This finding suggests that degradation of the 66 S RNA occurs within the virus at 37 °C in the growth medium. Breaks in the native RNA molecule may be hidden by its secondary structure.

In rate zonal sedimentation studies of visna virus RNA, species sedimenting between 4 and 7 S have also been observed (Lin and Thormar, 1971; Harter et al., 1971). A fraction of this 4–7 S RNA resists RNAase digestion and appears to be composed of both single-stranded RNA and RNA–DNA hybrid complexes when analyzed in cesium sulfate gradients (Harter et al., 1971). The base ratio of the low molecular weight RNA fraction differs from that of 63–67 S RNA and of bulk RNA extracted from uninfected SCP cells (Lin and Thormar, 1971). In electrophoretic analysis, visna virions were found to contain only a single low molecular weight RNA species migrating at 4 S (Haase et al., 1974b); 5, 7, 18 and 28 S species like those present in RSV were not detected. A small homogenous peak of 4 S RNA comprising about 0.4% of the 70 S RNA was also found after DMSO treatment of 70 S visna virus RNA. It may serve as an RNA primer molecule for the viral DNA polymerase.

Visna virus particles disrupted by exposure to 0.05 or 0.1% sodium dodecyl sulfate (SDS) release an internal nucleic acid component in the form of rings or short curvilinear rods (Friedmann et al., 1974). Additional exposure to DMSO causes uncoiling of the rings and produces a heterogenous population of single, unbranched filaments up to 9.3 μm long which are similar in size to the strands observed in 60–70 S visna virus RNA recovered from glycerol velocity gradients. Longer exposure to DMSO (30 min), results in complete denaturing of the virus RNA into short fragments that average 3.2 μm in length. Regions of double-strandedness or reduplication intercalated in the long strands or coils were not found. It is possible that junctions between genomic subunits may remain unrecognized by electron microscopic methods which were utilized. These observations are in keeping with the results of biochemical studies and indicate that the visna virus genome consists of a molecule 9.3 μm long which is composed of subunits and assumes a coiled configuration within the virus particles.

The nucleic acids of maedi and progressive pneumonia viruses have also been analyzed (Stone et al., 1971b; Lin and Thormar, 1972). Virus-infected cells were incubated in the presence of [³H]uridine for 12–24 h or for as long as 14 days. Rapidly sedimenting (62–65 S) and low molecular weight (6–13 S) RNAs were extracted from purified virions. Furthermore, a prominent species sedimenting at 33–36 S which was susceptible to RNAase digestion was also recovered. Attempts to demonstrate a subunit structure in high molecular weight RNA from these viruses have not been reported.

3.7.2 DNA polymerase activity (see also 5.2.3 and 4)

Visna, maedi and progressive pneumonia virus particles contain DNA polymerase activities similar to those found in oncogenic RNA viruses and primate syncytial virus (Scolnick et al., 1970; Lin and Thormar, 1970b, 1972; Schlom et al., 1971; Stone et al., 1971a, b; Filippi et al., 1972). The DNA product of the polymerase reaction using the viral RNA as template has been shown to be DNA by its resistance to RNAase and alkali digestion, its sensitivity to DNAase and its density after centrifugation in cesium sulfate solutions (Schlom et al., 1971). *N*-Methyl isatin β-thiosemicarbazone added to the polymerase mixture interferes with enzyme activity (Haase and Levinson, 1973).

DNA produced by the visna virus polymerase reaction is synthesized in a two-step reaction with the initial transcription of single-stranded DNA bound to template RNA followed by the formation of double-stranded DNA (Haase et al., 1974a). The second reaction is blocked by actinomycin D. The double-stranded DNA product of the visna virus polymerase reaction, like that of the avian RNA tumor viruses, represents only 5–10% of the viral genome as determined by reassociation kinetics. The single-stranded DNA synthesized in the presence of actinomycin D contains sequences complementary to the entire visna virus genome as shown by the ability of complementary viral DNA in 6-fold excess to protect 70 S RNA from

RNAase digestion (Haase et al., 1974a). The visna virus polymerase reaction, therefore, resembles that of the oncogenic RNA viruses.

The DNA polymerase of visna virus has been separated into three enzymatically active polypeptides by column chromatography on DEAE–cellulose (Lin et al., 1973) The three enzymes are different in their pH optimum, sensitivity to *N*-ethylmaleimide, rate of catalytic reaction and template preference. Polymerase I, which has a molecular weight of approximately 125,000, appears similar to the DNA polymerase purified from oncornaviruses.

Visna virus polymerase activity is not inhibited by antisera directed against the polymerase of avian myeloblastosis virus (AMV), the Schmidt–Rupin strain of RSV or Rauscher murine leukemia virus (Nowinski et al., 1972; Parks et al., 1972).

3.7.3 Viral proteins

Initial gel electrophoretic studies of SDS-extracted radioactively labeled visna virions revealed 11–14 proteins, including at least two glycoproteins (Mountcastle et al., 1972). The carbohydrate-containing proteins could be removed by treatment of the particles with bromelain, suggesting that they were surface components of the virus.

Subsequent analysis of the proteins of visna, maedi, progressive pneumonia and zwoegerziekte viruses indicated that the proteins of the four viruses were similar (Haase and Baringer, 1974). Fifteen polypeptides ranging in molecular weight from 14,000 to 140,000 were found. One polypeptide was phosphorylated and three glycosylated. Nine of the visna virus proteins co-migrated with 13 RSV polypeptides. Like RSV, the bulk of the total visna virion protein was in small polypeptides and in a major 25,000–30,000 dalton polypeptide. The pattern of glycosylation in visna virus particles, however, differed from that found in RSV virions.

Visna virus proteins have also been analyzed by agarose gel column chromatography in 6 M guanidine hydrochloride (Lin and Thormar, 1974). Ten polypeptides, two of which were glycopeptides, were found. Nonidet-disrupted visna virions have been resolved into two fractions by potassium tartrate density gradient centrifugation (Lin and Thormar, 1974). The heavier fraction (1.24 g/ml) contained 50% of the viral RNA, most of the endogenous DNA polymerase activity, and a major internal polypeptide with an estimated molecular weight of 28,000. The lighter fraction (1.08 g/ml) contained all the glycopeptides, 50% of the viral RNA, a part of each of the other viral protein components, and very little endogenous DNA polymerase activity.

3.8 Antigenic properties (see also 2.6.2)

Neutralizing, complement-fixing and precipitating antibodies have been demonstrated in serums from sheep with natural or experimentally-induced visna or maedi (Sigurdsson et al., 1960; Gudnadóttir and Kristinsdóttir, 1967; Terpstra and de Boer, 1973). In addition, antibodies detected by the fluorescence antibody technique are formed in sheep shortly after infection with visna virus before neutralizing antibodies are detectable (Thormar, 1969). The relationship of the appearance of these antibodies to the slow evolution of the desease is not clear (Thormar and Pálsson, 1967). Neutralizing antibodies are present chiefly in the IgG_1 class; low activity can be associated with immunoglobin (Ig)M, but no significant activity is noted in the IgG_2 class (Mehta and Thormar, 1974; see also 6.2).

Visna virus antibodies have also been detected using a passive hemagglutination test with tanned sheep erythrocytes (Karl and Thormar, 1971). Hyperimmunization of rabbits with purified visna virus by standard procedures provokes passive hemagglutinating, complement-fixing and immunofluorescence antibodies, but neutralizing antibodies cannot be detected (Karl and Thormar, 1971; see also 5.5).

Visna virus is also neutralized by human and bovine serums (Thormar and Sigurdardóttir, 1962; Thormar and von Magnus, 1963). The factor responsible for this neutralizing effect may represent a heat-stable, nonspecific inhibitor rather than an immunologically reactive substance.

3.9 Intracellular events in viral replication (see also 5.2.3 and 4)

Relatively little is known about the site and mode of synthesis of virusspecific components in cells infected with visna or maedi virus. Using immunofluorescence staining, viral antigen is detected first in the perinuclear cytoplasm (Harter et al., 1967; Thormar, 1969). Cytoplasmic fluorescence appears shortly before the formation of newly infective virus and later becomes particularly brilliant at the cell surface. At all times, it remains confined to the cytoplasm. The cytoplasm of cells infected with visna virus shows an intense red fluorescence when stained with acridine orange; this appears related to an increased RNA content (Thormar, 1966b).

5-Bromodeoxyuridine effectively inhibits visna and maedi virus multipli-

D. H. Harter

cation when added 1–2 h after infection (Thormar, 1965b). Actinomycin D
added to visna-infected cell cultures as late as 24 h after inoculation also
interferes with virus multiplication (Thormar, 1965b). The effect of these
inhibitors resembles that observed in cells infected with RNA tumor viruses
(Temin, 1963; Bader, 1964, 1965), and suggested the possibility that a
DNA intermediate may be involved in viral replication. Such a virus-specific
DNA intermediate was recently demonstrated by Haase and Varmus (1973).
DNA recovered from visna virus-infected SCP cells, which had first partici-
pated in network formation, accelerated the reassociation of the predominantly double-stranded DNA synthesized by the endogenous visna
virus polymerase reaction. In addition, 70 S visna virus RNA annealed to
DNA extracted from visna-infected cells in DNA excess. Thus, the visna
virus genome appears to be integrated into the host cell genome as a DNA
copy in a manner similar to that observed with RNA tumor viruses (Markham and Baluda, 1973; Varmus et al., 1973).

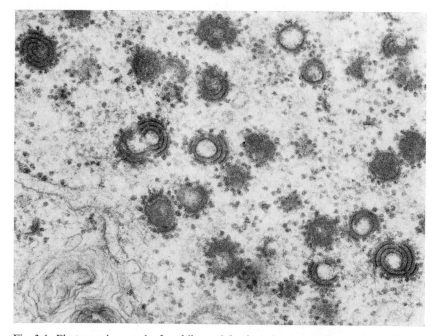

Fig. 3.4. Electron micrograph of multilayered, laminated structures in cytoplasm of visna
virus-infected sheep testes cells. Preparation fixed in glutaraldehyde, post-fixed in osmium
tetroxide, embedded in epoxy resin and stained with uranyl acetate and lead citrate.
× 66,000 (Courtesy of Dr. J. E. Coward).

Certain visna virus-infected cells contain intracytoplasmic multilayered spherical structures (Takemoto et al., 1971; Coward et al., 1972; Malmquist et al., 1972). This appears to be the only consistently observed intracellular ultrastructural alteration (other than the budding crescent-shaped surface forms) which has been recognized in visna virus-infected cells. The spiral layers are 13–14 nm thick and consist of a dense membranous component and a finely granular zone of lesser density (Fig. 3.4). The lamellae resemble the peripheral layer of buds arising from the cell membrane. At times, they are seen close to the plasma membrane; suggesting that some lamellar structures may differentiate at the cell surface. These structures are particularly striking in visna virus-infected sheep testes cells, but have also been observed in fetal sheep lung cells (Malmquist et al., 1972) and in bovine embryonic spleen cells infected with a visna-like syncytia-producing virus from cattle (Boothe and van der Maaten, 1974). The composition of these structures is still unknown.

3.10 Structural and serological similarities between visna, maedi and other related viruses

Visna and maedi viruses resemble one another in their mode of replication from infected tissue culture cells (Thormar, 1965a). A partial cross-reaction in neutralization studies with considerable quantitative strain variation has been found between visna and maedi viruses. All visna and maedi virus strains were neutralized by maedi antiserum, but some maedi strains were only partly neutralized by visna antiserums (Thormar and Helgadóttir, 1965). Visna and maedi viruses have identical antigens in gel diffusion and passive hemagglutination tests (Mehta and Thormar, 1975). Nucleic acid hybridization studies have also been unable to distinguish between visna and maedi virus nucleic acids (Harter et al., 1973).

Several other viral agents recovered from sheep with progressive interstitial pneumonia in various parts of the world also show serological cross-reactivity in neutralization tests with visna and maedi viruses (Thormar, 1966c). These include the progressive pneumonia of Montana sheep and zwoegerziekte, a well-described form of chronic pneumonia which occurs in Holland. The Montana isolate, progressive pneumonia virus, also cross-reacts with visna and maedi virus antiserums in immunofluorescence assays and resembles visna and maedi viruses in its fine structure and mode of development from tissue culture cells (Takemoto et al., 1971).

On the other hand, no serological cross-reactivity has been demonstrated

D. H. Harter

between visna and maedi viruses and several RNA tumor viruses. Anti-serums to Gross, Maloney and Friend murine leukemia virus, and to the Bryant strain of RSV fail to neutralize visna and maedi viruses (Thormar and Helgadóttir, 1965). Ether-treated visna virions do not give a precipitate in gel diffusion tests with group-specific antiserums prepared against avian leukosis–sarcoma, murine leukemia–sarcoma, hamster leukemia–sarcoma, feline leukemia, mouse mammary tumor or simian mammary tumor viruses (Nowinski et al., 1971). Furthermore, there appears to be no significant sequence homology between the nucleic acids of Rauscher murine leukemia virus and mouse mammary tumor virus and those of visna and maedi viruses (Harter et al., 1973). Therefore, although visna and maedi viruses may well utilize the same mode of intracellular replication as the oncogenic RNA viruses, they are not serologically related and their nucleic acid genomes appear individual and distinct.

References

BADER, J. P. (1964) Virology, 22, 462.

BADER, J. P. (1965) Virology, 26, 253.

BOOTHE, A. D. and VAN DER MAATEN, M. J. (1974) J. Virol., 13, 197.

BRAHIC, M., TAMALET, J. and CHIPPAUX-HYPPOLITE, C. (1971) C. R. Acad. Sci. (Paris), 272, 2115.

BRAHIC, M., TAMALET, J., FILIPPI, P. and DELBECCHI, L. (1973) Biochimie, 55, 885.

CHIPPAUX-HYPPOLITE, C., TARANGER, C., TAMALET, J., PAUTRAT, G. and BRAHIC, M. (1972) Ann. Inst. Pasteur, 123, 409.

COMPANS, R. W. (1974) J. Virol., 14, 1307.

COWARD, J. E., HARTER, D. H. and MORGAN, C. (1970) Virology, 40, 1030.

COWARD, J. E., HARTER, D. H., HSU, K. C. and MORGAN, C. (1972) Virology, 50, 925.

FILIPPI, P., BRAHIC, M., TAMALET, J. and DELBECCHI, L. (1972) C. R. Acad. Sci. (Paris), 275, 1567.

FRIEDMANN, A., COWARD, J. E., HARTER, D. H., LIPSET, J. S. and MORGAN, C. (1974) J. Gen. Virol., 25, 93.

FUCILLO, D. A., KURENT, J. E. and SEVER, J. L. (1974) Annu. Rev. Microbiol., 28, 231.

GILLESPIE, D., TAKEMOTO, K. K., ROBERT, M. and GALLO, R. C. (1973) Science, 179, 1328.

GUDNADÓTTIR, M. and KRISTINSDÓTTIR, K. (1967) J. Immunol., 98, 663.

GUDNADÓTTIR, M. and PÁLSSON, P. A. (1967) J. Infect. Dis., 117, 1.

HAASE, A. T. and LEVINSON, W. (1973) Biochem. Biophys. Res. Commun., 51, 875.

HAASE, A. T. and VARMUS, H. E. (1973) Nature, 254, 237.

HAASE, A. T. and BARINGER, J. R. (1974) Virology, 57, 238.

HAASE, A. T., GARAPIN, A. C., FARAS, A. J., VARMUS, H. E. and BISHOP, J. M. (1974a) Virology, 57, 251.

HAASE, A. T., GARAPIN, A. C., FARAS, A. J., TAYLOR, J. M. and BISHOP, J. M. (1974b) Virology, 57, 259.

HARTER, D. H. (1969) J. Gen. Virol., 5, 157.

HARTER, D. H. and CHOPPIN, P. W. (1967) Virology, 31, 176.

HARTER, D. H., HSU, K. C. and ROSE, H. M. (1967) J. Virol., 1, 1265.

HARTER, D. H., SCHLOM, J. and SPIEGELMAN, S. (1971) Biochim. Biophys. Acta, 240, 435.

HARTER, D. H., AXEL, R., BURNY, A., GULATI, S., SCHLOM, J. and SPIEGELMAN, S. (1973) Virology, 52, 287.

KARL, S. C. and THORMAR, H. (1971) Infect. Immun., 4, 715.

KREMZNER, L. T. and HARTER, D. H. (1970) Biochem. Pharmacol., 19, 2541.

LIN, F. H., GENOVESE, M. and THORMAR, H. (1973) Prep. Biochem., 3, 525.

LIN, F. H. and THORMAR, H. (1970a) Virology, 42, 1140.

LIN, F. H. and THORMAR, H. (1970b) J. Virol., 6, 702.

LIN, F. H. and THORMAR, H. (1971) J. Virol., 7, 582.

LIN, F. H. and THORMAR, H. (1972) J. Virol., 10, 228.

LIN, F. H. and THORMAR, H. (1974) J. Virol., 14, 782.

LOPEZ, C., EKLUND, C. M. and HADLOW, W. J. (1971) Proc. Soc. Exp. Biol. Med., 138, 1035.

MALMQUIST, W. A., KRAUSS, H. H., MOULTON, J. E. and WANDERA, J. G. (1972) Lab. Invest., 26, 528.

MARKHAM, P. D. and BALUDA, M. A. (1973) J. Virol., 12, 721.

MEHTA, P. D. and THORMAR, H. (1974) Infect. Immun., 10, 678.

MEHTA, P. D. and THORMAR, H. (1975) Infect. Immun., 11, 829.

MOUNTCASTLE, W. E., HARTER, D. H. and CHOPPIN, P. W. (1972) Virology, 47, 542.

NOWINSKI, R. C., EDYNAK, E. and SARKAR, N. H. (1971) Proc. Natl. Acad. Sci. U.S.A., 68, 1608.

NOWINSKI, R. C., WATSON, K. T., YANIV, A. and SPIEGELMAN, S. (1972) J. Virol., 10, 959.

PARKS, W. P., SCOLNICK, E. M., ROSS, J., TODARO, G. J. and AARONSON, S. A. (1972) J. Virol., 9, 110.

PAUTRAT, G., TAMALET, J., CHIPPAUX-HYPPOLITE, C. and BRAHIC, M. (1971) C. R. Acad. Sci. (Paris), 273, 653.

SCHLOM, J., HARTER, D. H., BURNY, A. and SPIEGELMAN, S. (1971) Proc. Natl. Acad. Sci. U.S.A., 68, 182.

SCOLNICK, E., RANDS, E., AARONSON, S. A. and TODARO, G. J. (1970) Proc. Natl. Acad. Sci. U.S.A. 67, 1789.

SIGURDARDÓTTIR, B. and THORMAR, H. (1964) J. Infect. Dis., 114, 55.

SIGURDSSON, B., THORMAR, H. and PÁLSSON, P. A. (1960) Arch. Ges. Virusforsch., 10, 368.

STONE, L. B., SCOLNICK, E., TAKEMOTO, K. K. and AARONSON, S. A. (1971a) Nature, 229, 257.

STONE, L. B., TAKEMOTO, K. K. and MARTIN, M. A. (1971b) J. Virol., 8, 573.

TAKEMOTO, K. K., MATTERN, C. F. T., STONE, L. B., COE, J. F. and LAVELLE, G. (1971) J. Virol., 7, 301.

TAKEMOTO, K. K., AOKI, T., GARON, C. and STURM, M. M. (1973) J. Natl. Cancer Inst., 50, 543.

TEMIN, H. (1963) Virology, 20, 577.

TERPSTRA, C. and DE BOER, G. F. (1973) Arch. Ges. Virusforsch., 43, 53.

THORMAR, H. (1960) Arch. Ges. Virusforsch., 10, 501.

THORMAR, H. (1961) Virology, 14, 463.

THORMAR, H. (1963a) Virology, 19, 273.

THORMAR, H. (1963b) J. Immunol., 90, 185.

THORMAR, H. (1965a) Res. Vet. Sci., 6, 117.

THORMAR, H. (1965b) Virology, 26, 36.

THORMAR, H. (1966a) In: (D. C. Gajdusek, C. J. Gibbs, Jr, and M. Alpers, Eds) Natl. Inst. Neurol. Dis. Blindness, Monogr. 2. Slow, Latent and Temperate Virus Infections. (U.S. Government Printing Office, Washington, D.C.) p. 335.

THORMAR, H. (1966b) Acta Pathol. Microbiol. Scand., 68, 54.

THORMAR, H. (1966c) In: (L. Severi, Ed.) Proc. Int. Conf. on Lung Tumours in Animals. (Perugia, Italy) p. 393.

THORMAR, H. (1969) Acta Pathol. Microbiol. Scand., 75, 296.

THORMAR, H. and CRUICKSHANK, J. G. (1965) Virology, 25, 145.

THORMAR, H. and HELGADÓTTIR, H. (1965) Res. Vet. Sci., 6, 456.

THORMAR, H. and MAGNUS, H. VON (1963) Acta Pathol. Microbiol. Scand., 57, 261.

THORMAR, H. and PÁLSSON, P. A. (1967) Perspect. Virol., 5, 291.

THORMAR, H. and PETERSEN, I. (1964) Acta Pathol. Microbiol. Scand., 62, 461.

THORMAR, H. and SIGURDADÓTTIR, B. (1962) Acta Pathol. Microbiol. Scand., 55, 180.

CHAPTER 4

The pathology of visna and maedi in sheep

Gudmundur GEORGSSON, Neal NATHANSON,
Páll A. PÁLSSON and Gudmundur PÉTURSSON

4.1 The pathology of visna

4.1.1 Introduction

The following description of visna will be confined to pathological changes observed in Icelandic sheep, since it is in this breed that the disease has been studied most extensively. Successful transmission of visna to other breeds of sheep has only been reported twice, and in both instances the experiments were done on a limited number of sheep (Pette et al., 1961; Narayan et al., 1974).

The first description of the pathology of visna, both natural and transmitted cases, was published in 1957 by Sigurdsson et al.; in the following year Sigurdsson and Pálsson (1958) added further details. This was followed in 1962 by the most extensive and detailed account of the pathology of visna (Sigurdsson et al., 1962). The following description will be based on these reports, especially the last one, as well as on our own cases from hitherto unpublished experiments. The histopathology of natural and experimental cases will be described together as, according to Sigurdsson et al. (1957, 1962), there is apparently no difference between the two. It should, however, be mentioned that the vast majority of examined cases

Original material presented in this report is drawn from a collaborative study between the Institute for Experimental Pathology, University of Iceland and the Johns Hopkins University, and is supported in part by Grant NS 11451 from the USPHS.

Slow virus diseases of animals and man, edited by R. H. Kimberlin
© *North-Holland Publishing Company 1976*

are from transmission experiments. This is simply due to the fact that visna
was eradicated in Iceland in 1951 when studies of this disease were in their
infancy (see 2.10).

Our cases are from recent experiments designed to study the pathogenesis
of visna which were started in the spring of 1973. The sheep were infected
by intracerebral inoculation by the method described by Sigurdsson et al.
(1957) using 0.3 ml of visna virus (about $10^{5.5}$ TCID$_{50}$) grown in sheep choroid
plexus cultures. In the first experiment, which was done to test the neuro-
virulence of a virus isolate, 5 sheep were inoculated with virus strain K1010.
This was a spinal fluid isolate from a sheep with clinical signs of visna that
had been infected with the strain K796. One of the group of 5 sheep, K1514,
developed acute clinical disease in about 6 weeks and was sacrificed one
week later. Virus isolated from the choroid plexus of this sheep was passed
twice in sheep choroid plexus cells and harvested and used for the subsequent experiments.

So far 26 sheep from these experiments have been sacrificed at intervals
from one month to 1½ years after inoculation. The left half of the brain
was processed for histological examination and 9 blocks were cut from each
brain, at standard levels, with some modification, based on an atlas (Yoshikawa, 1968). Specimens were also taken from three different levels of the
spinal cord, i.e. lower part of the cervical cord and middle of thoracic and
lumbar cord, and from the sciatic nerve on both sides and optic nerve at its
emergence from the ocular bulb on one side. Furthermore we examined
sections from internal organs and skeletal muscle in all cases. In 19 of the
26 cases definite histological lesions have been found in the central nervous
system and the results of the pathological examination in these cases will be
included in the following description. We will also present some observations
on ultrastructural changes, but these must be regarded as preliminary
findings as the analysis of the material embedded for electron microscopy
is in its initial stage.

4.1.2 Histopathology

4.1.2.1 Internal organs and skeletal muscle

The histological examination of our experimental cases has not revealed
any microscopic lesions in thymus, thyroid, adrenal, pancreas and small
intestine.

The mediastinal and mesenteric lymph nodes often showed non-specific
reactive hyperplasia, usually a follicular hyperplasia with active germinal

centers and/or histiocytic proliferation commonly accompanied by infiltration of eosinophilic leucocytes of varying degree. In the spleen we sometimes found hyperplasia of the Malphigian corpuscles with active germinal centers. However, similar changes were seen in the lymphoid tissue of uninoculated controls. We did not find any signs of a generalized reticuloendotheliosis in the lymphoid tissue, as observed by Pette et al. (1961) in German sheep experimentally infected with visna.

The kidneys often showed a discrete proliferative glomerulonephritis, which is a very common finding in apparently healthy Icelandic sheep as described in other breeds (Lerner and Dixon, 1966).

In the liver occasionally a slight periportal inflammation was found and in the cardiac and skeletal muscle sarcosporidia were regularly found and occasionally a mild inflammatory reaction. Such changes are commonly found in uninfected control sheep.

In the lungs we always found a mild inflammatory reaction which probably represented the early lesions of maedi (see 4.4.2), especially since visna virus was grown from the lungs in 58% of the cases tested. However, we would hesitate to classify all these slight changes as early maedi, because we always found some parasitic lesions in the lungs and these parasites, besides causing a nodular inflammation, apparently can provoke some interstitial inflammation in the adjacent tissue.

4.1.2.2 The nervous system

Meninges In our studies the leptomeninges of the brain always showed inflammation, whereas the spinal leptomeninges were spared in about 1/3 of the cases. Marked inflammatory reaction as shown in Fig. 4.1 is seldom encountered but tends to be located over the superior frontal gyrus, hippocampic fissure, lobus pyriformis, occipital lobe and cerebellar lingula. The inflammatory reaction of the spinal leptomeninges is usually mild but when marked there usually is an accentuation of the inflammatory process over the anterior median fissure (Fig. 4.6), posterior median sulcus and around the nerve roots, especially the posterior ones.

The inflammatory infiltrates of the meninges are composed mainly of lymphocytes and monocytes and/or macrophages but some plasma cells are also found, whereas polymorphonuclear leucocytes are not seen. A cytological study of the cells of the cerebrospinal fluid in a few cases with marked pleocytosis, showed a preponderance of macrophages with relatively few typical small lymphocytes (Fig. 4.2) and rare plasma cells. The macro-

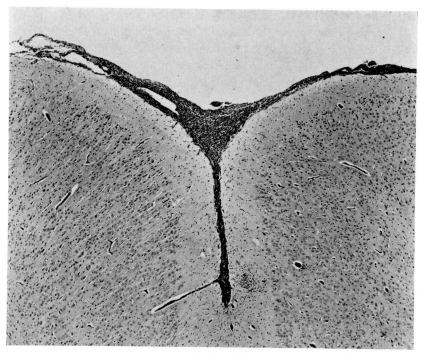

Fig. 4.1. Marked meningitis over the cerebral cortex. The inflammation extends into the cortex at the bottom of the sulcus. A small glial nodule is present in the molecular layer. Gallocyanin–Eosin. × 40.

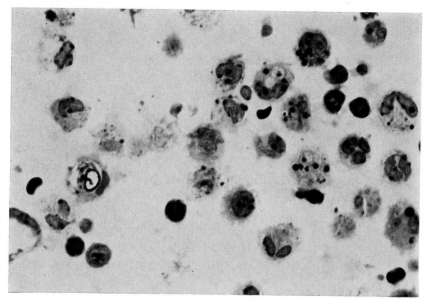

Fig. 4.2. Cells of the cerebrospinal fluid. Macrophages containing osmiophilic inclusions predominate, but small typical lymphocytes are also present. Epon, 0.75 μ section. Toluidine blue. × 1050.

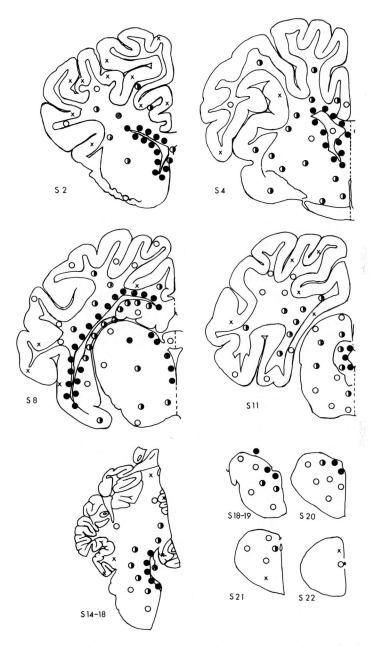

Fig. 4.3. Schematic representation of the frequency of lesions in different parts of the brain: ●, lesions found in 2/3 or more of the cases; ◑, lesions found in 1/3–2/3 of cases; ○, lesions found in less than 1/3 of cases; ×, solitary observation. The lesions of the choroid plexus of the lateral and third ventricles are represented by the sign inside the lateral ventricle in plane S4 and those of the fourth ventricle by the sign above plane S18-19. The S numbers refer to planes in the atlas by Yoshikawa (1968). For further details see text.

phages frequently contained ingested lipid droplets, possibly products of myelin breakdown, and often phagocytized cells or cell remnants, for example, lymphocytes. Mitotic figures were occasionally observed.

4.1.2.3 Brain and spinal cord

Topography of the lesions The schematic representation of brain lesions (Fig. 4.3) is based upon 19 cases sacrificed 1–6 months after infection. It shows the frequency of lesions in various areas but not the severity of the

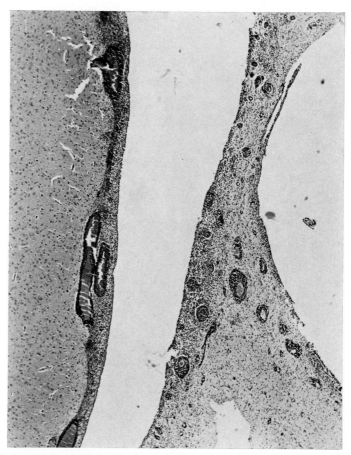

Fig. 4.4. Lateral ventricle. Confluent subependymal inflammation. Perivascular infiltrates extend into adjacent tissue. Caudate nucleus on the left, septum pellucidum on the right. Gallocyanin–Eosin. × 40.

process, although this usually coincides. This scheme is simplified so that lesions of the various nuclei as well as of the white matter are summarized in one or just a few signs. For example, the sign in the middle of the caudate nucleus at the S2 level represents lesions widely spread within this nucleus in this plane of section. Occasional cases showing extensive almost confluent lesions in the radiatio corporis callosi at the S2 level are included in the two signs drawn in this area. Lesions of the white matter of the cerebral gyri are generally represented by signs at the bases of the gyri, although the lesions are occasionally confluent or situated more peripherally.

Fig. 4.3 shows clearly that the sites of predilection for lesions are along the ventricular system. The frequency of lesions was similar in various periventricular sites. The relatively common involvement of the white matter of the occipital lobe is most likely due to extension of subependymal inflammation of the posterior horn of the lateral ventricle which was sometimes included in plane S11. Periventricular changes were always present although the extent and severity of the lesions varied from an almost confluent subependymal inflammation (Fig. 4.4) surrounding the entire ventricular system and extending into the spinal cord around the central canal, to only a few, scattered small foci of subependymal inflammation.

The choroid plexus of the lateral, third and fourth ventricles was frequently inflamed. There was apparently some correlation between the severity and extent of subependymal inflammation and the intensity of the inflammatory process in the choroid plexus.

The inflammation apparently spreads from the periventricular localization to the adjacent cerebral white and grey matter, thus the deep part of the cerebral white matter is more frequently involved than the white matter of the cerebral gyri (Fig. 4.3). The severity of the lesions of the deep cerebral white matter varied from a few, scattered, discrete perivascular infiltrates and/or small glial nodules to extensive confluent lesions, especially in the frontal lobe. In many cases the severe lesions accompanied an intense and extensive periventricular inflammation. Lesions were frequently found in the corpus callosum. The internal capsule was affected in half of the cases usually with few small foci but in one case multiple disseminated foci were observed (Fig. 4.5).

Lesions were commonly found in the basal ganglia, especially in the caudate nucleus and with decreasing frequency in the putamen and globus pallidus. The lesions in the basal ganglia were always in the form of small scattered foci. On the other hand, in the thalamus, more extensive inflammatory and sometimes necrotic foci were found, especially in the

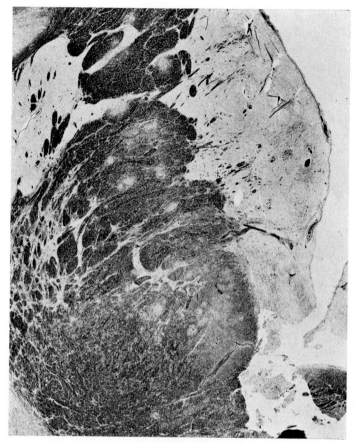

Fig. 4.5. Multiple, disseminated, demyelinated lesions in the internal capsule. Lateral ventricle on the right, section through caudate nucleus. Klüver–Barrera. × 10.

anterior thalamus often extending caudally into the pulvinar. The frequency and extent of lesions in this part of the thalamus is apparently related to the site of inoculation.

The cerebral cortex is seldom affected except by extension from overlying meningitis (Fig. 4.1) or from lesions in the underlying white matter. The cortical lesions are usually small, discrete and are irregularly distributed (Fig. 4.3). The amygdaloid was most frequently affected.

Lesions in the mesencephalon were most frequently found subependy-mally, extending to the peri-aqueductal grey matter, but occasionally more extensive, partly necrotic foci were noted either in the tectum or tegmentum.

In the cerebellum, pons and medulla the lesions were also most common subependymally, around the fourth ventricle whence they extended in an irregular fashion into the adjacent white matter. Again the lesions consisted of small inflammatory foci but occasionally larger areas with intensive inflammation sometimes with necrosis were observed in the cerebellar white matter. Only solitary lesions were found in the cerebellar cortex. The topography of the lesions in the cerebrum, brainstem and cerebellum in our series is very similar to the localization as described by Sigurdsson et al. (1962).

Lesions were found in the spinal cord in about 2/3 of our cases. The spinal cord was never involved unless the brain was also affected. When present, lesions were nearly always found at all three different levels examined, but were not always continuous. The topography of the histological changes in our series was of the so-called 'peri-ependymo-poliomyelitic' type (Sigurdsson et al., 1962). Subependymal lesions around the central canal were

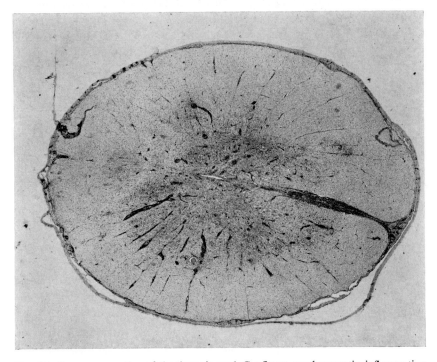

Fig. 4.6. Transverse section of the thoracic cord. Confluent, partly necrotic, inflammation of the entire grey matter extending into the white columns. Meningitis, especially prominent in the anterior median fissure (to the right). Gallocyanin–Eosin. × 14.

always present and usually lesions also involved the commissural grey matter and spread from there for a variable distance into the adjacent grey matter. Sometimes confluent inflammation of the entire grey matter was observed (Fig. 4.6). Commonly, inflammatory cells were seen migrating through the ependyma, which was occasionally disrupted, and in the central canal cellular and proteinaceous exudate was often found. Only in 3 of our cases were histological changes present in the white columns of the spinal cord (Fig. 4.6).

These findings are in contrast to those of Sigurdsson et al. (1962) who found prominent lesions in the white columns. Sometimes wedge or square-shaped areas were seen by these authors in the anterior column with a 'center of honeycomb' appearance bordered by inflammatory infiltrates. In other cases the whole hemi-cord was affected by confluent inflammatory foci apparently radiating from the central canal. Still another pattern of this focal or multifocal type consisted of rather small foci of varying size involving symmetrical regions of the spinal cord without respecting the limits between the grey and the white matter.

Character and evolution of the lesions The initial lesions of the neural

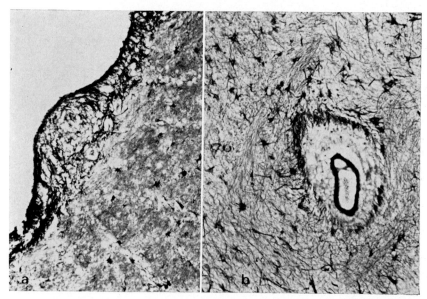

Fig. 4.7. (a) Subependymal inflammatory nodule with marked astrocytic proliferation bulges into the ventricle. Cajal. × 105. (b) Dense mat of astrocytic fibers surrounding a perivascular infiltrate. A distinct proliferation of astrocytes in the adjacent tissue. Cajal. × 105.

parenchyma are apparently small subependymal inflammatory foci, usually around a venule or a small vein. The overlying ependyma is often infiltrated by inflammatory cells and sometimes disrupted. These nodules are mainly composed of lymphocytes of varying size, monocytes and/or macrophages with plasma cells in variable numbers. Glial cells also participate in the formation of the foci, principally astrocytes (Fig. 4.7a) and microglia to a lesser degree. These subependymal foci then increase in number and often coalesce to form confluent subependymal inflammation (Fig. 4.4). With increasing severity of the subependymal inflammation, perivascular infiltrates are seen to spread into adjacent white or grey matter (Fig. 4.4).

The perivascular infiltrates may vary from only a few cells to broad cuffs composed of many layers of inflammatory cells (Fig. 4.8), and are frequently surrounded by a dense mat of astrocytic fibers (Fig. 4.7b). They are composed of variously sized lymphocytes, monocytes and/or macrophages and

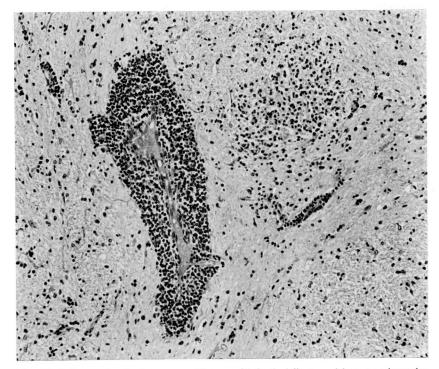

Fig. 4.8. Inflammation in the corpus striatum with both delicate and heavy perivascular cuffs. A small glial nodule on the upper right, with destruction of the white matter. Gallocyanin–Eosin. × 150.

plasma cells in varying numbers (Figs 4.9 and 10). Activated large lymphocytes with intense basophilia of the cytoplasm (Fig. 4.9) due to a rich content
of polyribosomes are often present (Figs 4.10 and 4.17). Mitotic figures are
common (Fig. 4.15). Cells staining positively with the Hortega method for
microglia are quite numerous. The endothelial cells are sometimes swollen
and there may be evidence of increased permeability with proteinaceous
exudate and sometimes discrete extravasation of erythrocytes into the perivascular space. The perivascular infiltrates, although often seen to respect
the glial membrane, may disrupt it and inflammatory cells then migrate into
the adjoining parenchyma. This is usually accompanied by glial reaction.
Glial nodules with occasional microcavitation and/or necrosis, are part of
the histological reaction in visna. They are usually small and irregularly
distributed both in grey and white matter (Figs 4.1 and 4.8) and often occur
apparently independent of perivascular inflammatory infiltrates.

With increasing severity of the inflammatory reaction the perivascular
infiltrates increase in size and number. Invasion of adjacent parenchyma by

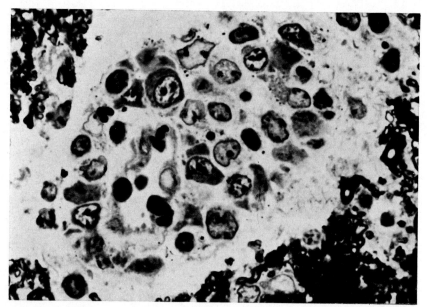

Fig. 4.9. Perivascular infiltrate. In the middle a venule with cells migrating through the
endothelium (lymphocytes). The infiltrate is composed of different types of lymphocytes,
monocytes, and/or macrophages and an occasional plasma cell. Epon. 0.75 μ. Toluidine
blue. × 1280.

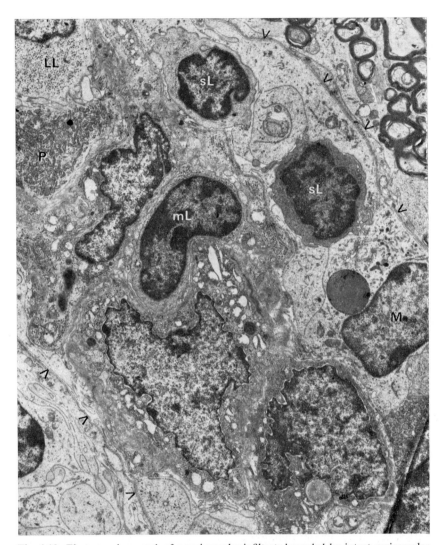

Fig. 4.10. Electron micrograph of a perivascular infiltrate bounded by intact perivascular glia limitans (arrowheads). The vessel is tangentially cut and the lumen not visible. ML, lymphocyte migrating through the endothelium; SL, small lymphocytes; LL, large lymphocyte with abundant polyribosomes; P, plasma cell; M, macrophage with lipid inclusion. Myelin and axons on the upper right well preserved. × 5200.

inflammatory cells becomes more prominent as do the concomitant glial nodules. The inflammatory foci then coalesce to form variable areas of semiconfluent or confluent infiltration (Fig. 4.11). The tissue in these areas

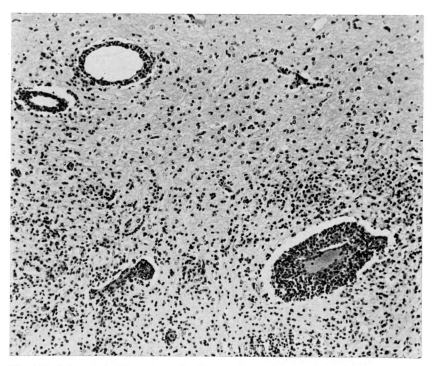

Fig. 4.11. Subcortical white matter. Confluent inflammation, with diffuse infiltration of the parenchyma accompanied by a marked glial reaction. Gallocyanin–Eosin. × 150.

may become oedematous and gradually liquefy. Thus finally liquefaction necrosis takes place centrally in areas with intense inflammatory reaction (Fig. 4.12). Necrotic lesions are preferentially found in the white matter and are sometimes very extensive and may lead to porencephaly after the necrotic tissue has been resorbed, as shown by Sigurdsson et al. (1962) in their Fig. 10. Necrotic lesions sometimes occurred in the grey matter in our series, especially in the thalamus. In the necrotic lesions microglia and/or macrophages are often numerous, both diffusely distributed and with perivascular accumulation. Occasional small clusters of phagocytic cells containing Scharlach R and Oil Red 0-positive material are present in the necrotic area. The breakdown of myelin in the necrotic foci is always accompanied by destruction of axons (Fig. 4.13). The necrotic areas are sometimes bordered by proliferation of astrocytes (Fig. 4.14). The astrocytic response may extend for some distance into adjacent tissue, but gradually fades off. Diffuse astrocytic response independent of inflammatory or

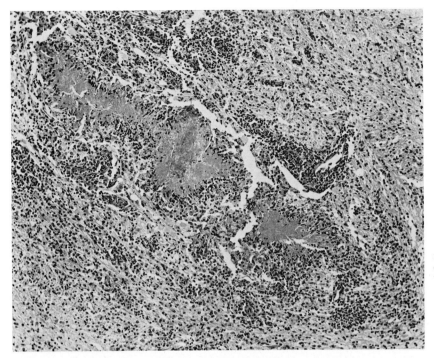

Fig. 4.12. Internal capsule. Intense inflammation with liquefaction necrosis and in the middle an irregular focus of coagulative necrosis bordered by cells arranged in a palisading fashion. Gallocyanin–Eosin. × 100.

necrotic lesions, as characteristically seen in scrapie (see Ch. 12), was not observed.

Besides liquefaction necrosis, small irregular foci of coagulative necrosis were found in two cases of our series. In one case only a solitary lesion was noted in the internal capsule, whereas in the other case multiple lesions were found, mainly in the internal capsule but also in the thalamus and cerebellar white matter. These lesions were usually found within areas with intense inflammatory reaction and/or liquefaction necrosis. They were bordered by astrocytes and macrophages arranged in a palisading fashion (Fig. 4.12).

Destruction of myelin seems to be a late event in the evolution of lesions. As mentioned above, the perivascular inflammatory infiltrates often respect the glial membrane. When they invade the adjacent white matter the inflammatory cells together with some proteinaceous exudate spread between the nerve fibers and in the early stages usually leave the axons with their

G. Georgsson et al.

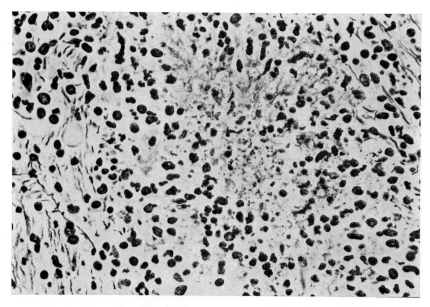

Fig. 4.13. Necrotic lesion with fragmentation and destruction of axons. Bodian. × 420.

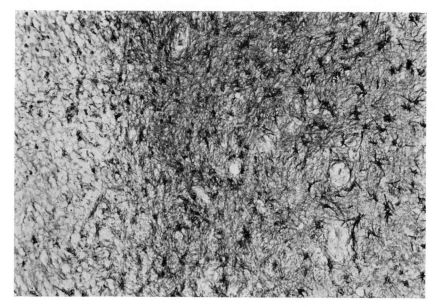

Fig. 4.14. Necrotic area (on the left) bordered by proliferated astrocytes, which form a dark zone next to the necrosis. Cajal. × 105.

myelin sheaths intact (Figs 4.10, 4.15 and 4.16). Myelinated nerve fibers with only minimal changes are sometimes observed running through inflammatory foci with oedema and incipient necrosis (Fig. 4.17). Accordingly, cells containing products of myelin breakdown are inconspicuous in the inflammation. The dissociation of myelinated nerve fibers by inflammatory exudate causes a rarifaction of the myelin (Fig. 4.15). Complete destruction of myelin is only observed in necrotic areas (Fig. 4.5).

Inflammation of the choroid plexus varies from a discrete sparse infiltration of lymphocytes, monocytes and/or macrophages intermingled with plasma cells to an intense inflammatory reaction often with massive proliferation of lymphoid tissue with active germinal centers (Fig. 4.18). Migration of inflammatory cells through the usually intact surface lining of the choroid plexus is frequently observed.

The nerve cells are seldom affected and then probably indirectly, perhaps by an ischaemic or toxic mechanism. Shrinkage and sclerosis of nerve cell bodies or swelling and/or chromatolysis, sometimes accompanied by satellitosis, is occasionally noted in inflammatory foci but may also occur independently of lesions. Complete destruction of nerve cells with neuro-

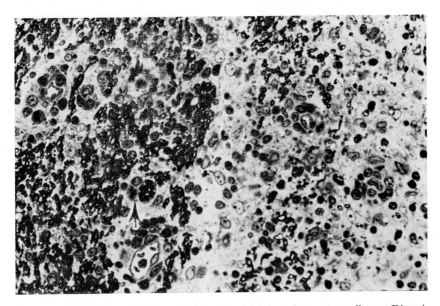

Fig. 4.15. Subependymal inflammation (on the right) invading corpus callosum. Dissociation of myelinated fibers by the cellular infiltrate with minimal breakdown of myelin. Note mitotic figure (arrow). Epon. 0.75 μ. Toluidine blue. × 450.

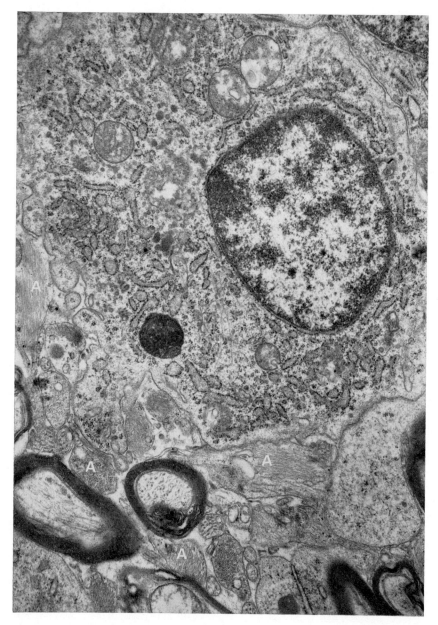

Fig. 4.16. Electron micrograph of a plasma cell in close contact with myelinated nerve fibers. Numerous astrocytic processes (A) with abundant microfilaments. × 22.000.

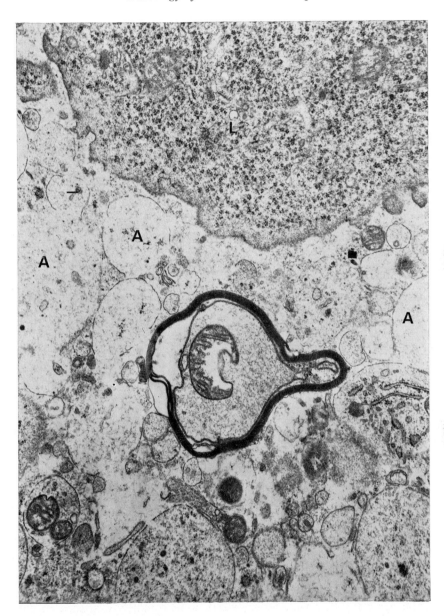

Fig. 4.17. Electron micrograph of a myelinated nerve fiber with intramyelinic oedema running through an area with oedema and incipient disintegration of the tissue. L, part of a large lymphocyte with abundant polyribosomes; A, swollen and partly disintegrating astrocytic processes. × 16,400.

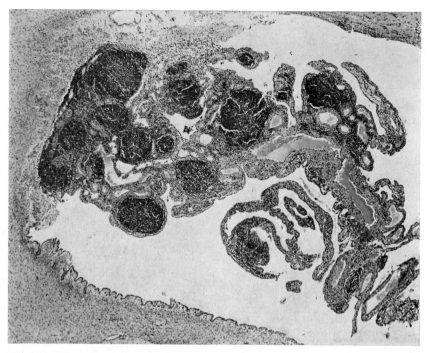

Fig. 4.18. Massive nodular lymphoid proliferation in the choroid plexus of the lateral ventricle. Gallocyanin–Eosin. × 40.

phagia was rarely observed. In general, neurons were well preserved, often remarkably so, even in areas with severe inflammation.

With the exception of swelling of endothelial cells of capillaries and venules commonly observed in inflammatory foci, the vessels rarely show changes. Occasionally slight intimal thickening in veins or arteries is present.

4.1.2.4 Peripheral nervous system

The peripheral nervous system was not affected in our series. Thus, we did not find inflammation in the nerve roots or spinal ganglia, whereas Sigurdsson et al. (1962) describe involvement of these structures by continuous extension of inflammatory infiltrates from the adjacent meninges. The sciatic nerves, optic nerve together with retina, which were routinely examined in our series, never showed lesions. But Sigurdsson et al. (1962) found isolated inflammatory foci in peripheral nerves at some distance from the nerve roots. In one of their cases granulomatous and diffuse interstitial inflammation composed of lymphocytes and plasma cells accompanied by degeneration of myelinated nerve fibers was found. But our observations suggest that involvement of the peripheral nervous system is probably rare.

4.1.3 Clinico-pathological correlation and progression of lesions

One of the outstanding features of visna is the long incubation period. In the field, clinical signs of visna were never observed in sheep under 2 years of age (Pálsson, 1972; see 2.2.2). In transmission experiments, clinical signs are usually first detected several months or years after intracerebral inoculation. The pathological lesions on the other hand may develop rather early in the course of infection.

Meningitis is apparently an early event, if not the first pathological lesion to develop in visna, as judged by pleocytosis of the cerebrospinal fluid, which has been observed as early as on the 15th day after infection (Sigurdsson et al., 1962). However, one must bear in mind that the pleocytosis of the cerebrospinal fluid does not only reflect infection of the meninges. Inflammation of the ependyma and of the choroid plexus, which commonly occurs in visna, can also contribute to the increased number of cells in the cerebrospinal fluid. In cases examined 16 and 23 days after infection both meningitis and subependymal inflammation was observed (Sigurdsson et al., 1962). According to Narayan et al. (1974) the initial lesion found 7 days after intracerebral inoculation of American foetal lambs is meningitis, whereas in experimentally infected newborn lambs examined after the same interval damage of the neural parenchyma was also observed. In our series, 8 out of 12 sheep examined 1 month after infection showed meningitis which was always accompanied by inflammation of the neural parenchyma.

The course of the meningitis is unpredictable according to both histological examination and cell counts of the cerebrospinal fluid. An initial rise in the number of cells may fall to normal levels which are then maintained (see Ch. 6, Fig. 6.1) or there may be repeated exacerbations as indicated by fluctuations in the cell count of the cerebrospinal fluid during the course of the disease (Gudnadóttir, 1974). In one of our experiments, groups of 4 sheep were examined histologically 1, 3 and 6 months after infection. The severity of the meningitis did not show any relation to the duration of infection.

In spite of the rather intensive inflammation found in the majority of the 19 histologically positive cases of our series examined 1–6 months after infection, only one sheep had shown clinical signs. The lack of overt clinical signs in spite of the distinct inflammatory reaction is easily explained by the predominant localization of the inflammation in the periventricular area, choroid plexus and meninges. Neurons and major fiber tracts are usually spared. Extension of the periventricular lesions may then affect the major fiber tracts and/or grey matter in cerebrum, cerebellum, brainstem and spinal cord and cause overt clinical signs, most frequently disturbance of motor function or cerebellar symptoms (see 2.2.2).

In our series, we did not find any relationship between the duration of the infection and severity of the lesions. Thus, the intensity of the inflammatory reaction observed in the group of sheep sacrificed 1 month after inoculation was comparable with the lesions found in the groups of sheep examined 3 and 6 months after inoculation. It appears that initially the pathological process develops quite rapidly, but then the tempo slows considerably. This finding is discussed further in Ch 6 (see 6.3.2). The results of our short-term experiments are in accord with the findings of Sigurdsson et al. (1962). They found typical lesions both with regard to severity and extent in sheep examined about 1 month after inoculation, whereas in cases of long standing, a meningo-ependymitis of average intensity was sometimes observed. This might be an expression of a different time of onset of pathological lesions, as indicated by the appearance of pleocytosis at variable intervals after infection, and also the remitting course of the disease, reflected in the often observed fluctuation of the pleocytosis and sometimes in the clinical course of the disease.

4.2 Comparison with visna-like diseases in other breeds of sheep and goats

Encephalomyelitis of sheep similar or apparently identical to visna has been reported from various countries (see 2.5.2). The detailed descriptions of the pathological lesions found in Dutch (Ressang et al., 1966) and German sheep (Schaltenbrand and Straub, 1972) as well as the short descriptions of an encephalitis observed in sheep concomitantly affected with maedi-like disease in Hungary (Süveges and Széky, 1973), India (Sharma, 1972) and Kenya (Wandera, 1970) are in accordance with the findings in visna of Icelandic sheep. Through the courtesy of Dr. R. Hoff-Jørgensen, who sent us histological sections, we have been able to confirm the resemblance of the lesions in sheep in Denmark to those of visna in the Icelandic sheep.

The studies by Pette et al. (1961) of experimental visna in German sheep showed a generalized enlargement of lymph nodes due to a diffuse and sinusoidal proliferation of reticulo-endothelial cells, also noted in the spleen. This together with reticulo-lymphocytic foci in the lungs was interpreted by these authors as evidence that the lesions in the brain in visna are a manifestation of a generalized reticulo-endotheliosis. The pathological findings in Icelandic sheep affected with visna do not support this view.

Narayan and co-workers (1974) found in transmission experiments lymphoid proliferation in the lungs and lymphoid hyperplasia in lymph nodes and spleen and suggested that visna virus infection, in addition to producing persistent infection of the brain, causes a lympho-proliferative disease; these authors also refer to the alleged oncogenic potential of the visna virus. Our pathological findings in visna of Icelandic sheep, as well as those of previous workers, do not support the idea that visna causes a malignant neoplasia of lymphoid tissue. Thus, we did not find a distinct enlargement of lymph nodes in visna. The histological changes were especially found in nodes draining organs that are frequently infected by either microorganisms or parasitic helminths, and were similar to reactive changes found in control sheep. According to the experience in Iceland, both with the extensive eradication programs and with experimental cases, there is no indication that visna causes tumours in Icelandic sheep (Pálsson, 1972).

Encephalomyelitis of goats similar to visna has been reported from Germany and the United States. The disease in German goats, called granulomatous encephalomyelitis, was first reported by Stavrou et al. (1969) and was later transmitted to goats with cell-free filtrates of organ suspensions from affected goats (Dahme et al., 1973). Serum antibodies to visna virus detected in affected goats, as well as the detection in tissue culture of a virus ultrastructurally very similar to visna virus (Weinhold, 1974), suggest that the causative agent is visna or a closely related virus. The pathological lesions in the CNS found in both spontaneously and experimentally infected goats resemble those of visna, especially the topographic distribution of lesions. The character of the lesions differs from those of visna in that the inflammatory foci in goats frequently showed a tendency to form granulomas with central coagulative necrosis, whereas this is very uncommon in visna-affected Icelandic sheep. This is possibly due to a difference in host response or to a biological difference in the causative agent.

The pathological lesions found in infectious leucoencephalomyelitis of young goats in the United States (Cork et al., 1974a, b) resemble the lesions of visna both with regard to the topography and character. Another interesting observation is the concomitant interstitial inflammation of the lungs accompanied by a pronounced lymphoid hyperplasia. Transmission experiments as well as failure to isolate bacteria or mycoplasma suggest a viral aetiology of the disease. But the causative agent has so far not been isolated or characterized and tests for serum antibodies to zwoegerziekte virus, a virus closely related if not identical to visna (de Boer, 1970b), were negative.

4.3 Comparative pathology of visna and diseases of the CNS of known or possible viral aetiology

The early inflammatory response in visna and the character of the evolving lesions resembles some diseases of the central nervous system in animals and humans of known or possible viral aetiology. On the other hand a distribution of lesions comparable to those of visna seems to be rare. As pointed out by Pette et al. (1961), the pathological lesions in visna – especially their topographic distribution – resemble those of a human disease known as primary reticuloendotheliosis of the brain or reticulohistiocytic granulomatous encephalitis (Cérvos-Navarro et al., 1960). A similar condition has been described in dogs under the term of granulomatous reticulosis (Fankhauser et al., 1972). The nature of these disorders is debated, but according to Fankhauser et al. they merge with the neoplasms of the sarcoma group. The aetiology is unknown, but there is some indication that the human disease may be caused by a virus (Iizuka and Spalke, 1972). A careful comparison of the pathological changes found in visna of Icelandic sheep with those of the above-mentioned conditions, reveals some significant differences both in character and distribution of lesions, that seem to distinguish visna from the group causing primary reticulosis of the brain. Thus, we do not find in visna of Icelandic sheep the characteristic reticulo-histiocytic cells in the inflammatory infiltrates as described in these disorders. Reticular fibers are sparse and glial nodules are common in visna in contrast to observations in the primary reticulosis of the brain. As far as the localization is concerned, the widespread diffuse subependymal distribution of inflammation commonly observed in visna is apparently not a feature of primary reticulosis of the brain.

Sigurdsson et al. (1957) drew attention to the similarity of visna with human demyelinating diseases. They pointed out that the intermittent progress often observed in visna was reminiscent of the course in multiple sclerosis, although the morphology of these diseases differed in that the inflammatory reaction dominated and preceded the demyelination in visna (Sigurdsson and Pálsson, 1958). The histological and ultrastructural findings in our transmission experiments support this concept. Demyelination is only found where an intense inflammatory reaction has progressed to necrosis and the axons are then destroyed together with myelin. Later, Sigurdsson et al. (1962) suggested that among the human demyelinating diseases the lesions found in post-infectious encephalitis particularly resemble those found in visna both with regard to the character and distribution of lesions. The inflammatory reaction in visna is, however, more intense, and confluent and perivenous sleeve-like demyelination is not observed in visna.

The striking resemblance of the lesions in visna to those of demyelinating canine distemper is interesting, although there are some differences in the character and distribution of lesions (Innes and Saunders, 1962). In this context it should be recorded that no inclusion bodies or giant cells are seen in visna-infected brains. This form of canine distemper has often been referred to as an animal model for acute multiple sclerosis and/or post-infectious perivenous encephalitis. Frauchiger and Fankhauser (1970), however, have questioned this view and pointed out, for example, some fundamental morphological differences between canine distemper and these human demyelinating diseases. Most of their arguments are also valid for visna.

On the other hand, there is little similarity between the lesions of visna and those observed in old dog encephalitis (Jubb and Kennedy, 1970). There is some evidence that this disease is another manifestation of infection with canine distemper virus (Lincoln et al., 1971, 1973). The same applies to the morphology of subacute sclerosing panencephalitis (Sourander and Haltia, 1973), which might be regarded as a human counterpart of old dog encephalitis.

The pathology of scrapie (see Ch. 12) and Aleutian mink disease (see Ch. 9) bears no resemblance to the pathology of visna. Scrapie is apparently identical to the disease known in Icelandic sheep as rida (Pálsson and Sigurdsson, 1958).

4.4 The pathology of maedi

4.4.1 Introduction

Few reports have dealt with the pathology of maedi in Icelandic sheep. The main pathological features have been described by Sigurdsson et al. (1952), Sigurdsson (1954) and Gíslason (1966); see also 2.3.1 for gross pathology of maedi. Later Georgsson and Pálsson (1971) gave a detailed

account of the histopathological changes in maedi. The following description
will be based on these reports, especially the last one.

4.4.2 Histopathology

The initial histopathological changes in maedi-affected lungs found in
experimental cases are scattered perivascular infiltrates, mainly composed
of lymphocytes and monocytes, accompanied later by some infiltration into
the alveoli and an apparently increased cellularity of the interalveolar septa.
Such lesions have been observed as early as 10 days after experimental infection (Sigurdsson et al., 1953). It is difficult to prove that such a minimal
and relatively non-specific response is actually caused by maedi virus,
especially in an organ normally harbouring various microorganisms and
parasites. The observed relationship between the duration of the infection
and the number of such foci supports the opinion that they are actually
caused by maedi virus.

  With further progression of the lesions the most characteristic histopathological feature in maedi becomes more prominent. This consists of a
diffuse thickening of the interalveolar septa, which encroaches upon the

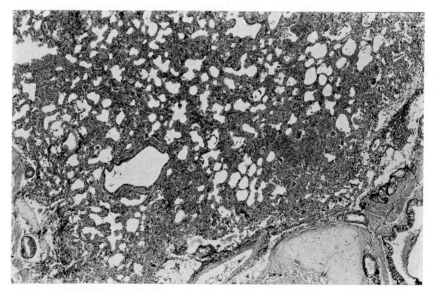

Fig. 4.19. Severe interstitial inflammation partly with total consolidation of the lung tissue.
Masson–Trichrome. × 42.

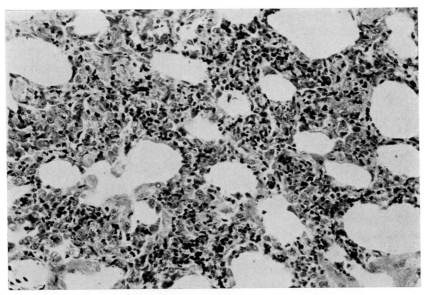

Fig. 4.20. Thickening of interalveolar septa by inflammatory infiltration composed mainly of lymphocytes and large mononuclear cells. Masson–Trichrome. × 245.

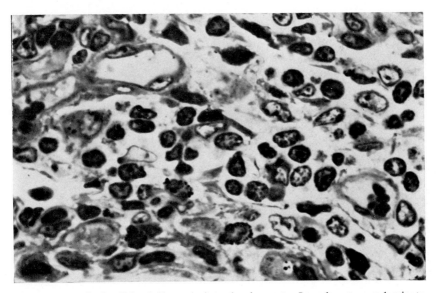

Fig. 4.21. Detail of cellular infiltrate in interalveolar septa. Lymphocytes predominate. Large mononuclear cells (monocytes and/or alveolar septal cells) are also numerous with occasional fibroblasts and mast cells. Epon, 0.75 μ section. Toluidine blue. × 980.

alveoli to a varying degree (Fig. 4.19). Thus, in some areas, an obliteration
of the alveoli with almost total consolidation of the tissue and few if any
patent alveoli is found, whereas in other areas the thickening is more
moderate. In totally consolidated areas atelectasis is commonly an additional
factor. The thickening of the interalveolar septa is mainly caused by a
cellular infiltration composed of lymphocytes, monocytes and/or macrophages in varying proportion (Figs 4.20 and 4.21). Plasma cells are also
found, but usually they are relatively sparse. Polymorphonuclear leucocytes
are not found in the inflammatory infiltrates in uncomplicated maedi.

Hypertrophy and/or hyperplasia of smooth muscles is of common occurrence and is frequently very prominent (Fig. 4.22). It is most commonly
observed at the level of the alveolar ducts and respiratory bronchioles and
frequently extends into adjacent interalveolar septa. Knob-like thickenings
of smooth muscles are often found in the interalveolar septa at the opening
of alveoli into alveolar ducts or respiratory bronchioles (Fig. 4.22). These
knobs are sometimes hyalinized with a dense network of reticular and elastic
fibers and may be lined by hyperplastic cuboidal epithelium. Occasionally
nodular thickenings of smooth muscles are observed in the bronchioles.

Reticular fibers are usually increased and form a dense network in the

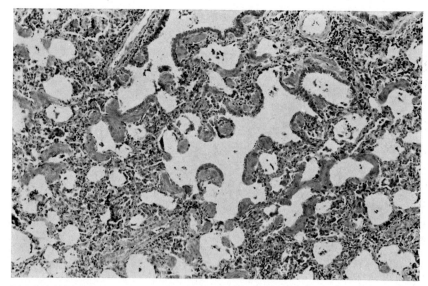

Fig. 4.22. Hypertrophy of smooth muscles in interalveolar septa and terminal air passages
with knob-like thickenings. Distinct interstitial inflammation is also present. Masson–
Trichrome. × 102.

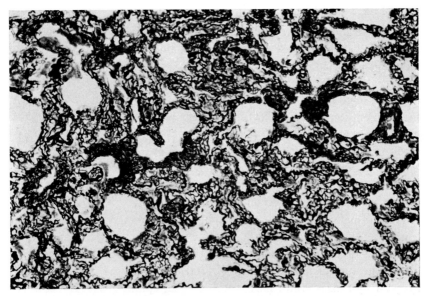

Fig. 4.23. Distinct increase of reticular fibers in the interalveolar septa. Gömöri. × 260.

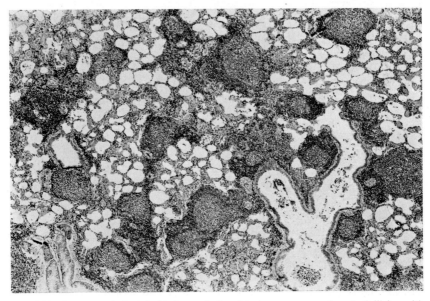

Fig. 4.24. Pronounced lymphoid hyperplasia showing numerous lymph follicles with germinal centers. Moderate interstitial inflammation. Masson–Trichrome. × 42.

interalveolar septa (Fig. 4.23), whereas elastic fibers in the septa are apparently sparse and often fragmented. Fibrosis is on the whole inconspicuous. Even in far advanced cases with prominent thickening of the interalveolar septa there is usually only a slight increase of collagenous fibers.

A regular finding is accumulation of lymphoid tissue. In some cases only a few scattered small accumulates of lymphocytes are found. In other cases the proliferation of lymphoid tissue is very prominent (Fig. 4.24) and frequently forms regular lymph follicles with active germinal centers showing many mitotic figures. The lymphoid tissue is predominantly peribronchial but partly perivascular. Peribronchial lymphoid tissue is usually external to the smooth musculature, but sometimes discrete infiltration into the lamina propria of the bronchial epithelium is noted.

Transformation of the squamous epithelium of the alveoli into a cuboidal epithelium, i.e., epithelialization, is sometimes noted, especially where the thickening of the interalveolar septa is most pronounced. In some areas the architecture of the lung tissue is totally disorganized and replaced by irregular cyst-like cavities lined with cuboidal epithelium (Fig. 4.25). The cyst-like cavities are partly disorganized bronchi and bronchioles and partly

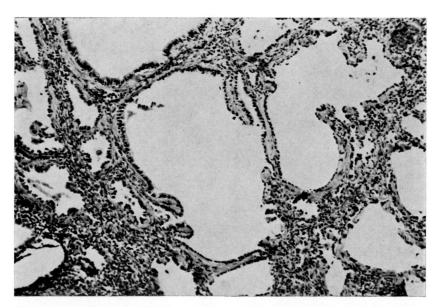

Fig. 4.25. Cyst-like cavities lined by a simple cuboidal epithelium and separated by fibrous tissue with chronic inflammation. Masson–Trichrome. × 130. (From Georgsson and Pálsson (1971) by courtesy of S. Karger.)

distended alveoli with epithelialization. These areas may bear a superficial resemblance to lesions found in pulmonary adenomatosis (but see 4.5).

In the alveoli desquamated macrophages and sparse proteinaceous exudate is occasionally noted. Bi- or tri-nucleated alveolar macrophages are sometimes found, whereas multinucleated giant cells of similar appearance as those typical for the cytopathic effect of maedi virus in tissue culture (Sigurdardóttir and Thormar, 1964; see also Ch. 5) are not found.

The bronchial lumen often contains sparse mucus. The epithelium of smaller bronchi and bronchioles is frequently hyperplastic with distinct mucoid metaplasia. The bronchial glands are sometimes widened and filled with mucus. Occasionally small intraluminal polyps are found in the bronchioles with a fibrous core infiltrated by mononuclear cells and covered by hyperplastic epithelium.

In far advanced cases some hyperplasia of the media of arterioles and small arteries with splitting of elastic lamina occurs, as well as slight thickening of the intima in larger arteries and veins. But on the whole vascular changes are inconspicuous.

In Giemsa-stained smears from maedi lungs Sigurdsson et al. (1952) always found greyish-blue cytoplasmic inclusions in large mononuclear cells. The significance of this finding is uncertain. Inclusion bodies have neither been demonstrated in sections from maedi lungs nor are they found in tissue culture infected with maedi virus (Sigurdardóttir and Thormar, 1964; see also Ch. 5).

The histological changes observed in maedi are usually complicated by lesions caused by parasites. These are minimal in experimental cases kept indoors but often pronounced in field cases. Furthermore, signs of terminal acute bronchopneumonia are often found.

The tracheo-bronchial and mediastinal lymph nodes show a diffuse lymphoid hyperplasia. Infiltration of eosinophilic leucocytes to a varying degree is commonly present and sometimes partly calcified granulomatous foci are found.

In the kidneys proliferative glomerulonephritis is commonly observed, a common finding in uninfected sheep, (see 4.1.2), but in other organs there are no constant histological changes.

4.5 Comparison with maedi-like diseases in other breeds of sheep

Pulmonary diseases of sheep similar or apparently identical to maedi in Icelandic sheep have been reported under different terms in many countries.

In Denmark, Hungary, India, Israel and Norway the Icelandic term maedi has been used. In the United States a comparable disease has been known as 'Montana progressive pneumonia', whereas a similar disease of sheep in South Africa is named Graaff–Reinet disease, in Holland 'zwoegerziekte' and in France 'la bouhite'. In Kenya, Germany, Bulgaria and apparently also in Kirgizia, the term 'progressive interstitial pneumonia' has been used for a lung infection similar to maedi. The histopathological features of all these diseases are generally very similar to those already described for maedi in Icelandic sheep (see 2.5.1 for further details and for references).

One of the most thoroughly studied diseases related to maedi is zwoeger-ziekte in Holland which is caused by a virus closely related to the maedi–visna virus (de Boer, 1970a). The description of the pathological lesions in this disease by Ressang et al. (1968) conforms essentially to the pathological findings in maedi, i.e., chronic interstitial inflammation with thickening of the interalveolar septa, peribronchial and perivascular lymphoid hyperplasia and hypertrophy of smooth muscles. The main difference is in the degree of fibrosis, which is slight in maedi, and the knob-like thickenings on the tips of the interalveolar septa seen in maedi and maedi-like disease in Kenya (Roach, R. W., personal communication) but not found in zwoegerziekte. This feature seems also to be lacking in Montana progressive pneumonia, although in other respects the pathological lesions are strikingly similar to maedi (Marsh, 1923, 1966). This lack of knobs on the interalveolar septa in zwoegerziekte and Montana progressive pneumonia may be related to different response of the musculature, which we tend to regard as a secondary phenomenon.

The hypertrophy of the smooth muscles is not pathognomonic for maedi, because, for example, infestation with lungworms can provoke a similar response (Li, 1946) and it also occurs as a compensatory response to diminished elastic recoil of the lungs in various chronic pulmonary diseases of man, e.g., emphysema (Liebow et al., 1953). Such a mechanical concept to explain the hypertrophy of the musculature of the lungs might also be valid for maedi, where thickening of the interalveolar septa and fragmenta-tion and destruction of elastic fibers would certainly hamper the elastic recoil of the lungs.

Lymphoid hyperplasia, one of the characteristic pathological lesions in maedi, is also present in zwoegerziekte and Montana progressive pneumonia. In zwoegerziekte as in maedi this is apparently one of the earliest lesions observed. In maedi the lymphoid hyperplasia is variable, often discrete but sometimes very pronounced, as shown in our Fig. 4.24. In 'la bouhite', where the pathological lesions are in general similar to those found in maedi, lymphoid hyperplasia is apparently very prominent and was regarded by Lucam (1942) as a neoplastic process. In maedi there is no indication that the lymphoid hyperplasia is neoplastic and we agree with the general opinion that the lymphoid hyperplasia in the lungs in maedi and related conditions is a reactive response to infection. In chronic inflammation due to various agents it is relatively common to see lymph follicles in tissues where normally there is no lymphoid tissue. In other infectious diseases of lungs, e.g. enzootic pneumonia of pigs and atypical pneumonia of calves, lymphoid hyperplasia is prominent. According to Jericho's (1966) study on hysterectomy-produced and colostrum-deprived piglets, an increase of lymphoid tissue of the lungs with age is directly related to challenge with various agents or even irritating substances. Thus, the lymphoid hyperplasia in the lungs in maedi and related conditions, although apparently provoked by infection with maedi virus, is not a pathognomonic feature.

Hyperplasia of the epithelium in small bronchi and bronchioles together with alveolar epithelialization is common to maedi, zwoegerziekte and Montana progressive pneumonia. These changes are not specific for infection with maedi virus and like Marsh (1966) we found such lesions most prominent in cases complicated either by bacterial infection or parasitic infestation (Georgsson and Pálsson, 1971). In this context it is perhaps appropriate to emphasize that maedi and pulmonary adenomatosis are different disease entities (see Ch. 2), although the epithelial hyperplasia, together with disorganization of the lung tissue sometimes observed in far advanced cases of maedi, may bear a superficial resemblance to the lesions found in pulmonary adenomatosis. However, there is a distinct difference in the pathological lesions found in these two diseases (Marsh, 1966; Georgsson and Pálsson, 1971).

The different features of the histopathology of maedi are not pathognomonic and can be provoked by various agents as already mentioned. Although the overall histological picture is rather characteristic it is not sufficient to make a definite diagnosis of maedi. This is perhaps best illustrated by a pulmonary disease of sheep in Scotland, atypical pneumonia (Stamp and Nisbet, 1963), which has often been referred to as related to maedi because of the similarity of the histological lesions to those of maedi. However, the clinical course as well as the macroscopic appearance with predilection for apical and cardiac lobes differs from maedi. The aetiology of atypical pneumonia has not been ascertained and it has been questioned whether it is a specific entity. Lesions resembling those found in Scottish

atypical pneumonia have been produced in lambs experimentally infected
with either parainfluenza 3 virus or Bedsonia organisms (Stevenson, 1969).
In this context it is of interest that in Australia a proliferative interstitial
pneumonia of sheep with macroscopic and microscopic appearance similar
to Scottish atypical pneumonia has been described. Aetiological studies
have shown that this lung disease is caused by mycoplasma (Sullivan et al.,
1973). These examples serve to emphasize that the diagnosis of maedi can
not be based on histopathological examination alone. The macroscopic
appearance and the clinical course must also be considered, but the final
proof can only be obtained by virological and serological methods.

4.6 Comparison with human viral and chronic interstitial pneumonias

The histopathology of maedi shows some features that are generally present
in viral pneumonias, especially those with a chronic course. Thus, the
inflammatory infiltrates found in various viral pneumonias of man consist
mainly of mononuclear cells, the inflammation is particularly interstitial
and epithelial hyperplasia is of common occurrence. On the other hand
muscular hypertrophy and lymphoid hyperplasia, constant findings in
maedi, are apparently rare in human viral pneumonias. In measles pneumonia, however, considerable hyperplasia of peribronchial lymphoid tissue
may occur. But in spite of some general histological features shared by
maedi and viral pneumonias, an overall histological picture comparable
to that of maedi is not found in human viral pneumonias (Spencer, 1963).
    Chronic interstitial pneumonia (Liebow, 1968) has been reported under
various terms, e.g., Hamman–Rich syndrome. These pneumonias may
possibly represent a late stage in the evolution of viral pneumonias, although
mycoplasma or physical and chemical factors may apparently be involved
(Liebow, 1968). In about half of the cases the aetiology remains unknown
(Solliday et al., 1973). The pathological lesions found in the chronic interstitial pneumonias may show some similarities with those found in maedi;
the lungs are usually diffusely affected and the histopathological lesions
include chronic interstitial inflammation, alveolar epithelialization and
hypertrophy of the smooth musculature of the lungs. The inflammatory
reaction is usually more intense in maedi than in chronic interstitial pneumonia. But the main difference is in the degree of interstitial fibrosis that
usually predominates in chronic interstitial pneumonia, whereas it is
generally slight in maedi. Occasionally, however, fibrosis and disorganization

of the lung tissue in maedi may simulate the so-called 'honey-combing' often observed in the chronic interstitial pneumonias.

In cases of chronic interstitial pneumonia of unknown aetiology, the association with possible autoimmune diseases, (e.g., rheumatoid arthritis or scleroderma) as well as the demonstration of immune complexes in affected lungs, have suggested that autoimmune reaction might be operative in the pathogenesis (Nagaya et al., 1973). This is of some interest as the possibility has been suggested that immune mechanisms might play a part not only in the pathogenesis of visna (see 6.3.3) but also in maedi.

Acknowledgements

We are indebted to Mrs. Elsa Benediktsdóttir, Mrs. Ragnhildur Jóhannesdóttir and Mr. Chester Reather for expert technical assistance.

References

DE BOER, G. F. (1970a) Thesis, Utrecht.

DE BOER, G. F. (1970b) T. Diergeneesk., 95, 725.

CÉRVOS-NAVARRO, J., HÜBNER, G., PUCHSTEIN, G. and STAMMLER, A. (1960) Frankfurt. Z. Pathol., 70, 458.

CORK, L. C., HADLOW, W. J., CRAWFORD, T. B., GORHAM, J. R. and PIPER, R. C. (1974a) J. Infect. Dis., 129, 134.

CORK, L. C., HADLOW, W. J., GORHAM, J. R., PIPER, R. C. and CRAWFORD, T. B. (1974b) Acta Neuropathol. (Berl.), 29, 281.

DAHME, E., STAVROU, D., DEUTSCHLÄNDER, N., ARNOLD, W. and KAISER, E. (1973) Acta Neuropathol. (Berl.), 23, 59.

FANKHAUSER, R., FATZER, R., LUGINBÜHL, H. and MCGRATH, J. T. (1972) Adv. Vet. Sci. Comp. Med., 16, 35.

FRAUCHIGER, E. and FANKHAUSER, R. (1970) In: (P. J. Vinken and G. W. Bruyn Eds) Handbook of Clinical Neurology. Vol. 9 (North-Holland Publishing Company, Amsterdam) p. 664.

GEORGSSON, G. and PÁLSSON, P. A. (1971) Vet. Pathol., 8, 63.

GÍSLASON, G. (1966) In: (T. Dalling, Ed.) Int. Encycl. Vet. Med. Vol. 3 (Green, Edinburgh) p. 1780.

GUDNADÓTTIR, M. (1974) Progr. Med. Virol., 18, 336.

IIZUKA, R. and SPALKE, G. (1972) Acta Neuropathol. (Berl.), 21, 39.

INNES, J. R. M. and SAUNDERS, L. Z. (1962) In: (J. R. M. Innes and L. Z. Saunders, Eds) Comparative Neuropathology. (Academic Press, London) p. 373.

JERICHO, K. W. (1966) Vet. Bull., 38, 687.

JUBB, K. V. F. and KENNEDY, P. C. (1970) Pathology of Domestic Animals. 2nd edn., Vol. 2 (Academic Press, London) p. 426.

LERNER, R. A. and DIXON, F. J. (1966) Lab. Invest., 15, 1279.

LI, P. L. (1946) J. Pathol. Bact., 58, 373.

LIEBOW, A. A. (1968) In: (Liebow and Smith, Eds) The lung. (Williams & Wilkins, Baltimore) p. 332.

LIEBOW, A. A., LORING, W. E. and FELTON, W. L. (1953) Am. J. Pathol., 29, 885.

LINCOLN, S. D., GORHAM, J. R., OTT, R. L. and HEGREBERG, G. A. (1971) Vet. Pathol., 8, 1.

LINCOLN, S. D., GORHAM, J. R., DAVIS, W. C. and OTT, R. L. (1973) Vet. Pathol., 10, 124.

LUCAM, F. (1942) Réc. Méd. Vét., 118, 273.

MARSH, H. (1923) J. Am. Vet. Med. Assoc., 62, 458.

MARSH, H. (1966) In: (L. Severi, Ed.) Proc. Int. Conf. on Lung Tumors in Animals. (Perugia, Italy) p. 285.

NAGAYA, H., ELMORE, M. and FORD, C. D. (1973) Am. Rev. Resp. Dis., 107, 826.

NARAYAN, O., SILVERSTEIN, A. M., PRICE, D. and JOHNSON, R. T. (1974) Science, 183, 1202.

PÁLSSON, P. A. (1972) J. Clin. Pathol. 25, suppl. (R. Coll. Path.), 6, 115.

PÁLSSON, P. A. and SIGURDSSON, B. (1958) VIII Nordiske Veterinärmötet – Sektion A, rapport 8.

PETTE, E., MANNWEILER, K. and PALACIOS, O. (1961) Dtsch. Z. Nervenheilk., 182, 635.

RESSANG, A. A., STAM, F. C. and DE BOER, G. F. (1966) Pathol. Vet., 3, 401.

RESSANG, A. A., DE BOER, G. F. and DE WIJN, G. C. (1968) Pathol. Vet., 5, 353.

SCHALTENBRAND, G. and STRAUB, O. C. (1972) Dtsch. Tierärztl. Wochensch., 79, 10.

SHARMA, D. N. (1972) Agra Univ. J. Res. (Sci.), 21, 85.

SIGURDARDÓTTIR, B. and THORMAR, H. (1964) J. Infect. Dis., 114, 55.

SIGURDSSON, B. (1954) Br. Vet. J., 110, 255.

SIGURDSSON, B. and PÁLSSON, P. A. (1958) Br. J. Exp. Pathol., 39, 519.

SIGURDSSON, B., GRÍMSSON, H. and PÁLSSON, P. A. (1952) J. Infect. Dis., 90, 233.

SIGURDSSON, B., PÁLSSON, P. A. and VAN BOGAERT, L. (1962) Acta Neuropathol. (Berl.), 1, 343.

SIGURDSSON, B., PÁLSSON, P. A. and GRÍMSSON, H. (1957) J. Neuropathol. Exp. Neurol., 16, 389.

SIGURDSSON, B., PÁLSSON, P. A. and TRYGGVADÓTTIR, A. (1953) J. Infect. Dis. 93, 166.

SOLLIDAY, N. H., WILLIAMS, J. A., GAENSLER, E. A., COUTU, R. E. and CARRINGTON, C. B. (1973) Am. Rev. Resp. Dis., 108, 193.

SOURANDER, P. and HALTIA, M. (1973) Ann. Clin. Res., 5, 298.

SPENCER, H. (1963) In: (H. Spencer, Ed.) Pathology of the Lung. (Pergamon Press, Oxford) p. 143.

STAMP, J. T. and NISBET, D. J. (1963) J. Comp. Pathol., 73, 319.

STAVROU, D., DEUTSCHLÄNDER, N. and DAHME, E. (1969) J. Comp. Pathol., 79, 393.

STEVENSON, R. G. (1969) Vet. Bull., 39, 747.

SULLIVAN, N. D., ST. GEORGE, T. D. and HORSFALL, N. (1973) Aust. Vet. J., 49, 57.

SÜVEGES, T. and SZÉKY, A. (1973) Acta Vet. Acad. Sci. Hung., 23, 205.

WANDERA, J. G. (1970) Vet. Rec., 86, 434.

WEINHOLD, E. (1974) Zentralbl. Vet. Med. B, 21, 32.

YOSHIKAWA, T. (1968) Atlas of the Brains of Domestic Animals. (University of Tokyo Press, Tokyo.)

CHAPTER 5

Visna-maedi virus infection in cell cultures and in laboratory animals

Halldor THORMAR

5.1 Introduction

Since the first tissue culture isolations of visna and maedi virus were made in Iceland in 1957–1958, related viruses have been isolated from sheep in many other countries such as zwoegerziekte virus in Holland and progressive pneumonia virus (PPV) in the U.S. (de Boer, 1969; Lopez et al., 1971). Studies of these virus isolates in tissue culture have shown that most of their properties are identical (de Boer, 1969; Takemoto, 1972; Thormar et al., 1974). The visna virus strain, first isolated from the brain of Icelandic sheep with visna, has been most thoroughly studied in tissue culture. It will therefore be used as the prototype for this group of related viruses, except where significant differences have been observed.

5.2 Cell–virus interaction in sheep cell cultures

5.2.1 Cytopathic effect

Most in vitro studies of visna virus have been performed in confluent monolayers of cells derived from sheep choroid plexus (SCP) and maintained in medium containing low serum concentration. In this type of culture a characteristic cytopathic effect (CPE) appears at various times after inoculation depending on the virus concentration (Sigurdsson et al., 1960). As first shown by Harter and Choppin (1967), fusion from without occurs as early as 30–60 min after inoculation with visna virus at a multipli-

Slow virus diseases of animals and man, edited by R. H. Kimberlin
© *North-Holland Publishing Company 1976*

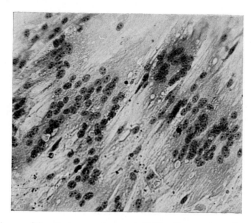

Fig. 5.1. Fusion from without. A monolayer of SCP cells was inoculated with 20 $TCID_{50}$ of visna virus per cell and incubated at 37°C for 10 h. Most of the cells have fused and formed large multinucleated syncytia. Formation of new virus is not detectable at this time of infection. Giemsa staining. × 135.

city of 28 median tissue culture infective doses ($TCID_{50}$) per cell. By 6 h the entire cell layer has undergone fusion into giant syncytia containing clusters of nuclei (Fig. 5.1). At lower virus concentrations the fusion process occurs less rapidly and is less complete. However, even at input multiplicities as low as 1 $TCID_{50}$ per cell approximately 10% of all cells fuse to form syncytia at 6 h after inoculation. The cell-fusing activity was shown to be associated with the visna virus particle and not to be dependent on infectivity or viral multiplication (Harter and Choppin, 1967; de Boer, 1969; Lopez et al., 1971).

In addition to the direct effect on cell membranes that leads to fusion and disintegration of the syncytia within a few hours, visna virus causes a CPE that is dependent on viral infectivity. This CPE, which can be observed by light microscopy of live cell layers, correlates with the multiplication of virus in the cell culture. It is characterized by the formation of refractile mononucleated spindle-shaped cells with dendritic processes and of giant multinucleated stellate cells (Fig. 5.2). As the infection progresses most of the cells are seen to undergo cytopathological changes leading to cytolysis. Cell layers fixed and stained with hematoxylin and eosin show that infected cells have a deeply basophilic cytoplasm and normal looking nuclei (Harter et al., 1967; de Boer, 1969; Lopez et al., 1971). Electron microscopic studies of infected SCP cells have not revealed any specific changes in either the cytoplasm or the nuclei. Arrays of intracytoplasmic particles and multi-

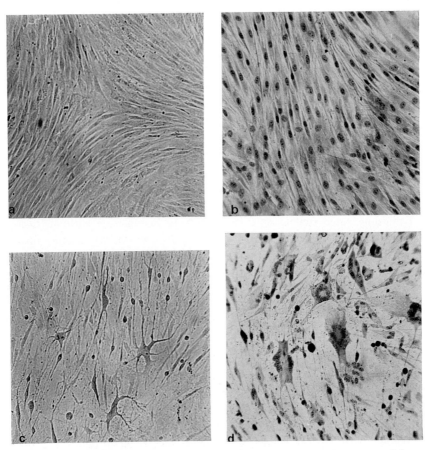

Fig. 5.2. (a) A monolayer of uninfected SCP cells. Live × 69. (b) The same cell layer stained with Giemsa. × 115. (c) A SCP monolayer inoculated with 0.5–1 $TCID_{50}$ of visna virus per cell and showing pronounced CPE after incubation at 37°C for 4 days. Note refractile mononucleated spindle-shaped cells and large multinucleated stellate cells. Live. × 57. (d) The same cell layer stained with Giemsa. × 95.

laminated structures have been observed in the cytoplasm of infected cells (Coward et al., 1970, 1972; Takemoto et al., 1971). However they do not resemble the extracellular virus particles and their relationship to the viral infection is obscure. The same is true for a variety of structural changes observed by Macintyre et al. (1973).

It is known that the viral particles are formed by budding of the cell membrane (Thormar, 1961). Staining with specific ferritin-conjugated antibodies (Coward et al., 1972) has shown that buds in all stages of development contain viral antigens on their surface. It is noteworthy that other areas of the cell membrane were free of ferritin-labeling. There was no intracellular staining by ferritin-labeled antibodies, but Macintyre et al. (1973) observed staining of paranuclear cysternae of the granular endoplasmic reticulum by visna virus-specific peroxidase–labeled antibodies. The electron microscope studies are in agreement with fluorescent-antibody studies which have shown diffuse staining of the cytoplasm of infected cells, heavy staining of the cell membranes in late stages of infection and no staining of the nuclei (Harter et al., 1967; de Boer, 1969; Thormar, 1969; Lopez et al., 1971; Takemoto et al., 1971).

The hallmark of the cytopathic effect of visna virus in vitro seems to be the alteration of cell membranes which results in cell fusion and formation of multinucleated stellate cells. In contrast to fusion from without, fusion from within depends on new viral synthesis. It is not known whether fusion from within only requires intracellular synthesis of virus-specific proteins or is dependent on the release of mature virus particles from the cell membrane.

5.2.2 Viral growth cycle

The one-step growth cycles of visna–maedi and zwoegerziekte viruses in SCP and of progressive pneumonia virus (PPV) in normal sheep lung (NSL) and sheep testis (ST) cultures have been studied by a number of laboratories. A latent period of 16–24 h has been reported for visna virus and PPV (Thormar, 1963b, 1965a; Harter and Choppin, 1967; Lopez et al., 1971; Takemoto et al., 1971) whereas a somewhat shorter latent period (12–14 h) has recently been observed by Trowbridge (unpublished results). A latent period of 25–30 h has been found in growth cycles of zwoegerziekte virus and maedi virus (Thormar, 1965a; de Boer, 1969). The latent period is followed by an exponential increase in infective virus lasting for 20–30 h. This correlates with a gradual increase in the CPE and with accumulation of viral antigen in the cytoplasm of infected cells (Harter et al., 1967; de Boer, 1969; Thormar, 1969; Lopez et al., 1971; Takemoto et al., 1971), but both continue to increase after the rate of viral production has leveled off. At this time the budding activity of the cellular membrane is at its maximum.

The adsorption of visna virus to SCP cells has been shown to be completed in about 2 h (Harter, 1969). Electron microscope studies indicate that the virus enters the cells after fusion of the viral and cellular membranes (Chippaux-Hyppolite et al., 1972).

5.2.3 Conditions affecting cell–virus interaction

The visna virus infection has been studied most in confluent monolayers of secondary sheep cell cultures maintained in medium with low serum concentration at 37 °C and at neutral or slightly alkaline pH. Few studies have been made to determine how changes in these conditions affect the infection. Since CPE and viral production occur in dense monolayers of non-dividing cells which have been maintained for days without serum it has been concluded that visna virus synthesis and CPE are not dependent on prior division of the host cell. This has been confirmed in experiments where host cell nucleic acid synthesis was blocked by UV-irradiation or by treatment with drugs at the time of inoculation with visna virus (Trowbridge et al., 1975). Somewhat larger yields of virus are obtained in confluent SCP cultures if they are maintained in medium with high lamb serum concentration rather than in medium without serum. On the other hand, the rate of virus synthesis was temporarily decreased in actively dividing non-confluent cultures maintained in high serum concentration (Trowbridge, unpublished).

Little is known of how visna virus infection affects the physiological state of the host cells, for instance whether the infection inhibits, stimulates or is without an effect on cell proliferation. Preliminary observations indicate that there is no significant effect on cell division, one way or the other (Thormar, unpublished results). High serum concentration in the medium is found to enhance the cell-fusing activity of the virus (fusion from within).

We know little about the effect of temperature and pH on visna virus CPE and proliferation. Preliminary studies have indicated that both are inhibited at 35–36 °C, whereas they are enhanced at 39–40 °C, as compared with 37 °C.

Some studies have been made on the influence of various compounds, such as drugs, on visna virus–host cell interactions. Compounds that inhibit DNA synthesis, namely 5-fluorodeoxyuridine (FUDR), 5-iododeoxyuridine (IUDR) and 5-bromodeoxyuridine (BUDR) were found to inhibit visna virus synthesis to a varying degree. BUDR was most active, but only if it was applied to tissue cultures less than 8 h after inoculation with the virus (Thormar, 1965b). The effect of BUDR was largely reversed by addition of thymidine. Furthermore, actinomycin D, which is known to block DNA-dependent RNA synthesis, was found to inhibit visna virus formation throughout the growth cycle. These findings suggested that the proliferation of visna virus depends on synthesis and function of DNA. The role of DNA in the formation of visna virus was further substantiated by the demonstration of RNA-dependent DNA polymerase (reverse transcriptase) in the visna virion (Lin and Thormar, 1970; Schlom et al., 1971) and of a visna DNA provirus in infected cells (Haase and Varmus, 1973). The drug thiosemi-carbazone was found to inhibit viral reverse transcriptase and viral infectivity to the same extent (Haase and Levinson, 1973), indicating that the enzyme activity was required for viral proliferation and cytopathic effect.

Miracil D and mitomycin C do not inhibit visna virus synthesis, although both are potent inhibitors of host cell DNA synthesis and mitomycin C also inhibits RNA synthesis of SCP cells (Trowbridge et al., 1975). It is concluded that prior host cell nucleic acid synthesis and division are not required in order that transcription of the visna DNA provirus into viral RNA can begin.

5.2.4 Biochemical characteristics (see also 3.7)

A great deal of information has been obtained about the biochemical composition of the visna virion and some data are becoming available on the biochemical changes caused by the virus in infected sheep cells.

Visna virus (Brahic et al., 1971; Harter et al., 1971; Lin and Thormar, 1971), PPV (Stone et al., 1971) and maedi virus (Lin and Thormar, 1972a) all contain a single-stranded RNA, predominantly of the 60–70 S type. Brahic and Vigne (1975) have shown that visna virus harvested a few minutes after its formation and release from the cell membrane contains only RNA subunits of 30–40 S. These subunits mature within 30–60 min into the 60–70 S RNA species. The RNA of mature visna virus has been most thoroughly studied by Haase et al. (1974a) who confirmed the presence of only two RNA species, namely 70 S and 4 S. Each molecule of 70 S RNA, with a molecular weight of $10–12 \times 10^6$, dissociated by heating into 3 to 4 subunits with a molecular weight of about 3×10^6. In contrast to the 70 S RNA species of Rous sarcoma virus (RSV) the visna 70 S species contained only minute amounts of 4 S component, comprising about 0.4% of the 70 S molecule. The visna virus RNA was found to be similar to RSV RNA in genetic complexity and to be 3–4 times more complex than the poliovirus genome. On the other hand, the secondary structure of the 70 S RNA of visna virus seems to differ from that of RSV 70 S RNA as judged from their relative migration rates in polyacrylamide gels and sedimentation rates in buffers of low ionic strength in which the visna RNA unfolds to a greater extent than the RSV RNA. This difference was considered to be related to the very low contents of 4 S RNA species in the 70 S molecule of visna virus as compared with RSV (Haase et al., 1974a).

The 70 S visna RNA functions as a template for visna virus reverse transcriptase. The DNA product of the in vitro enzymatic reaction has been characterized by Haase et al. (1974b). The initial product was found to be a single-stranded DNA bound to the RNA template. The formation of the single-stranded DNA was followed in 30–60 min by the appearance of double-stranded DNA. Both DNA products have a low molecular weight and sedimentation coefficients of approximately 4 S. It was estimated that the double-stranded DNA only represented 5–10% of the viral genome, although hybridization experiments with the single-stranded DNA product showed it to contain sequences complementary to the entire viral RNA.

A two-step reaction resulting in the formation of double-stranded DNA is indicated by this study (Haase et al., 1974b). It is suggested that the first step synthesizes single-stranded DNA using viral RNA as template, whereas the second step utilizes the initial DNA product as template to synthesize double-stranded DNA. The question of whether or not the two steps are catalyzed by more than one DNA polymerase (Haase et al., 1974b) has been raised by Lin (Lin and Thormar, 1972b; Lin et al., 1973) who found that the visna DNA polymerase is composed of at least two enzymatically active proteins. The two proteins showed differences in kinetics, pH optimum, sensitivity to inhibitors and template specificity. In a more recent study, Lin (unpublished) has further purified the DNA polymerase and demonstrated the presence of three components, 5.5, 7 and 11 S. The 5.5 S polymerase component shows a strong preference for visna virus RNA as a template, compared with native calf thymus DNA. In contrast, the 7 S component responds equally well to these two natural nucleic acids. It is suggested that the 5.5 S enzyme is the reverse transcriptase, and that the 7 S enzyme is a DNA-dependent DNA polymerase. The 11 S component represents a small fraction of the total enzyme. It seems to function in the absence of template and has tentatively been ascribed a terminal deoxynucleotidyl transferase activity.

The studies of the biochemical characteristics of visna virus in a cell-free system are now being extended to virus-infected cells. The first study by Haase and Varmus (1973) demonstrated that the entire 70 S viral RNA is transcribed into DNA during the latent phase of viral growth in SCP cells and at least some of the DNA becomes covalently associated with the genome of the host cell. The integration of visna virus DNA into host cell genome during infection is consistent with the DNA provirus hypothesis proposed

by Temin (1967) for RNA tumour viruses. A more recent time sequence
study performed by Haase and Brahic (Brahic, personal communication)
has shown that the RNA of the parent virus remained in the cytoplasm of
infected SCP cells for at least 7 h after infection but was greatly diminished
at 14 h after infection. The synthesis of virus-specific DNA in the cytoplasm
of infected cells begins almost immediately after infection and is completed
in about 7 h. Somewhat later, at about 20 h after infection, viral DNA is
present in the nucleus of the host cell in an integrated form. At about the
same time or a little later, progeny virus RNA is found in the cytoplasm.
The results of these elegant biochemical studies agree well with earlier
biological studies on the effect of inhibitors applied to SCP cell cultures at
various times after infection with visna virus (Thormar, 1965b).

5.2.5 *Persistently infected cultures*

Primary tissue cultures from choroid plexus and lungs of sheep affected
with visna have been found to produce small amounts of visna virus for
long periods of time without showing much CPE or cell lysis. A low serum
concentration in the medium was found to enhance the CPE, whereas a
high serum concentration stimulated growth and regeneration of the tissue
culture (Sigurdsson et al., 1960; Thormar, 1967). After the virus had been
passed a few times in tissue culture its cytopathogenicity was found to
increase and to result in an almost complete lysis and destruction of infected
cell layers regardless of whether or not the culture medium contained low
or high concentrations of serum. By maintaining infected SCP cultures in
fluid medium with a high serum concentration for long periods of time after
the majority of the cells had been lysed, Trowbridge (Trowbridge and
Thormar, 1972; Trowbridge, 1974a, 1975) was able to stimulate cell growth
that resulted in repopulation of the culture vessel and the establishment of
a carrier culture for visna virus. The growth became detectable by the
formation of colonies of surviving cells on the 15th day after infection,
when visna virus was still present at a high titer. On the 45th day after
infection a confluent cell layer had formed. When subcultured, the carrier
cultures required more time to become confluent than comparable cultures
of uninfected cells. Their viability was also reduced as determined by their
relative efficiency of colony formation. This was less than 0.02% as compared with 20–30% in uninfected SCP cultures. No CPE was observed
in the carrier cultures, except for a slightly increased frequency of cells
with two or more nuclei. The carrier cultures maintained virus titers at

considerably high levels for over one year, particularly if subcultured at long intervals. Cultures which were subcultured more frequently produced much less free virus detectable in the fluid medium. However, in both cultures the majority of the cells apparently contained visna-specific antigen as determined by immunofluorescent staining. Both cultures also contained significant activities of visna-specific RNA-dependent DNA polymerase.

The carrier cultures are resistant to superinfection with visna virus but are susceptible to infection by vesicular stomatitis virus (VSV) and vaccinia virus. This indicates that interferon is not responsible for the persistence of visna virus in the carrier cultures (Trowbridge, 1974a, 1975). In another experiment it was shown that uninfected SCP cells could be made resistant to VSV infection by treatment with an interferon-inducing agent (poly(I) · poly(C)) whereas they were fully susceptible to visna virus infection. The above results suggest that visna virus has a very low degree of susceptibility to interferon produced by SCP cultures and that the virus is a poor interferon-inducing agent, at least in persistently infected SCP cultures.

5.2.6 Lymphocytes, macrophages and fetal cells

Few studies have been made of the effect of visna virus on sheep cells other than SCP, ST and NSL cell cultures. De Boer (1969) found that zwoeger-ziekte virus was associated with lymphocytes from sheep with viremia but nothing is known about the lymphocyte–virus interaction. Preliminary experiments indicate that macrophages from sheep brain and spleen are susceptible to visna virus (Thormar, unpublished results). Fetal lung and brain cultures support the propagation of visna virus and undergo CPE, but it has not been determined whether they are more or less susceptible to the virus than cultures from adult sheep.

5.3 Infection of cells from species other than sheep

5.3.1 Human cells

Visna virus has been found to persist at a low titer in kidney and choroid plexus cell cultures from human embryos for more than two months (Thormar and Sigurdardóttir, 1962). The cultures which were maintained at a low serum concentration showed little CPE except for a few multinucleated and stellate cells. More recently Macintyre et al. (1972a, b, 1973, 1974a, b) have studied the effect of visna virus on a permanent line of human astrocytes derived from a brain tumour. The virus was propagated by the human cell cultures which were maintained in 10% serum, and produced a CPE similar to that observed in SCP cultures, namely stellate cells and multinucleated giant cells. The characteristic CPE and a virus titer of 10^5–10^6 TCID$_{50}$ per ml persisted in the human astrocyte cultures for at least 24 months and during numerous subcultivations. When inoculated into SCP cultures the human-passaged visna virus caused the characteristic cytopathic changes resulting in complete cytolysis.

In one group of persistently infected human astrocyte cultures, designated XV-4, morphological changes were observed one year after initial infection (Macintyre et al., 1974a, b). The XV-4 cell line was characterized by the appearance of layers of round cells attached to a base layer showing the usual CPE of multinucleated and stellate cells. It produced virus, designated XV visna virus, at a titer of 10^5–10^6 TCID$_{50}$ per ml, and visna virus-specific cytoplasmic antigen and typical extracellular virions were present. Subculturing of the rounded cells or passage of the XV virus into fresh cultures of human astrocytes resulted in the formation of the characteristic XV-4 cultures. If the XV virus was passed into SCP cells maintained in 20% lamb serum it behaved differently from the parent visna virus maintained under the same conditions. Instead of a complete cytolysis, only about 70% of the cells cytolysed. The surviving cells proliferated and formed a carrier culture which showed syncytia and produced virus-specific cytoplasmic antigen and free XV virus to a titer of 10^3–10^4 TCID$_{50}$ per ml. The authors speculate that XV might be either a visna-associated virus or a variant of visna virus and raise the question of its possible oncogenic capacity.

5.3.2 Mouse and hamster cells

Takemoto and Stone (1971) have reported transformation of AL/N and BALB/C mouse cell lines after inoculation with visna virus and PPV. Foci of morphologically altered, spindle-shaped cells appeared 3 weeks after inoculation. In addition to change in morphology, the cells showed loss of contact inhibition and rapid growth. Visna virus could not be recovered from subcultures of the transformed cells except by co-cultivation with ST cells. The transformed cells produced tumours in X-irradiated young adult mice one month after subcutaneous inoculation. It was concluded that the transformation was visna virus-induced and that the viral genome was incorporated into the genome of the transformed cells. Tumour-specific transplantation antigens were demonstrated in the transformed mouse cells in vivo (Law and Takemoto, 1973), but there was no cross-reaction between the various 'visna-or PPV-induced' tumours. This is in contrast to common virus-specific transplantation antigens, which are cross-reactive for various tumours induced by the same virus. No cross-reactivity was demonstrated between visna and PPV viral antigens and the transplantation antigens of transformed cells.

In a recent study, Brown and Thormar (1975) inoculated AL/N mouse embryo cultures, 3T3 and BALB/3T3 mouse cell lines and hamster embryo cultures with a high multiplicity of visna virus. The AL/N and hamster embryo cultures produced small amounts of virus for at least 100 days with little or no CPE. The 3T3 mouse cell line was found to contain visna virus for at least 20 cell passages and after a final dilution of the inoculated cells of 10^{20}-fold. The virus could be demonstrated only by co-cultivation of in-fected 3T3 cells with susceptible sheep cells, but was not detectable in the fluid medium or in freeze-thawed cells. The cell–virus relationship was, therefore, similar to that observed by Takemoto and Stone (1971) in their AL/N cell line. There was no CPE in the infected 3T3 cultures and no in-dication of cell transformation in any of the infected mouse or hamster embryo cell cultures.

5.3.3 Other species

Primary cultures of choroid plexus cells from a number of species were found to produce low amounts of virus for 1–3 months (Thormar and Sigurdardóttir, 1962). Only cultures of bovine cells propagated the virus to high titers and showed characteristic CPE. The susceptibility of bovine cells to visna virus was confirmed by Harter et al. (1968) in a continuous cell line of bovine origin. Most continuous cell lines from other species do not support infectious visna virus replication although BHK21 cells and a number of monkey kidney cell lines (August and Harter, 1974) are susceptible to fusion from without by the virus.

5.4 Variants of visna virus

A direct plaque assay for visna–maedi viruses was developed by Trowbridge (1974b) using an overlay medium containing DEAE-dextran and staining with neutral red 10–18 days after infection. Plaque size heterogeneity was observed in plates infected with stocks of visna and maedi viruses. Large plaque-forming (Lpf) and small plaque-forming (Spf) variants of visna virus were isolated and cloned and found to produce only the respective size plaques (Torchio et al., 1973). The Spf variant is less cytopathogenic than the Lpf variant in SCP cells, and less virus is produced. The difference in infectivity is even more pronounced when infected SCP cultures are incubated at 42 °C rather than at 37 °C. Synthesis of the Lpf variant was found to be depressed in SCP cultures infected 48 h earlier with the Spf variant, indicating interference (Trowbridge et al., 1974). The significance of these variants in relation to the disease in sheep is not known.

A variant of visna virus, the XV virus, has been isolated by Macintyre et al. (1974a, b), as mentioned earlier (5.3.1). This variant shows a characteristic CPE in human astrocyte cultures and is less productive in SCP cultures than the parent strain. Otherwise it has not been well characterized and its purity is unknown.

5.5 Infection of animals other than sheep

Numerous attempts to infect small laboratory animals with visna virus have failed, even after inoculation of large quantities of virus (Thormar, unpublished). In these early experiments no immune response against the virus was detectable. However, rabbits immunized with repeated injections of concentrated and purified visna virus responded with the formation of visna antibodies demonstrable by the passive hemagglutination (PHA), complement fixation (CF) and immunofluorescent tests (Karl and Thormar, 1971). No neutralizing antibodies could be detected. By giving one intravenous injection with purified virus, followed one week later by an intramuscular injection with virus and complete Freund's adjuvant, Karl (unpublished) was able to produce neutralizing antibodies in rabbits. In contrast to the PHA and CF antibodies that appeared 2–3 weeks after immunization, the neutralizing antibodies were not detectable until 2–3 months later and increased for at least 6 months to a titer of 1:128 to 1:256. Visna virus could not be isolated from organs of immunized rabbits at the time of sacrifice and no signs of disease were observed.

In a recent study (Mehta and Thormar, 1975) rabbits were immunized with purified visna and maedi viruses using complete Freund's adjuvant and injection into footpad and intramuscular sites. The PHA and CF antibody activities were associated mainly with the IgG class, but significant activities were also found in the IgM class. Precipitating and neutralizing antibodies were found only in the IgG class. Visna and maedi viruses were shown by this study to have identical antigens when compared in gel diffusion and PHA tests.

A recent publication (Dwivedi, 1974) reports typical histopathological changes in lungs of guinea pigs inoculated intrapulmonarily with suspensions of maedi lungs. No virological or serological studies were done on these animals.

Although visna virus has been shown to multiply in bovine and human tissue cultures, there is no indication that the virus is infectious in these species in vivo. Neutralizing activities against visna virus, which are found in low titers in human sera and in high titers in bovine sera (Thormar and Sigurdardóttir, 1962; Thormar and von Magnus, 1963; Thormar, unpublished), do not have the properties of antibodies and probably represent non-specific inhibitors.

Visna virus is widespread in goats in India, where it causes both interstitial pneumonia (maedi) and leukoencephalitis resembling visna (Dwivedi, 1974). The virus has been isolated from lungs (Hajela et al., 1974) and from choroid plexus (Weinhold, 1974) of goats showing signs of disease.

TABLE 5.1

Properties of visna and maedi virus in SCP cell cultures.

Visna virus	Maedi virus
1. Causes fusion from without	Not studied. (N. St.)
2. CPE: Refractile spindle-shaped cells with dendritic processes. Multinucleated giant cells. No inclusion bodies.	Refractile spindle-shaped cells most pronounced.
3. Diffuse staining of the cytoplasm by fluorescent visna antibodies. No fluorescent staining of the nuclei.	
4. Virions are released by budding from the plasma membrane.	
5. Growth cycle about 20 h. Possibly less.	Growth cycle 25–30 h. Possibly less.
6. Virus synthesis depends on prior formation of a DNA provirus.	
7. Virions carry reverse transcriptase and possibly a DNA-dependent DNA polymerase.	
8. Virus synthesis and CPE are not dependent on prior host cell division.	(N. St.)
9. Preliminary observations indicate no effect of viral infection on host cell division.	(N. St.)
10. Persists in low titers in carrier cultures.	(N. St.)
11. Little or no interferon production. Very low susceptibility to interferon.	(N. St.)

5.6 Relevance of tissue culture and small animal studies to the pathogenesis of visna and maedi in sheep

Although visna virus is highly cytopathogenic in sheep cell cultures there is no evidence for this in vivo. Specifically, multinucleated giant cells or syncytia formation have not been observed in brain or lung tissues from infected animals (see Ch. 4). Since the cell-fusing activity is a distinct property of the visna virus particle in vitro the question arises of whether or not this property plays any role in the pathogenesis of the disease. It has been suggested that the membrane-fusing activity of the virus is the cause of demyelination observed in visna (Harter and Choppin, 1967), but there is no direct evidence in support of this. Macintyre et al. (1972b) have speculated that multinucleated syncytia might be detectable in infected tissues only in the earliest stages of the disease process, before an immune response has developed. After the development of humoral or cellular immunity the syncytia could be destroyed immediately upon formation by either a cytolytic effect of visna antibodies in the presence of complement or by sensitized lymphocytes. No studies of these mechanisms in visna have been reported, either in vitro or in vivo. Since visna virus is released by budding, viral antigens are present on the outer surface of the cell membrane, but it is noteworthy that only the actual viral buds stain with specific ferritin-labeled antibodies (Coward et al., 1972). Further studies of the formation of visna virus antigens on the surface of infected cells in vitro and in vivo would be of interest. An interaction of such surface antigens with specific antibodies or with sensitized lymphocytes might play a role in the pathogenesis of the disease and could initially be studied in tissue culture. Virus-producing carrier cultures with the majority of the cells infected but showing little or no CPE might be a good system in which to study this aspect of virus–cell interaction.

Whether or not carrier cultures are a useful model system for studying other aspects of the visna virus–cell interaction is not known. In any case cultures derived directly from tissues of infected animals would be preferable to regenerated lysed cultures. However, studies of the latter type of carrier cultures have already rendered information of some general interest. For example, in such a system visna virus does not induce detectable interferon production and does not stimulate cell proliferation or cause transformation of cells.

The question of the oncogenicity of visna virus is important in view of its striking morphological and biochemical similarities to oncornaviruses (see Ch. 3). The absence of any known oncogenicity of the virus in sheep or in sheep tissue cultures must be weighed against the one still unconfirmed report of visna virus-induced transformation in mouse cell lines (Takemoto and Stone, 1971). The failure of visna virus to induce common cross-reacting transplantation antigens in various 'visna virus-induced' tumours is in contrast to the viral-specificity of transplantation antigens induced by oncornaviruses. Although it may seem unlikely that the transformation observed in visna-infected mouse cells was of spontaneous origin this must be considered a possibility. A confirmation of this observation under conditions where spontaneous transformation is ruled out is needed before visna virus can be accepted as an oncogenic agent. Preliminary results also indicate that visna virus–cell interaction differs from that of oncornaviruses in other respects. Thus, in contrast to oncornavirus synthesis (Temin, 1967) synthesis of infectious visna virus is not significantly inhibited in cell cultures in which both cell division and cellular DNA synthesis are blocked at the time of infection (Trowbridge et al., 1975).

Although visna virus has been found to grow in cultures of human astrocytes, no attempts have been made to grow the virus in sheep astrocytes or in any other type of glial cells from sheep brain. Visna virus infection should be studied by immunofluorescent and electron microscopic techniques in myelinated brain cultures from sheep, if such cultures can be maintained in a satisfactory condition. A comparison with immunofluorescent and electron microscope studies of brain tissues from infected sheep might be of value in elucidating the demyelinating process in visna.

Cells of the lymphoid system seem to be of importance in the pathogenesis of visna (6.2). There is evidence for proliferation of the virus in lymphocytes and macrophages in vitro, but the cell–virus interactions have not been studied. It is unknown whether lymphocyte stimulation in the presence of mitogens or different antigens is suppressed by visna virus; neither do we know whether or not the virus induces proliferation of lymphocytes from normal or infected sheep under the proper conditions. It would be surprising if visna virus had no effect on lymphocytes.

The cause of the extreme slowness of the visna and maedi disease in sheep is still obscure. The virus has the capacity for rapid proliferation in vitro under optimal conditions. It is doubtful, however, that this capacity is utilized in cells of the infected animal. Whether or not the DNA visna provirus plays a role in the slowness of the disease is another question (see 6.3.1). This may be related to the question of why the host defense mechanisms consistently fail to overcome the visna–maedi infection and eliminate the virus. Techniques are now available to study the role of the provirus in the pathogenesis of the slow disease.

The inability to eliminate the infection may be caused, at least partly, by the ineffectiveness of antibody formation. Not only do neutralizing antibodies appear late, i.e. months after infection (see 6.2), but they also seem to react slowly with the virus (Thormar, 1963a). Whether or not infectious virus–antibody complexes are of any importance in visna is unknown. The demonstration of infectious virus in the blood of sheep in the presence of neutralizing antibodies is probably due to virus-producing leukocytes in the blood. As discussed earlier, the effect of cytotoxic antibodies or sensitized lymphocytes on visna-infected cells has not yet been studied, and the role of cell-mediated immune response in visna is unknown.

It would obviously be helpful if visna could be studied in small laboratory animals. Except for one observation of lung infection in inoculated guinea pigs, all attempts to infect small animals with visna–maedi virus have failed. It is noteworthy that in immunized rabbits, like in infected sheep, neutralizing antibodies appear much later than other types of antibody. The reason for this is unknown.

Acknowledgement

The assistance of Mrs. E. M. Riedel in the preparation of this manuscript is gratefully acknowledged.

References

AUGUST, M. J. and HARTER, D. H. (1974) Arch. Ges. Virusforsch., 44, 92.

BRAHIC, M., TAMALET, J. and CHIPPAUX-HYPPOLITE, C. (1971) C. R. Acad. Sci. Paris, 272, 2115.

BRAHIC, M. and VIGNE, R. (1975) J. Virol., 18, 1222.

BROWN, H. R. and THORMAR, H. (1975) Microbios, in press.

CHIPPAUX-HYPPOLITE, C., TARANGER, C., TAMALET, J., PAUTRAT, G. and BRAHIC, M. (1972) Ann. Inst. Pasteur, 123, 409.

COWARD, J. E., HARTER, D. H. and MORGAN, C. (1970) Virology, 40, 1030.

COWARD, J. E., HARTER, D. H., HSU, K. C. and MORGAN, C. (1972) Virology, 50, 925.

DE BOER, G. F. (1969) Thesis, Utrecht.

DWIVEDI, J. N. (1974) U. P. Vet. Coll., Mathura, India.

HAASE, A. T. and LEVINSON, W. (1973) Biochem. Biophys. Res. Commun., 51, 875.

HAASE, A. T. and VARMUS, H. E. (1973) Nature, New Biol., 245, 237.

HAASE, A. T., GARAPIN, A. C., FARAS, A. J., TAYLOR, J. M. and BISHOP, J. M. (1974a) Virology, 57, 259.

HAASE, A. T., GARAPIN, A. C., FARAS, A. J., VARMUS, H. E. and BISHOP, J. M. (1974b) Virology, 57, 251.

HAJELA, S. K., SHARMA, D. N. and DWIVEDI, J. N. (1974) Ind. J. Anim. Sci., 44, 10.

HARTER, D. H. (1969) J. Gen. Virol., 5, 157.

HARTER, D. H. and CHOPPIN, P. W. (1967) Virology, 31, 279.

HARTER, D. H., HSU, K. C. and ROSE, H. M. (1967) J. Virol., 1, 1265.

HARTER, D. H., HSU, K. C. and ROSE, H. M. (1968) Proc. Soc. Exp. Biol. Med., 129, 295.

HARTER, D. H., SCHLOM, J. and SPIEGELMAN, S. (1971) Biochim. Biophys. Acta, 240, 435.

KARL, S. C. and THORMAR, H. (1971) Infect. Immun., 4, 715.

LÄW, L. W. and TAKEMOTO, K. K. (1973) J. Natl. Cancer Inst., 50, 1075.

LIN, F. H., GENOVESE, M. and THORMAR, H. (1973) Prep. Biochem., 3, 525.

LIN, F. H. and THORMAR, H. (1970) J. Virol., 6, 702.

LIN, F. H. and THORMAR, H. (1971) J. Virol., 7, 582.

LIN, F. H. and THORMAR, H. (1972a) J. Virol., 10, 228.

LIN, F. H. and THORMAR, H. (1972b) Abstr. 72nd Annu. Meeting Am. Soc. Microbiol., p. 219.

LOPEZ, C., EKLUND, C. M. and HADLOW, W. J. (1971) Proc. Soc. Exp. Biol. Med., 138, 1035.

MACINTYRE, E. H., WINTERSGILL, C. J. and THORMAR, H. (1972a) Nature, New Biol., 237, 111.

MACINTYRE, E. H., WINTERSGILL, C. J. and THORMAR, H. (1972b) Med. Res. Eng. 11, 7.

MACINTYRE, E. H., WINTERSGILL, C. J. and VATTER, A. E. (1973) J. Cell Sci., 13, 173.

MACINTYRE, E. H., WINTERSGILL, C. J. and VATTER, A. E. (1974a) Am. J. Vet. Res., 35, 1161.

MACINTYRE, E. H., WINTERSGILL, C. J. and VATTER, A. E. (1974b) Beitr. Pathol. Bd., 152, 163.

MEHTA, P. D. and THORMAR, H. (1975) Infect. Immun., 11, 829.

SCHLOM, J., HARTER, D. H., BURNY, A. and SPIEGELMAN, S. (1971) Proc. Natl. Acad. Sci. U.S.A., 68, 182.

SIGURDSSON, B., THORMAR, H. and PÁLSSON, P. A. (1960) Arch. Ges. Virusforsch., 10, 368.

STONE, L. B., TAKEMOTO, K. K. and MARTIN, M. A. (1971) J. Virol., 8, 573

TAKEMOTO, K. K. (1972) In (F. Woltgram, Ed.) Multiple Sclerosis: Immunology, Virology and Ultrastructure. (Academic Press, New York) p. 211.

TAKEMOTO, K. K., MATTERN, C. F. T., STONE, L. B., COE, J. E. and LAVELLE, G. (1971) J. Virol., 7, 301.

TAKEMOTO, K. K. and STONE, L. B. (1971) J. Virol., 7, 770.

TEMIN, H. M. (1967) J. Cell. Physiol., 69, 53.

THORMAR, H. (1961) Virology, 14, 463.

THORMAR, H. (1963a) J. Immunol., 90, 185.

THORMAR, H. (1963b) Virology, 19, 273.

THORMAR, H. (1965a) Res. Vet. Sci., 6, 117.

THORMAR, H. (1965b) Virology, 26, 36.

THORMAR, H. (1967) Curr. Topics Microbiol. Immunol., 40, 22.

THORMAR, H. (1969) Acta Pathol. Microbiol. Scand., 75, 296.

THORMAR, H., LIN, F. H. and TROWBRIDGE, R. S. (1974) Progr. Med. Virol., 18, 323.

THORMAR, H. and MAGNUS, H. VON (1963) Acta Pathol. Microbiol. Scand., 57, 261.

THORMAR, H. and SIGURDARDÓTTIR, B. (1962) Acta Pathol. Microbiol. Scand., 55, 180.

TORCHIO, C., TROWBRIDGE, R. S. and THORMAR, H. (1973) Abstr. 73rd Ann. Meeting Am. Soc. Microbiol., p. 208.

TROWBRIDGE, R. S. (1974a) Abstr. 74th Annu. Meeting Am. Soc. Microbiol., p. 263.

TROWBRIDGE, R. S. (1974b) Appl. Microbiol., 28, 366.

TROWBRIDGE, R. S. (1975) Infect. Immun., 11, 862.

TROWBRIDGE, R. S. and THORMAR, H. (1972) Abstr. 72nd Annu. Meeting Am. Soc. Microbiol., p. 232.

TROWBRIDGE, R. S., TORCHIO, C. and THORMAR, H. (1974) Abstr. 74th Annu. Meeting Am. Soc. Microbiol., p. 239.

TROWBRIDGE, R. S., TORCHIO, C., MAZZELLA, M. M. and BROPHY, P. A. (1975) Abstr. 75th Annu. Meeting Am. Soc. Microbiol.

WEINHOLD, E. (1974) Zentralbl. Vet. Med. B., 21, 32.

CHAPTER 6

Pathogenesis of visna: review and speculation*

Neal NATHANSON, Hillel PANITCH and
Gudmundur PÉTURSSON

6.1 Introduction

Visna and maedi are chronic and progressive sheep diseases, probably best considered as different manifestations of infection with a single agent. Not only are they one of the prototypes used by Sigurdsson to delineate the concept of slow infection (Sigurdsson, 1954; see Ch. 1), but they continue to attract attention because understanding their pathogenesis remains an unresolved challenge. An additional source of interest is the accumulating evidence which has firmly established the close relationship of visna–maedi to the RNA tumor viruses, thus confirming the early intuition of Sigurdsson and Thormar (see Chs 3 and 5). Current information on the molecular anatomy, mode of replication, and morphogenesis of visna (Haase et al., 1973, 1974a, b, c; see also Ch. 3) underlies many of the speculations which follow below.

The salient questions which the pathogenesis of visna–maedi present to the investigator are: (1) The 'slowness' of the disease. Incubation periods to clinical disease are highly irregular, ranging from 2 months to over 10 years in experimentally infected sheep, and it remains somewhat a moot point whether incubation may exceed the natural lifespan in some instances. (2)

* The original data in this report are drawn from a collaborative study undertaken by a group whose members are G. Georgsson, N. Nathanson, P. Pálsson, and G. Petursson. Supported in part by USPHS grant NS 11451.

The pathology of the disease. Lesions are confined to central nervous system (CNS) and lungs, and present a characteristic picture in which the accumulation of lymphoid and mononuclear cells is prominent (see Ch. 4). (3) The role of the immune response. Naturally and experimentally infected sheep regularly develop serum antibody to the virus, but this is characteristically slow to appear. Furthermore, the immune response fails to suppress the infection or to contain the evolving pathological lesions. Finally, the possibility must be considered that the pathological lesions are immunologically mediated.

We have embarked on a collaborative study of the pathogenesis of visna, and our recent observations will be summarized in the next section, to set the stage for a more speculative discussion of the questions set forth above. Where pertinent, comparisons with other viral infections will be made. Much of the relevant data on visna and maedi are presented in the foregoing chapters of this book and in reviews by Gudnadóttir (1974), Pálsson (1972), and Thormar and Pálsson (1967).

6.2 Visna in sheep: description of the evolving infection

Two groups of Icelandic sheep were inoculated intracerebrally with a highly neurovirulent strain of visna virus (1514). In the first group (Exp. I) 20 sheep were observed for signs of clinical disease while serial blood and cerebrospinal fluid (CSF) samples were obtained for cell counts and viral isolations. The second group (Exp. II) contained 12 sheep which were killed at 1, 3 and 6 months. Serial studies on these animals are summarized in Fig. 6.1 and Tables 6.1–3.

Isolations of virus were carried out by explantation of tissues; by inoculation of sheep choroid plexus (SCP) cells with serum, CSF, or cell-free homogenates of tissues; and by co-cultivation of explants with SCP cells. Cultures were maintained in roller tubes and observed for cytopathic effect at 7 and 14 days. All negative cultures were then passed into tubes of fresh SCP cells which were observed for an additional 14 days.

Virus was isolated from the spinal fluid (CSF) in most sheep tested 30 days after infection, but the frequency of isolation consistently declined thereafter, so that no isolations were made at 7 or 9 months (Fig. 6.1). Viremia did not appear until about 3 weeks after infection but virus was isolated from about 20% of sheep at each subsequent bleeding (studied to 7 months). All sheep were viremic but only intermittently, thus accounting for the low frequency of recoveries. Virus in the blood was associated with

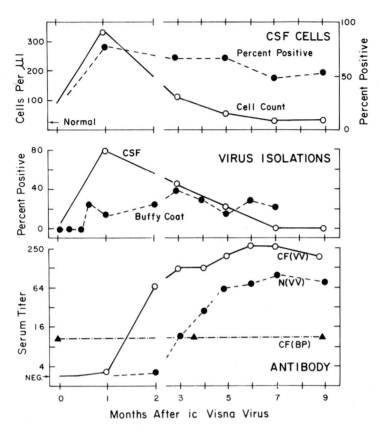

Fig. 6.1 Twenty Icelandic ewes, age about 6 months, were inoculated intracerebrally with $10^{5.5}$ TCD$_{50}$ of visna virus, strain 1514, and studied sequentially for CSF cell count (median), virus in blood and CSF (percent of tested specimens yielding virus), and serum antibody (geometric mean titer) (Exp. I). Antibody: complement fixation and neutralizing titers to visna virus, and complement fixation to basic protein (see 6.3.3).

the leukocyte fraction (buffy coat) and was never isolated from plasma (Table 6.1). In sheep killed at 1, 3 and 6 months, virus was isolated with high frequency from choroid plexus, and less often from selected areas of the neuroparenchyma. The association of infection with lymphoid tissues was indicated by high isolation rates from spleen and lymph nodes. It was consistently observed that virus was present at only minimal levels (Table 6.2). Thus, even a 10-fold dilution of buffy coat or of tissue homogenates reduced the frequency of isolation. Also, blind passage was required for virus recovery in many instances.

TABLE 6.1

Recovery of visna virus from tissues of Icelandic sheep 1–7 months following i.c. inoculation of $10^{5.5}$ TCD_{50} of strain 1514. Exp. 1: repeated tests, 20 days to 7 months after
infection of 20 sheep. Exp. II: tests on 12 sheep sacrificed 1, 3 or 6 months after infection.

Exp.	Tissue	Specimens tested	Positive (no.)	Positive (%)
I	Plasma	132	0	0
	Buffy coat	132	32	24
	CSF	63	25	40
II	Lymph nodes	12	10	83
	Spleen	12	10	83
	Bone marrow	12	2	17
	Choroid plexus	12	11	92
	Cerebellum	12	4	33
	Medulla	12	8	67
	Spinal cord	12	3	25
	Lung	12	7	58

TABLE 6.2

Recovery of visna virus from tissues of Icelandic sheep 20 days to 7 months following i.c.
inoculation of $10^{5.5}$ TCD_{50} of strain 1514. Influence of inoculum size, technique, and
blind passage. Data include Exp. I (Isolations from CSF and buffy coat) and Exp. II
(isolations from tissues). Tissues are those listed in Table 6.1. V/W: volume or weight
tested per roller tube. For buffy coat volumes refer to dilutions of 0.1 ml sample containing 3×10^6 WBC. Tissue weights refer to wet weight. 0, 1, 2 P: number of blind
passages required for isolation. Ex + SCP: explants co-cultivated with sheep choroid
plexus cells.

Specimen	V/W		Specimens tested	Isolations		Positive (%)
				0 P	1 or 2 P	
CSF	0.1	ml	67	18	5	34
Buffy coat	0.1	ml	152	5	22	18
	0.01	ml	152	1	8	6
	0.001	ml	152	3	7	7
Tissues:						
Homogenates	0.01	g	95	16	16	34
Explants	0.1	g	157	3	29	20
Ex + SCP	0.1	g	114	5	24	25

TABLE 6.3

Histological lesions of visna in 12 Icelandic sheep sacrificed 1–6 months following i.c. injection with strain 1514. Grading scale: 1-minimal inflammation; 2-moderate inflammation; 3-marked inflammation, some necrosis; 4-severe inflammation and considerable necrosis. The severity of pathology was similar at 1, 3 and 6 months (Exp. II) (see Ch. 4).

CNS area	Median	Range
Cerebral cortex	0	0–1
Cerebral white matter	1	0–2
Basal ganglia	1	0–4
Midbrain	1.5	0–4
Pons-medulla	1	0–2
Cerebellar cortex	0	0–1
Cerebellar white matter	0	0–1
Cord	1	0–2
Periventricular areas	3	0–4
Leptomeninges	1	0–3

The infection appeared to be cell-associated to a certain extent, as suggested by the ability to isolate virus from buffy coat but never from plasma, as well as frequent isolations from explants when homogenates were negative (20 of 64 isolations). On the other hand, homogenates were sometimes positive when explants were negative (16 of 64). In addition, free virus was occasionally present in the CSF since isolations could be made from supernatants of centrifuged samples in 8 of 20 instances.

The antibody response was slow but very regular (Fig. 6.1), CF antibody appearing between 1 and 2 months and neutralizing antibody between 2 and 4 months. All sheep raised both antibodies, albeit at slightly variable rates.

In order to characterize the antibodies, immunoglobulins from visna-infected sheep were fractionated according to molecular size and charge by various methods (gel filtration, ultracentrifugation, ion-exchange chromatography and others). These studies will be published in detail elsewhere (Pétursson et al., unpublished) but the main findings indicate that neutralizing antibodies are highly charged and can be clearly separated from complement-fixing antibodies by DEAE–cellulose ion-exchange chromatography. Initially, the high charge of the neutralizing antibodies led to the belief that they were not of the IgG classes but perhaps belonged to IgA (Pétursson, 1970). Further studies using specific antisera to the different sheep immunoglobulin classes have not shown any evidence of neutralizing activity associated with IgA. These antibodies seem to be associated with the IgG_1 class but are more highly charged than the average immunoglobulin molecule of this class. Complement-fixing antibodies, however, seem to have a charge distribution similar to the general population of IgG_1 immunoglobulin molecules.

We have not seen any antibody activity in the IgM class and have therefore been unable to confirm the findings of Mehta and Thormar (1974) who observed low titers of neutralizing antibodies to visna associated with the IgM class. A lack of an early IgM response to infection with visna virus if confirmed by further research may indicate an incomplete or defective immune response to this virus in sheep.

The pathological response (see 4.1.3) was reflected in the CSF cell counts which were elevated in 80% of animals at 30 days (Fig. 6.1). Counts dropped thereafter, reaching low levels at 7 and 9 months. Lesions were present in the CNS of all sheep examined at 1, 3 and 6 months and were quite severe at 1 month with no obvious increase by 6 months (Table 6.3). The character of the CNS lesions has been detailed by Georgsson et al. in Ch. 4, but attention should be called to several features. The early lesions were mainly of an inflammatory type, consisting of perivascular and parenchymal infiltrates of lymphocytes and monocytes, with a scattering of plasma cells, while polymorphonuclear leukocytes were absent. Lesions were most severe and often confined to the inoculation site and periventricular areas. Choroid plexus was likewise frequently involved and extensive infiltrates resembling germinal centers were not uncommon.

Many of these results are consistent with previous studies of visna. Sigurdsson et al. (1957, 1958, 1962) described the transmissibility of visna, the characteristic CSF pleocytosis, and the pathological lesions. Both these early reports and our present observations indicate that destruction of myelin follows an intense inflammatory reaction. There appears to be a concomitant destruction of axons and myelin sheaths with little evidence of primary demyelination.

Although only 2 of 24 sheep followed for 1 year developed clinical disease, all sheep appeared to be infected using either elevation of CSF cell count, recovery of virus, or development of antibody as criteria.

Studies of Gudnadóttir and Pálsson (1965a) and de Boer (1970) described sheep observed for periods of several years, and again the findings with respect to CSF, recovery of virus, and histopathology were similar to our own. In addition these investigators found that the majority of sheep inoculated with visna virus intracerebrally developed the pulmonary lesions of maedi, while a few animals inoculated by the intrapulmonary route developed lesions in the CNS as well as the lungs. These observations were interpreted as being consistent with the hypothesis that both visna and maedi were caused by the same agent (de Boer, 1970). In more recent studies Gudnadóttir and Pálsson (Gudnadóttir, 1974) have followed the course of infection for up to 11 years, and have contributed much important long-term data on the pathogenesis of the disease. Almost all sheep, whether inoculated by the intracerebral or intrapulmonary routes, developed CSF pleocytosis, produced virus from choroid plexus, had lesions typical of both visna and maedi, and eventually died of clinical visna. CSF cell counts were almost invariably elevated, and virus could be recovered from blood and CSF during the first few months after inoculation. In individual animals these parameters fluctuated greatly over the course of years, although the animals remained persistently infected until the time of death. An additional interesting observation was that virus isolated from individual sheep late in the course of disease appeared to differ antigenically from virus isolated earlier, thus escaping the full effect of neutralizing antibody and perhaps providing a mechanism for viral persistence.

As these reports show, although the clinical disease may be very protracted, many reliable parameters of infection appear quite early. Our results, though confined to the first year of infection, have confirmed most of these findings and thus provide a useful model for relatively short term studies of viral and immunological mechanisms in visna.

6.3 Speculations upon the pathogenesis of visna

6.3.1 The 'slowness' of infection

A notable feature of visna and maedi is the very minimal levels of virus found, which remain at consistent trace levels. The requirement for blind passage and for explantation (with or without co-cultivation) techniques for maximal isolation indicates that infected cells are producing only trace amounts of free infectious virus. If it is assumed that the slowness of infection is due to these minimal levels of infectivity, there are several possible mechanisms which require comment.

(1) Immunological damping of infection. Although infected sheep produce antibody capable of neutralizing virus, there is little evidence that antibody could play a role. Thus, virus levels are low from the outset of

infection prior to advent of the immune response. Furthermore, thym-
ectomized antilymphocytic serum (ALS)-treated fetal sheep yield the same
minimal amounts of virus (Narayan et al., 1974, and unpublished) indicating
that cell-mediated immunity does not suppress infection. Later in infection
there is some evidence for an effect of antibody indicated by the disappearance of virus from CSF (Fig. 6.1) or blood (de Boer, 1970), but this only
occurs 3–6 months after inoculation.

(2) Requirement for DNA synthesis in host cells. A round of DNA
synthesis may be required for replication of visna in permissive cells, as
for other C-type viruses (Humphries and Temin, 1974) and it could be
argued that the few cell types permissive for visna virus rarely divide in adult
sheep. However, the finding that the same minimal virus levels occur in fetal
sheep (Narayan et al., 1974), where most cell populations are rapidly
proliferating, makes this mechanism unlikely. It is of interest that the
addition of a mitogen (concanavalin A) to buffy coat cultures does not
enhance the isolation of virus therefrom. Finally visna replicates well in
stationary cultures of ovine cells (Haase and Varmus, 1973).

(3) Defectiveness of the virus. In certain slowly progressive CNS infections, particularly subacute sclerosing panencephalitis and progressive
multifocal leukoencephalopathy, the causal viruses can only be isolated with
difficulty, and even when present in tissue explants it may be difficult to
induce production of infectious or rapidly replicating virus (Payne et al.,
1969; Padgett et al., 1971; Weiner et al., 1972). By contrast, the evidence does
not indicate that visna virus is defective, since all isolates quickly grow
to high titer and readily infect ovine cell cultures. Furthermore, titration
data fail to suggest the presence of a helper virus in permissive cell cultures
(Haase, unpublished).

Another possibility is the synthesis in vivo of defective interfering virus
which might effectively retard the spread of infection. Although there is no
direct evidence for this mechanism, recent work with vesicular stomatitis
virus (Doyle and Holland, 1973) suggests that this should be considered
in future studies.

(4) Limited permissiveness of sheep cells in vivo. The minimal amounts
of free infectious virus in infected sheep suggest a restriction at some step
in the replicative cycle. A precedent for this postulated semipermissiveness
is provided by early observations of Thormar and Sigurdardóttir (1962) who
found that primary cultures of human, canine, or porcine origin would yield
only low levels of visna virus over a 1–2 month period (see 5.3.3). A similar

finding was reported by Harter et al. (1968) for a line of bovine kidney (MDBK) cells.

If one turns to other C-type viruses, it is clear that host restriction can occur at different points in the replicative cycle. Thus, the host range of Rous sarcoma viruses in chicken fibroblasts is determined by surface receptors; a nonpermissive cell type will bind viruses of restricted subgroups but subsequent steps in infection fail to proceed. In certain nonpermissive chicken cells infection is quantitatively reduced but not absolutely excluded (Duff and Vogt, 1969). Murine leukemia viruses (MuLV) provide another example of semipermissiveness. Many MuLV are either N- or B-tropic; thus, N-type mouse fibroblasts are at least 100-fold less sensitive to B-tropic than to N-tropic MuLV (Hartley et al., 1970). The nature of the restriction is not well understood, but apparently occurs after virus penetration (Pincus et al., 1971; Krontiris et al., 1973).

It is clear that, regardless of the molecular mechanism, if sheep cells in vivo are very poorly permissive, the infectious process may proceed at a slow pace. This hypothesis implies that there is a major difference between the behavior of ovine cells in vivo and in vitro, but there are precedents for such differences, notably the classic example (Holland, 1961) of the monkey renal cell which becomes permissive for poliovirus in culture, due to the appearance or unmasking of specific receptors.

(5) The occurrence of an integrated provirus in visna replication can also be invoked to explain 'slowness'. In this view many cells may be latently infected and occasional activation of virus could then produce either virus-mediated cell destruction or limited amounts of viral antigen which trigger foci of immunological damage.

6.3.2 Visna infection: a hypothetical reconstruction

It is possible to construct a hypothetical description of the events which follow intracerebral infection with visna virus, based on the considerations already discussed. Following infection, the inoculum is partially deposited focally in the neuroparenchyma and a considerable proportion probably enters the CSF by regurgitation along the needle track into the ventricles and subarachnoid space. The distribution of the inoculum may play a particularly important role in determining initial cellular sites of replication, because secondary amplification is so retarded due to the in vivo restriction of virus replication. It is also conceivable that ependymal cells of the ventricle and choroid plexus are relatively more permissive for virus. This may

TABLE 6.4

Virus isolations and pathological lesions 30 days following i.c. inoculation of $10^{5.5}$ TCD_{50} of visna virus, strain 1514. To illustrate the different course of infection in 2 groups of 4 Icelandic sheep inoculated under identical conditions. Two of the sheep in each group received a course of ALS which had no apparent effect. CSF: range of counts. Isolations: multiple specimens of CNS and lymphoreticular tissues were tested.

	'Early' lesions	No 'early' lesions
No. of sheep	4	4
CNS pathology:		
choroiditis	4/4	0/4
periventriculitis	4/4	0/4
leptomeningitis	4/4	0/4
neuroparenchyma	4/4	0/4
CSF cell count	44–440	10–36
Virus isolations:		
CSF	3/4	1/4
CNS	13/16	2/16
lymphoreticular	12/16	1/16
buffy coat	2/2	0/4

account for the striking periventricular localization of lesions, which is perhaps most notable in the spinal cord where a rather unique lesion is seen, i.e., a focus of inflammation in grey matter around the central canal (Fig. 4.6).

Observations from a recent experiment (Table 6.4) support the foregoing reconstruction. A group of 8 sheep was inoculated intracerebrally and killed one month later. Two different outcomes were observed. One group exhibited 'early' lesions, with elevated cell count, severe choroiditis and periventricular cuffs, and a high frequency of virus isolations from CSF and CNS tissues. The other animals had few if any lesions and few CNS samples yielded virus. It is suggested that this marked dichotomy may be determined by the localization of the initial inoculum, so that heavy seeding of the intra-ventricular CSF produces widespread cellular infection and resultant early lesions.

An instructive comparison may be made with several acute viral infections of the CNS. Certain non-neuroadapted strains of influenza are capable of infecting ependymal cells of the mouse brain after intracerebral injection, but undergo an abortive cycle of replication without production of infective progeny (Mims, 1960). The resulting acute pathology is confined to the inoculation site, meninges and choroid plexus. Lymphocytic choriomeningitis (LCM) virus replicates first in ependymal and pial cells after brain injection of the adult mouse, again with acute choriomeningitis. In immunosuppressed animals, where the virus persists indefinitely, infection gradually spreads (over 1–3 months) to selected glial elements, particularly of the cerebellum. The disease which follows immunological reconstitution of these persistently infected animals likewise includes necrotic lesions of the cerebellar cortex (Gilden et al., 1972).

These examples suggest by analogy that a distribution of lesions reminiscent of visna can be a consequence of intracerebral infection of non- or semi-permissive hosts. Another phenomenon which may be related is the observation in our recent studies that rather marked pathological lesions are seen one month after infection and that these then progress quite slowly. The high frequency of isolation of cell-free virus from CSF one month after injection suggests early infection of cells lining the ventricle while the low titers again reflect the limited potential for virus amplification.

Although early lesions of the choroid plexus and periventricular zones do not indicate any predilection for white matter, it is clear that subsequent extension into the neuroparenchyma has some tendency to concentrate in major myelinated tracts. This in turn suggests a possible preferential replication in glial elements which are more common in white matter.

The immune response to visna, as reflected in serum antibody, is raised only slowly but eventually becomes marked. The 1–2 month interval before appearance may reflect the tardy and gradual presentation of antigen to responsive lymphoid cells (i.e., viremia is not seen until 3 weeks after inoculation) rather than any 'weakness' of antigenicity or hyporesponsiveness of the host. It is evident that the immune response cannot cope with the infection, perhaps seen most clearly in the blood, where the frequency of isolation from buffy coat continues at the same frequency in the face of rising neutralizing antibody titers. De Boer (1970) has observed, however, that after antibody reached high titers, he could increase the frequency of isolations by washing leukocytes prior to culturing. Likewise, it seems probable that a local production of antibody occurs in the CNS, in view of the participation of plasma cells in the leukocytic infiltrate (see 4.1). This could well account for the gradual disappearance of infectivity from the CSF, since antibody is quite capable of neutralizing cell-free virus. In turn, sterilization of the CSF might favor a movement of the evolving infection away from the ventricles with the passage of time.

6.3.3 Pathological lesions: their nature and mechanism

Visna virus produces a destructive effect upon permissive cells in vitro. Although this has not been studied in detail, cytological observations (Thormar and Sigurdardóttir, 1962) indicate that cell fusion and polykaryo-cytosis are prominent components of the cytopathic effect (see 5.2.1). Furthermore, it has been convincingly shown (Harter and Choppin, 1967) that non-infectious UV-inactivated virus can produce fusion and lysis, demonstrating that a destructive effect may indeed be initiated by the fusion of cell membranes. Among C type viruses there are precedents (Klement et al., 1969) for the view that viral biogenesis is not innately destructive, except for the occurrence of syncytia. In fact, Harter et al. (1968) described a porcine cell line in which visna virus replicated vigorously without a cytopathic effect.

The mechanism of visna-induced CNS lesions may be considered in the light of the foregoing observations. It seems apparent that the characteristic evolving lesion has none of the stigmata that might be expected of a virus which destroys through a fusion effect. Parenthetically, it may be noted that virus-induced syncytia in the CNS can be readily detected following infection with certain enveloped viruses such as measles (Baringer and Griffith, 1970).

The early lesions produced by visna virus are leukocytic infiltrates, at first indistinguishable in character from those induced by numerous other viruses. In our current studies it has been possible to look in further detail at the participating cell types (see 4.1.2). The majority are mononuclear cells, both lymphocytes and monocytes, with essentially no polymorpho-nuclear leukocytes. An infiltrate of this character would be typical of other cell-mediated immune responses in the CNS, such as those seen in lympho-cytic choriomeningitis and experimental allergic encephalitis, both of which are T cell-dependent immunopathologies.

Taken together, these points suggest that the initial lesions produced by visna virus in the CNS may be mediated by an immunologic mechanism. As the disease progresses inflammatory lesions become extensive in myelinated tracts and foci of white matter destruction appear. At this stage it is difficult to dissect the chain of events from morphological observations, which show invading mononuclear cells interspersed among apparently normal myelinated fibers. The picture is reminiscent of EAE (Prineas et al., 1969) and of demyelinating distemper (Innes and Saunders, 1962; McCullough et al., 1974), and consistent with an immunological process.

Although the evidence that an immunological mechanism underlies visna is not yet at hand, it is reasonable to consider which antigens might be incriminated. Viral envelope glycoproteins are inevitably present on the cell membrane during the budding process by which C-types viruses undergo morphogenesis. An immune response directed against these membrane antigens could initiate the pathological events described. This hypothesis is attractive because it also could account for the simultaneous occurrence of similar lesions in the lung (see 4.4).

An additional possibility is that, during the course of chronic CNS disease, visna virus-infected sheep are sensitized to their own myelin antigens, so that an autoimmune process plays a role in the evolving white matter destruction. The possible role of an anti-myelin immune response is suggested

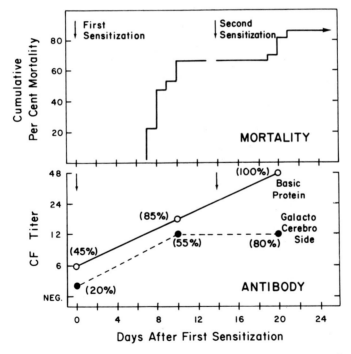

Fig. 6.2. Experimental allergic encephalitis in 23 American sheep (Hampshire breed) following sensitization with whole brain and complete Freund's adjuvant. Animals surviving 14 days after first injection were resensitized. All sheep which developed signs became moribund within 2–3 days. CF antibody titers are geometric means, and the percentages indicate the proportion with detectable antibody ($\geq$ 1:6).

by the similarities between the histological appearance of EAE and visna, noted above.

As part of our investigation of visna, a study of EAE in sheep was therefore undertaken, and selected data are summarized in Fig. 6.2. First, it was established that intradermal and intramuscular inoculation of whole sheep brain emulsified in complete Freund's adjuvant (WB-CFA) was capable of eliciting EAE. About 70% of animals developed acute disease 9–11 days after a single injection and about 20% responded to a second injection given 14 days later, with a shorter interval (5–7 days). A residual 10–15% failed to show symptoms after two injections. The disease in sheep was somewhat unusual in that all animals which developed clinical signs experienced a fulminating course, becoming moribund within 2–3 days after onset, while unaffected sensitized sheep had no apparent illness and the few examined showed no lesions in the CNS. Also, it appears that the high inducibility of EAE may be more characteristic of the Hampshire breed, while Icelandic sheep are less susceptible (3 of 6 developed signs or pathology after WB and CFA). This may explain the discrepancy between our success and the relatively limited inducibility reported by Innes (1951).

Serological studies have shown that sensitized sheep develop rapidly rising CF antibodies to basic protein and to galactocerebroside, the only two well-characterized myelin antigens (Fig. 6.2). In addition, an extensive effort to develop an assay for cell-mediated immunity has indicated that a minimal lymphocyte transformation response to basic protein (1.5–2.5-fold) appears in sheep with acute EAE.

Application of these findings to visna-infected sheep has yielded several preliminary observations. (1) A high proportion of Icelandic sheep carry antibody to myelin antigens, particularly basic protein. (2) There is no evidence that these antibody levels change during visna infection (Fig. 6.1). (3) Blast transformation has revealed no evidence that visna-infected sheep have a cell-mediated response to basic protein, but a better indicator of this response must be developed before the evidence can be considered conclusive. Macrophage migration inhibition for example may correlate better with cell-mediated immunity to basic protein than lymphocytic transformation (Vandenbark and Hinrichs, 1974).

6.3.4 Directions for the future

The views developed in the foregoing sections clearly point to a number of experimental approaches which require exploration. These include:

(1) The effect of immunosuppression upon visna lesions. A preliminary experiment, using antilymphoid serum, failed to suppress early lesions, but ancillary data suggested that a more vigorous approach will be required to achieve adequate immunosuppression. (2) The intimate details of the evolving lesion. Material is now at hand for ultrastructural studies of the visna lesion, but our search for visna antigens awaits production of a satisfactory immunofluorescent conjugate. (3) The cell-mediated immune response in visna. Techniques (blast transformation, MIF production, or cytolytic assay) which will indicate the cell-mediated response to myelin and visna antigens are under development. Their application should provide further insights regarding the immunopathogenesis of visna.

6.4 Comparative pathogenesis of visna

6.4.1 Maedi

Maedi and visna are probably best considered different aspects of the same infection, as was indicated above. Several factors appear to have an influence over the relative degree of involvement of brain or lung, including the route of infection and passage history of the infecting strain. Under natural conditions infections are transmitted horizontally, presumably by the respiratory route, and this tends to favor pulmonary involvement, maedi being more common than visna (see Ch. 2). However, it is clear that either CNS or pulmonary inoculation of lung or CNS isolates can cause both maedi and visna (Sigurdsson et al., 1953; Gudnadóttir and Pálsson, 1965a, b, 1967; Gudnadóttir, 1974).

The replication of visna–maedi virus in the lung is subject to the same restrictions seen in other tissues, and the same mechanisms undoubtedly are operative. The lesions of maedi (see 4.4) consist mainly of a round cell infiltrate into the interstitial tissue, and perivascular cuffing and sheets of lymphoid cells are common. As discussed above, these features suggest that the mechanism of disease production is similar to that operative in visna. However, the pathophysiology may be different, since the accumulating infiltrate in the lungs appears gradually to compromise alveolar exchange and pulmonary elasticity rather than produce any striking destruction of tissue.

6.4.2 Aleutian mink disease and the spongiform encephalopathies

The three slow infections (visna–maedi, Aleutian mink disease, and scrapie) considered in this book are strikingly different, both as to their etiological agents and pathogenesis. The spongiform encephalopathies produce no immunological response yet detectable (see Ch. 14). Aleutian mink disease involve primarily B cells which undergo clonal proliferation with production of vast amounts of specific antibody and an immune complex-initiated pathology (see Ch. 9). Visna–maedi causes a lymphoid infiltrate which resembles cell-mediated immunopathologies and which may be T cell-dependent. Thus, the three entities illustrate the wide diversity of slow infections and analysis of each is an independent undertaking which casts little light on the others.

Acknowledgements

Albert Lossinsky, Roger Lutley and Kolbrun Kristinsdóttir have contributed expert technical assistance. Robert Kibler kindly made the sheep basic protein for us and Ashley Haase contributed helpful suggestions on the manuscript.

References

BARINGER, J. R. and GRIFFITH, J. R. (1970) Lab. Invest., 23, 335.
DE BOER, G. F. (1970) Thesis, Utrecht.
DOYLE, M. and HOLLAND, J. J. (1973) Proc. Natl. Acad. Sci. U.S.A., 70, 2105.
DUFF, R. G. and VOGT, P. K. (1969) Virology, 39, 18.
GILDEN, D. G., COLE, G. A. and NATHANSON, N. (1972) J. Exp. Med., 135, 874.
GUDNADÓTTIR, M. (1974) Prog. Med. Virol., 18, 336.
GUDNADÓTTIR, M. and PÁLSSON, P. A. (1965a) J. Immunol., 95, 1116.
GUDNADÓTTIR, M. and PÁLSSON, P. A. (1965b) J. Infect. Dis., 115, 217.
GUDNADÓTTIR, M. and PÁLSSON, P. A. (1967) J. Infect. Dis. 117, 1.
HAASE, A. T. and VARMUS, H. E. (1973) Nature, New Biol., 245, 237.
HAASE, A. T. and BARINGER, J. R. (1974a) Virology, 57, 238.
HAASE, A. T., GARAPIN, A. C., FARAS, A. J., VARMUS, H. E. and BISHOP, J. M. (1974b) Virology, 57, 251.
HAASE, A. T., GARAPIN, A. C., FARAS, A. J., TAYLOR, J. M. and BISHOP, J. M. (1974c) Virology, 57, 259.
HARTER, D. G. and CHOPPIN, P. (1967) Virology, 31, 279.
HARTER, D. G., HSU, K. C. and ROSE, H. M. (1968) Proc. Soc. Exp. Biol. Med., 129, 295.
HARTLEY, J. W., ROWE, W. P. and HEUBNER, R. J. (1970) J. Virol., 5, 221.
HOLLAND, J. J. (1961) Virology, 15, 312.
HUMPHRIES, E. G. and TEMIN, H. M. (1974) J. Virol., 14, 531.
INNES, J. R. M. (1951) J. Comp. Pathol., 61, 241.
INNES, J. R. M. and SAUNDERS, L. Z. (1962) Comparative Neuropathology. (Academic Press, New York.)

KLEMENT, V., ROWE, W. P., HARTLEY, J. W. and PUGH, W. E. (1969) Proc. Natl. Acad. Sci. U.S.A., 63, 753.

KRONTIRIS, T. G., SUEIRO, R. and FIELDS, R. B. (1973) Proc. Natl. Acad. Sci. U.S.A., 70, 2549.

MCCULLOUGH, B., KRAKOWKA, S., KOESTNER, A. and SHADDUCK, J. (1974) J. Infect. Dis., 130, 343.

MIMS, C. A. (1960) Br. J. Exp. Pathol., 41, 586.

MEHTA, P. D. and THORMAR, H. (1974) Infect. Immun. 10, 678.

NARAYAN, O. T., SILVERSTEIN, A. M., PRICE, D. L. and JOHNSON, R. T. (1974) Science, 183, 1202.

PADGETT, B. L., WALKER, D. L., ZURHEIN, G. M. and ECKROADE, R. J. (1971) Lancet, 1, 1257.

PÁLSSON, P. A. (1972) J. Clin. Pathol. 25 (suppl. 6), 115.

PAYNE, F. E., BAUBLIS, J. V. AND ITABASHI, H. H. (1969) N. Engl. J. Med., 281, 585.

PÉTURSSON, G. (1970) Proc. 6th Int. Congr. Neuropathol., 831.

PINCUS, T., HARTLEY, J. W. and ROWE, W. P. (1971) J. Exp. Med., 133, 1219.

PRINEAS, J., RAINE, C. S. and WISNIEWSKI, H. (1969) Lab. Invest., 21, 472.

SIGURDSSON, B. (1954) Br. Vet. J., 110, 255.

SIGURDSSON, B. and PÁLSSON, P. A. (1958) Br. J. Exp. Pathol., 39, 519.

SIGURDSSON, B., PÁLSSON, P. A. and GRIMSSON, H. (1957) J. Neuropathol. Exp. Neurol., 16, 389.

SIGURDSSON, B., PÁLSSON, P. A. and TRYGGVADOTTIR, A. (1953) J. Infect. Dis., 93, 166.

SIGURDSSON, B., PÁLSSON, P. A. and VAN BOGAERT, L. (1962) Acta. Neuropathol., 1, 343.

THORMAR, H. and PÁLSSON, P. A. (1967) Perspect. Virol., 5, 291.

THORMAR, H. and SIGURDARDÓTTIR, B. (1962) Acta Pathol. Microbiol. Scand., 55, 180.

VANDENBARK, A. A. and HINRICHS, D. J. (1974) Cell Immunol., 12, 85.

WEINER, L. P., HERNDON, R. M., NARAYAN, O., JOHNSON, R. T., SHAH, K., RUBINSTEIN, L. J., PREZIOSI, T. J. and CONLEY, F. K. (1972) N. Engl. J. Med., 286, 385.

Part III

Aleutian disease of mink

The disease

The virus

Virus–host interactions

CHAPTER 7

The epizootiology of Aleutian disease

J. R. GORHAM, J. B. HENSON, T. B. CRAWFORD and G. A. PADGETT

7.1 Introduction

Aleutian disease (AD) is the most severe infectious disease of farm-raised mink. The mortality rate is frequently very high in mink of the Aleutian genotype *(aa)*. Although death is less common in non-Aleutian mink *(AA, Aa)*, subtle losses in pelt value and decreased production have resulted in annual losses of millions of dollars. No other disease problem has contributed more to the mink farmer's final decision to discontinue mink raising (Hartsough, 1975). This slow viral disease has been reported in all commercial mink producing countries of the world.

Scandinavian veterinarians have the most accurate estimate of AD prevalence in populations of farm-raised mink. Helgebostad (1970) reported that when 133,000 mink were tested in Norway by means of the iodine agglutination test (IAT), with its inherent limitations, 17% reacted to the test. Danish veterinarians, using the same criterion of infection, determined that prevalence of AD was about 20% in Denmark in the winter of 1973/1974 (Hansen, personal communication). They tested about 250,000 of a total of one million breeder mink. Surveys of this magnitude are not available from other countries.

The present review pivots on a previous publication (Gorham et al., 1965). We will attempt to record established facts in the development of several central themes. Hopefully, our dogmatic deductions will be attacked and stimulate a more critical evaluation.

Supported in part by grants AI 06477, AI 06591 and RR 00515 from the National Institutes of Health and the Mink Farmers Research Foundation.

Slow virus diseases of animals and man, edited by R. H. Kimberlin
© *North-Holland Publishing Company 1976*

Fig. 7.1. Aleutian mink genotype *aa* are also called blue mink. All mink of this genotype have the Chediak–Higashi Syndrome (CHS).

7.2 Mink raising

Because of its specialized nature, a brief account of a commercial mink farming operation is helpful in understanding the epizootiologic features of AD. Female mink (Fig. 7.1) are bred in March and have their young in May. The kittens are weaned in July and are usually separated into individual woven-wire pens to prevent fighting and pelt damage. They have a fur molt in early fall and are killed when their pelt is prime in November. Highly desirable kits and some adults are maintained for the next breeding season at a ratio of one male to 5 adult females. Other than at breeding time and before the time the kittens are weaned and separated, all mink are usually housed in individual wire pens. Individual wire pens are placed in long open sheds for convenience in feeding and handling. Frequently, as many as 2000 mink are crowded onto 4000 square meters of land. Because of the concentration of the mink, the distemper virus that uses the respiratory route and mink virus enteritis that uses an intestinal–oral route spread easily through unvaccinated herds.

7.3 The host

7.3.1 Early history

In 1941, two gun-metal colored mink kittens were noticed in a litter of standard dark, wild-type mink. The dam of these mink was the descendant of wild mink trapped in Northwest Oregon, and the sire was the descendant of mink obtained from the Yukon area of Alaska. The 'off-colored' mink were bred, and the results showed that the coat color was inherited as an autosomal recessive trait. They were called Aleutian mink because their pelage resembled the coat color of the Aleutian blue fox. These mink and their progeny did not succumb to AD, but the owner (Waris, personal communication) recognized a 'weakness' in that they died of a variety of other causes, including heat exhaustion.

The recessive trait was designated *a* (Shackelford, 1950). When Aleutian mink were bred with other recessive color phase mutations, several recessive mutations were developed and collectively called 'blue mink'. High pelt prices resulted in a ready sale of breeding stock. Enterprising mink farmers distributed the Aleutian genotype *a* (blue mink) throughout the mink-raising areas of the world. Soon disturbing field reports indicated that raising Aleutian mink was difficult because they were affected by a condition

J. R. Gorham et al.

characterized by a slowly progressive downward course that extended over a few months and always terminated in death.

Because the condition was seemingly confined to Aleutian mink, it was thought to be hereditary and was called Aleutian disease (AD). Interestingly, AD fits the pattern of other slow virus diseases (scrapie and kuru) in that they were considered at one time to be genetic diseases.

7.3.2 Chediak–Higashi syndrome

Padgett et al. (1964, 1968) pointed out that the Aleutian mutation was actually an example of the Chediak–Higashi syndrome (CHS). Although the mechanism was not apparent, we had an explanation for the remarkable responsiveness of the Aleutian mink to AD virus. CHS is an autosomal recessive disease syndrome first described in man (Beguez-Cesar, 1943) but also found in cattle (Padgett et al., 1964), in mice (Lutzner et al., 1967), in a killer whale (Taylor and Farrell, 1973) and in a cat (Kramer et al., 1975). The primary features of the CHS are pigmentary dilution (partial oculo-cutaneous albinism), repeated and severe bacterial infections and abnormally enlarged granules considered to be lysosomes. The fundamental defect that leads to increased susceptibility to pathogenic bacteria is likely due to impaired function of peripheral blood granulocytes. We believe that the abnormal lysosomes of CHS mink have an impaired ability to catabolize immune complexes that deposit in the kidney in AD and evoke glomerular disease (see Ch. 9).

7.3.3 Aleutian disease in CHS mink

All ages and either sex are susceptible to AD. Clinical signs include anorexia, weight loss, polydipsia, lassitude, pale mucous membranes and foot pads (anemia), rare nervous signs and coma that precedes death. A progressive, diffuse, immune complex glomerulonephritis (Henson et al., 1969; Porter et al., 1969; Pan et al., 1970) with elevated levels of blood urea nitrogen, anemia, thrombocytopenia (Eklund et al., 1968) and hypergammaglobulinemia typifies the unremitting course. We feel that all CHS mink infected with the Pullman or Utah 1 isolates are viremic for life.

7.3.4 Aleutian disease in non-CHS mink

Hartsough and Gorham (1956) recognized that AD occurred in non-CHS

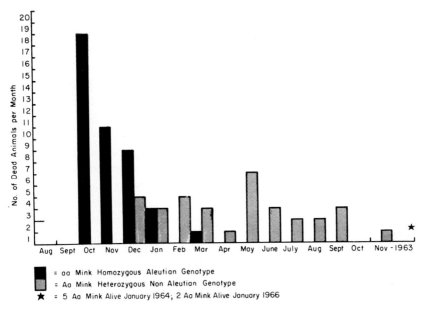

Fig. 7.2. Time of death of 40 CHS *aa* and 40 non-CHS *Aa* after intraperitoneal inoculation of 10^4 ID$_{50}$ of the Pullman AD virus isolate. The mean death-time of the CHS mink
was 102 days post-inoculation, whilst in the non-CHS mink it was 241 days, excluding
the 5 mink alive in January, 1964, (from Padgett et al., 1968).

mink. In contrast to the 100% case/fatality rate that North American observers have recorded in CHS mink, the response of non-CHS mink to AD
virus varies, and the prognosis is unpredictable. Padgett et al. (1968) injected
40 CHS and 40 non-CHS mink with a high dose of AD virus (Fig. 7.2).
Eighteen of the CHS mink died within 70 days after injection, 28 were dead
after 100 days and at that time all of the non-CHS mink were still alive.
Five non-CHS mink survived for more than 2½ years and 2 survived for
4½ years. Quantitive virologic studies by Eklund et al. (1968) demonstrated
that CHS mink are much more responsive to the Pullman isolate than non-
CHS mink.

Some non-CHS mink show all of the clinical features and lesions as described for CHS mink, but the disease is more protracted. Other non-CHS
mink seemingly recover – at least clinically. These non-CHS mink show only
a short period of viremia of about 1 month, with a transient rise in serum
gamma globulin (Padgett, 1969). Occasionally, non-CHS mink are not
hypergammaglobulinemic when infected with the Pullman isolate (Hadlow,

W. J., personal communication). There is no doubt that virus strains play a role, along with the mink genotype, in determining the nature and severity of the disease. Such studies are underway at the Rocky Mountain Laboratory, Hamilton, Montana.

Larsen and Porter (1975) injected a group of pastel mink (non-CHS) to determine whether the mink were 'non-persistently infected'. These were mink that developed a transient elevation of serum gamma globulin and had markedly lower specific AD antibody titers than 'persistently infected' mink. No AD lesions were found in the non-persistently infected mink. They bred the mink that were non-persistently infected and challenged their offspring with $10^5 ID_{50}$ of the same passage of AD virus used to infect their parents and found that 21 % of the offspring were non-persistently infected. Because 29 % of their unselected parents were non-persistently infected, Larsen and Porter felt that the ability to develop non-persistent AD infection was not genetically determined.

Before definitive work can be done on the influence of the mink genotype in determining something more than a lack of clinical response or recovery, we need more information on the kinetics of replication of different AD virus isolates in non-CHS mink. Recently, Hadlow (personal communication) found the Pullman virus isolate in the mesenteric lymph nodes of non-CHS mink 40 months after infection.

Hemmingsen and Heje (1964) had previously studied the role of genetic factors in controlled breeding of mink on a farm where AD was enzootic. They concluded that the disposition for AD does not depend solely on a single gene but on a polygenic system.

7.3.5 The ferret

AD is a natural occurring disease in commercial and experimental ferret colonies. Ferrets may exhibit hypergammaglobulinemia but not glomerular disease. Occasionally, affected ferrets may succumb after a long protracted course. They are much less responsive to the Pullman isolate of AD than non-CHS mink. Kenyon et al. (1966) described AD in ferrets and found that the virus persisted for as long as 136 days after injection. Even though the disease occurs in the ferret, the ferret is not very likely to be a natural reservoir of the virus.

7.3.6 Other carnivores

Serums from other wild carnivores were found to contain specific antibody by countercurrent electrophoresis (CEP). Ingram and Cho (1974) found positive samples in 2 of 100 foxes, 1 of 27 raccoons and 128 of 196 skunks. The importance of these animals in maintaining the virus in nature is an enigma. The skunk, particularly, deserves further investigation. We have experimentally injected AD virus into dogs and cats and have observed a transient positive serologic response with CEP. The dogs and cats appeared normal during the 6-month observation period.

7.3.7 Man

Epidemiologically, there has been no causal association between mink farming and a human disease that resembled AD. Porter and Larsen (1974) feel that the purported instances of human AD were probably cases of lymphoma. We agree with them when they state that it would seem prudent to be cautious when handling AD virus, particularly highly concentrated laboratory preparations of the agent. We have evidence of development of antibody that reacts with AD virus in workers in our laboratory. The length of persistence of antibody appears to be variable. The nature, specificity and duration of these responses is under investigation.

7.3.8 Age incidence

Disease signs in naturally occurring cases are rarely seen in mink less than 4 months of age. The onset of clinical disease is insidious in any genotype and usually not apparent in CHS mink until 1 month to 6 weeks after the mink becomes hypergammaglobulinemic. However, some CHS mink die after a short course from massive brain hemorrhage or from acute brain swelling with coning of the cerebellar vermis into the foramen magnum (Eklund et al., 1968). Padgett (unpublished observations) recorded the incidence in mink kittens by the IAT from 2 separate groups of CHS females in one herd in succeeding years. Whereas the IAT was used for diagnosis during the field study, all positive mink were autopsied after pelting and tissues collected for histopathology. The correlation between AD lesions and IAT positivity was greater than 90% in this totally CHS herd. Data in Table 7.1 revealed that the percentage of new cases per month (summer and fall) in kittens varied between 5.1 and 16.7. By the time the kittens were ready to be pelted in November, the prevalence of hypergammaglobulinemia among them was 30% in the 1964 group and 51% in the 1965 group. The relatively high (13.8%), incidence in November 1965, is probably related

J. R. Gorham et al.

TABLE 7.1

Incidence of Aleutian disease by iodine agglutination test in two groups of female mink and their young, 1964 and 1965.

Year	Adult breeder females negative to IAT in March	Kittens	Incidence of IAT positivity of kittens					Total IAT positivity in November	
			July	August	September	October	November	Kittens	Adult females
1964	85	396	26(6.6%)	24(6.1%)	20(5.1%)	22(5.6%)	29(7.3%)	121(30.5%)	38(44.8%)
1965	91	450	75(16.7%)	35(7.8%)	43(9.6%)	16(3.6%)	62(13.8%)	231(51.3%)	55(60.4%)

to the time of testing; i.e., the kit populations was tested early in October, but late in November. Furthermore, of the adult breeding females that had a negative IAT test the preceeding March, 44.8 % (1964) and 60.4 % (1965), were positive in November. These females were infected by contact.

Mink farming operations preclude the determination of accurate age incidence and mortality rate data on a year-round basis because all clinically ill and unthrifty mink are pelted each November. The age incidence data for infection should be based on results of sequential CEP or other tests for specific antibody.

7.4 *The virus*

7.4.1 *Early history*

AD probably was present in wild mink populations for untold generations. That most isolates of AD are not highly virulent for wild mink suggests long-term coexistence. Because mink lead a solitary life, the virus was not easily transmitted in nature by direct and indirect contact. But vertical transmission of the virus from the wild female to her kittens insured maintenance of the virus from one generation of mink to the next. Ingram and Cho (1974) demonstrated AD antibody in 16 of 29 sera collected from trapped wild mink. Many mink escape from mink farms; thus, one cannot be certain that their survey represented a valid wild population. However, this level of positivity in wild mink is a strong argument for the occurrence of AD in wild mink populations before commercial mink raising.

Thus, we suspect that AD was smoldering in the dark mink herds of North America from the time they were first raised in captivity. If this assumption is valid, then the scattered AD losses apparently did not attract the attention of the owner or diagnostician. Many early commercial mink herds were probably free of the virus, because the purchase and exchange of breeding stock was not a widespread practice until the forties.

Then the CHS mutation suddenly entered the picture. To this day, this highly responsive, albeit expensive, genotype serves as the sentinal signifying new foci of infection. Although the CHS mink have been blamed for spread of the disease, more likely the non-CHS mink with their proclivity for inapparent infections and longer disease course are the most important reservoir and source of the virus.

7.4.2 *Virus isolates*
There are at least 4 commonly used North American virus isolates of AD.

For convenience in this publication, we will designate them as to the location of the research group where the isolation was made: Ontario (Canada), Pullman (Washington), Utah-1 and Connecticut. These isolates have not been compared by classical virologic procedures because of animal expense and the lack of in vitro assay systems. Recently, Porter et al. (1975) reported the growth of AD virus in feline kidney cells maintained at 31.8 °C. If the utility of the system is confirmed, it should aid in the differentiation of possible strains, assuming specific antibody neutralizes AD virus in vitro.

A few observations suggest the likelihood of different AD virus strains. The case/fatality rate of all North American isolates that we are aware of for CHS mink must approach 100%. We know of no CHS mink that has recovered from the malady. However, Haagsma (personal communication) has a Netherlands isolate (designated 70261) that is definitely less virulent for CHS mink than his other isolates. After experimental infection with the 70261 isolate, CHS mink have remained clinically normal for more than 4 years. The incubation period based on positivity of the IAT varied from 2 to 40 months. The use of Haagsma's isolate in experimental studies should be helpful in elucidating not only the epizootiology of AD, but also the pathogenetic mechanisms.

Most experienced observers have noticed the rare fulminant AD epizootic in non-CHS mink that behaved similarly to AD in CHS herds. Although there may be gradations in the resistance of non-CHS mink, the probability of highly virulent AD strains seems more likely.

Hadlow and Eklund (personal communication) have shown that the Utah-1 isolate is more virulent for non-CHS mink than the Pullman isolate. Hartsough (personal communication), a careful observer with considerable field experience, has noted the possibility of AD strains. He has observed several herds that had an annual and expected low incidence of AD. New breeding stock that were unknowingly carriers of AD were introduced onto these ranches. Within a few months, the enzootic nature of the disease of the herd changed to that of an epizootic. Hartsough feels that a new strain of AD virus might have been introduced into a susceptible population.

Finally, one would hazard the guess that the persistent viremia encountered in mink with AD would not discourage the appearance of viral mutants. Japanese workers have reported that the equine infectious anemia virus may undergo antigenic shifts during the course of the disease in an individual horse. All horses are continually viremic for life (Kono et al., 1973).

7.4.3 Dual infections

Field impressions have led most observers to agree that subclinical or overt AD in all genotypes predispose to increases in the severity of secondary infections. Because AD-free CHS are themselves more susceptible to bacterial diseases, we will confine the present discussion to suprainfection of AD-infected non-CHS mink (Padgett et al., 1968).

That AD infection renders mink more responsive to the distemper virus was observed on Danish farms by Hansen (personal communication). This observation prompted him to expose healthy 6-month-old non-CHS pastel and dark mink to the distemper virus. Fifty percent of them died, whereas 80% of the AD infected non-CHS mink of the same genotype succumbed to distemper.

Hansen (personal communication) also found that it was more difficult to immunize AD-infected non-CHS mink against distemper. When AD infected mink were vaccinated and later challenged, 56% succumbed to distemper, whereas 28% of the mink that were normal at the time of vaccination died following the same distemper virus challenge.

Larsen and Gorham (1975) reported a 'new' viral enteritis of mink. Both investigators independently recorded that mink affected with AD were more likely to exhibit a more severe enteritis and die.

It is not unreasonable to expect that dual infections would most likely result in severe disease. First, AD-affected mink are likely to be anemic and in some degree of renal dysfunction. Second, several studies have shown impairment in the ability of AD-affected mink to form antibody against a variety of antigens. The antigens included *Brucella abortus* (Kenyon, 1966), keyhole limpet hemocyanin and goat erythrocytes (Porter et al., 1965; Lodmell et al., 1970, 1971) and horseradish peroxidase (Trautwein et al., 1974). Porter and Larsen (1974) feel that these results can be explained by antigenic competition with AD virus antigens that appear to have pre-empted the immune response to other antigens.

The study of Perryman et al. (1975) has been concerned with the role of cell-mediated immunity in AD. They determined the ability of lymphocytes from AD-infected mink to respond to phytolectins in a lymphocyte transformation test. Infected mink had a significantly lower response to phyto-hemagglutinin-P and concanavalin-A than normal mink. These findings suggest a depression of cell-mediated immunity in mink with AD. Thus, it is not difficult to understand why AD-infected mink are particularly prone to severe secondary infections since both humoral and cellular immune functions appear impaired (see Ch. 9).

7.5 Transmission

AD is not a highly contagious disease. Non-infected mink are often penned with or have been near infected animals for months, and the disease has not been transmitted. The slow propagation of the disease among animals in such circumstances suggests that a threshold dosage is necessary to effect viral transfer to a susceptible mink. To prevent fighting and fur damage, mink are usually penned separately throughout the year, which limits spread by contact.

7.5.1 Sources of virus

Kenyon et al. (1963a) were the first to report a portal of exit for the virus when they demonstrated the agent in the urine. This work has been confirmed (Gorham et al., 1964; Larsen, 1966). We also found AD virus in the saliva and feces but pointed out that urine contamination of the feces could not be excluded. Larsen (1966) collected fresh feces on sterile swabs from affected mink and prepared suspensions for intraperitoneal injection. Three of 7 test mink became positive within a 2-month period.

Haagsma (personal communication) demonstrated AD virus in the feces and saliva 10 days after experimental infection. He found virus in the urine at 140 days, but not at 10 or 20 days post-exposure.

With virus in the urine, feces and saliva, transmission per os is likely to be an important means of transmission. Gorham et al. (1964) showed that mink could be infected after the oral dosing of feces or splenic material that contained the virus. The results of oral titrations with the Pullman virus has revealed that a large dose of virus, in a bolus of food, is required to cause transmission (Hadlow, personal communication). Larsen (1966) also transferred the disease by the oral route. He felt that for effective transmission, large doses of virus were necessary or the animal had to have a prolonged exposure to smaller dosages. Recently, we inadvertently transmitted AD to 35 of 40 mink after the per os administration of ground intestines and spleens in an attempt to reproduce a viral-caused enteritis.

Porter and Larsen (1964) made a field observation that suggested that AD was not likely to be transmitted by food. Three mink farms had a common food supply. On two farms, less than 1% of the mink examined had AD;

but on the third farm, 75% of the mink had AD. However, farmers often redistribute uneaten food among mink pens. Conceivably uneaten food might become contaminated with virus and serve as a source of virus.

We detected AD virus in the saliva of experimental mink infected by the airborne route with high doses of aerosolized virus under highly artificial circumstances (Gorham et al., 1964). We agree with Larsen (1966) and Haagsma (1969a) that the saliva–aerosol–respiratory tract circuit is possible, but it plays a minor role under farm conditions. Quantitative data are lacking on the level of virus in the saliva.

Saliva containing the virus can apparently be inoculated into a female, by male mink biting the neck of a female during breeding. Larsen (1966) found by breeding IAT-positive males to negative females that over 60% of the females became positive. On the other hand, positive females rarely transmitted AD to negative males during breeding. These data also indicated to Larsen (1966) that the disease was not spread by copulation.

Several investigations have demonstrated the virus in blood or serum. Thus, the possibility of transmission by blood-contaminated syringes during routine vaccinations was investigated by Larsen (1966). He reported that there was essentially no difference in the prevalence of AD in mink after vaccination with either an automatic syringe (many animals vaccinated with the same syringe and needle) or one syringe and needle per animal or in an appropriate unvaccinated control group. Larsen felt that transmission did not occur because the dose of contaminating virus was too low to effect transmission.

7.5.2 *Biological products*

Inadequately formalized biological products containing residual live AD virus have caused devastating epizootics of AD. In 1949, farmers in the state of Oregon, U.S.A., lost over a thousand mink after the use of an autogenous tissue vaccine prepared from distemper-infected mink that were unknowingly infected with AD virus. The formalin concentration was sufficient to inactivate the distemper virus, but the safety test procedures did not detect AD. Presently, we have evidence that AD in crude suspension will survive in a final formalin concentration of 0.3% at 5°C for 2 weeks (Henson et al., 1962b; see 8.4.4). However, at the time (1949) AD was not considered to be infectious. Indeed, it was this vaccine accident and the epizootic reported by Helmboldt and Jungherr (1958) caused by a homemade vaccine that furnished the first solid evidence that the disease is not hereditary but is caused by an infectious agent. Interestingly, these vaccine accidents involving AD are reminiscent of the contaminated louping ill vaccines that revealed or suggested the viral etiology of scrapie (Gordon et al., 1939; see 10.4.1).

7.5.3 Insect transmission

The continuing viremia stimulated Haagsma (1969a) and Shen et al. (1973) to investigate insect transmission. Haagsma collected fleas from AD-affected animals and prepared suspensions that, when injected into test mink, failed to evoke AD.

Mosquitoes *(Aedes fitchii)* were maintained for as long as 35 days after a blood meal on AD-infected mink. When these mosquitoes were homogenized and injected into CHS test mink, AD was produced. Even though a vector–pathogen relationship was suggested by the above and other studies, Shen et al. (1973) felt that the prospect of *A. fitchii* serving as a natural vector was speculative because they were unable to effect transmission from mink to mink by this mosquito. We agree with Porter and Larsen (1964) that blood-sucking insects are not important in the transmission of AD.

7.5.4 Indirect contact

The virus exhibits remarkable stability in crude tissue suspensions (8.4). If this stability is indicative of the stability of the virus in urine, feces or saliva on fomites, indirect transmission is a ready possibility.

Mink invariably chew on mink handling gloves; thus, these gloves are an excellent candidate for virus indirectly contacted. Larsen (1966) purposely contaminated gloves with an AD spleen–kidney extract. After intervals as long as 20 min, mink were allowed to chew on the gloves. He readily transmitted the AD virus by this simple experiment. The use of disinfectants on the gloves reduced the efficacy of transmission. Haagsma (personal communication) intentionally contaminated gloves with AD virus. Then, mink were handled with these gloves, but he was unable to effect transmission by this procedure.

Field observations suggest that wire pens and wooden nestboxes that have housed AD-infected mink can harbor virus for a time after the infected animal has been removed. Haagsma (personal communication) removed AD-infected mink from pens. He allowed the contaminated pens to remain empty for a day. Susceptible mink that were confined in the pens eventually died of AD. Although there is no research that would confirm or deny these findings, Johansen's (1974) description of mink pens as a probable source of virus typifies field reports. A Danish mink farmer purchased mink pens that had housed AD-infected mink previously. Part of his herd was introduced into the newly acquired pens and the rest were kept in his original pens. No information was provided regarding the selection of animals for the original and new pens, but later IAT testing showed 14% reactors in his original pens and 40% in the AD-contaminated pens.

7.5.5 Vertical transmission

Mink farmers were the first to point out the familial occurrence of AD. A spot map of AD-infected and dead animals in a CHS population often revealed infected kits in the litter if the female was affected. That their observation was valid was shown by the IAT (Henson et al., 1963). They found that 45% of the kits from infected dams were positive at pelting time, whereas only 19% of the kits from non-infected dams were positive. Harang (1966) reported that 45% of the kittens from IAT-positive mothers were positive to the IAT at the time of pelting. Kittens from negative dams housed in disinfected cages on the same farm remained IAT negative during the same period.

There is good evidence that the virus crosses the placenta and causes intra-uterine infection and fetal death. Padgett et al. (1967) collected fetuses from the uterus of AD-infected CHS and non-CHS dams. Of a total of 87 kittens, 34 (40%) were dead. Fifty-three live fetuses were tested for AD virus by injecting fetal material into non-infected mink, and 32 (60%) contained AD virus. Haagsma (1969a) reported that AD was a cause of prenatal deaths, stillborn fetuses, and the death of kits during lactation. He was able to detect AD virus in a stillborn kit.

Crawford (1972) used a direct immunofluorescent test and demonstrated AD antigen in the ovary, placenta, mammary gland and most fetal tissues. Porter et al. (1972b) injected pregnant female mink with AD virus 2 weeks before parturition. At 3 months of age, the kittens exhibited mild lesions of AD. Porter and his co-workers suggested that the fetal mink had decreased responsiveness to viral antigens. In a somewhat similiar trial, we failed to induce tolerance by injecting individual fetuses through the wall of the uterus. The injections were in mid-to-late gestation. By the third month after parturition, 17 of 21 kittens were hypergammaglobulinemic and showing clinical signs, 18 of 21 had gross and microscopic lesions and 11 of 12 had glomerular deposition of gammaglobulin and complement.

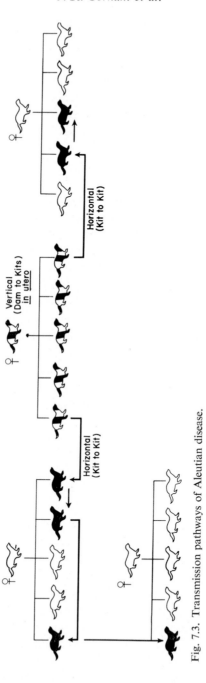

Fig. 7.3. Transmission pathways of Aleutian disease.

Our trials failed to detect AD virus in the colostral milk of infected mink (Gorham et al., 1964). On the other hand, Haagsma (personal communication) detected AD virus in the milk of an AD-infected female mink about a month after parturition.

Whether vertical or horizontal transmission is the most efficient is difficult to state (Fig. 7.3). One cannot disagree with Porter and Larsen (1974) when they state that although vertical transmission would seem ideal for maintaining the virus, most mink in the advanced stage of the disease do not become pregnant or if they do, abort or resorb the fetuses. Moreover, many kittens from chronically infected dams are likely to die during the first 2 weeks of life.

We mildly disagree with Porter and Larsen (1974), albeit on tenuous field evidence, that horizontal transmission probably accounts for most AD infections. Padgett (1965) recorded the pattern of IAT positivity on a farm with only CHS mink. The cases were plotted at monthly intervals on a geographic map on which each adult and progeny was shown. The distribution of onsets of positivity and spacial clustering of cases suggested that roughly one-half the cases in the young were the result of vertical transmission from an infected dam.

7.5.6 Infectious period

The time at which AD begins to be shed and the end of shedding of virus sufficient for transmission is of considerable epizootiologic importance. That the virus was demonstrated in urine, feces, blood, serum and a wide variety of tissues has been pointed out elsewhere (7.5.1). For the most part, these were random samples and were not collected sequentially to ascertain the infectious period. Until subjected to experimentation, we can assume that CHS mink are infectious for life. The infectiousness of non-CHS mink is open to question but will probably be related to the outcome of the disease; i.e., non-CHS mink that eventually die from AD probably excrete the virus for life. The epizootiologic significance of AD virus persisting in the mesenteric lymph nodes of non-CHS mink (7.3.4) will eventually have to be resolved.

7.6 Environment

CHS herds with a low prevalence or non-CHS herds with a prevalence of less than 15% (IAT) usually are not markedly affected by conditions of climate and stress. At least the mink farmers are not usually cognizant of losses. Environmental factors have a marked effect on the monthly incidence of AD only if a sufficient number of chronically affected animals comprise the farm population.

Note that virtually all clinically affected mink eventually die from AD. However, if they survive long enough, they are killed when their pelts are prime. Clinical disease occurs at all times of the year but most often in animals older than 6 months. The intrusion of adverse environmental factors only hastens their death and directly influences epizootic 'clustering' in time and space.

During the fall, winter and early spring, sharp drops in the temperature are accompanied by an increase in the number of fatalities. Exposure coupled with frozen food and water causes a flurry of losses. Because of impaired kidney function, affected mink are extremely thirsty and become dehydrated. Sick mink are especially prone to die in hot, humid weather. The stress of pregnancy, whelping and nursing in the spring and early summer and fur molting in the early fall are physiologic factors that can be correlated with the monthly incidence of AD. Hansen (personal communication) feels on the basis of field observations, that the presence of nitrite in the drinking water may increase the severity of AD in infected herds.

7.7 Diagnosis and control

By 1960 evidence showed that pelting of clinically affected mink would never control the malady. The disease was an important factor in determining whether some farmers continued to raise CHS mink. Farms that had a low prevalence in the 1950s had an increasing annual increase in the mortality rate. We recorded mortality rates of 500 to 850 per 5000 CHS mink on different farms.

7.7.1 Non-specific tests

Henson et al. (1961, 1962b) and Kenyon et al. (1963b) found that AD-infected mink have altered patterns of serum proteins. These changes were a significant increase in total serum protein and gamma globulin and decrease in albumin. These findings provided the basis for a rapid, simple field test for detecting subclinical disease; the methods described by Mallen et al. (1950) were used. This IAT is used extensively in all mink-raising countries (Henson et al., 1962a). It was emphasized that the IAT was not specific for AD and that other processes, particularly suppurative, resulted in serum protein changes that evoked a positive IAT. Tuberculosis also evokes a positive IAT (Hadlow, personal communication).

Nevertheless, the IAT test is effective, particularly in CHS herds, because almost all animals have altered serum protein patterns until death supervenes. Its use on CHS farms decreased the AD losses by 50% or more on some farms within 2 years. We have found that the IAT test, along with vigorous disinfection procedures, can control the disease on CHS farms. The inherent limitations of the IAT test, i.e., the 3–5 week period before the test converts to positive after infection and its limited false positivity, are not serious detriments in CHS herds if the mink are tested regularly.

The Utah workers used serum electrophoresis to detect mink with abnormally high levels of gamma globulin. On the basis of thousands of tests, Larsen (1965) concluded that the percentage of gamma globulin in normal mink serum was 5–14 in normal animals, whereas the percentage in AD mink was 15–60. Although serum electrophoresis is more precise than the IAT, the test is more difficult to conduct.

Hadlow (personal communication) found that the percentage of gamma globulin was higher than 15% in some apparently normal CHS females. He felt that the increase might be age related in that older mink were more likely to be involved.

Sandholm and Kangas (1973) developed a semi-quantitative method for the demonstration of increased globulin levels in mink blood. The method is based on the polymerization of protein NH_2 groups with glutaraldehyde, which can be observed as the formation of a clot in a capillary tube.

All workers realized that the alteration in serum proteins after infection of non-CHS mink was not predictable. Some non-CHS mink exhibited a sustained hypergammaglobulinemia and a progressive but extended downward course until death supervened. Other non-CHS mink showed a transient gamma globulin rise, and others showed no hypergammaglobulinemia, evidence that the virus strain may play a role in directing the response of non-CHS mink (Hadlow, personal communication).

7.7.2 Specific serologic tests (see also 8.3)

Specific serologic tests were urgently needed because non-specific tests would never eradicate or even control AD in non-CHS herds. Porter et al. (1969) reported that high levels of AD antibody were shown by indirect immunofluorescence when liver sections from infected mink early in the course of the disease were used as a source of antigen. The serums of all mink collected at 10 days or more after infection contained demonstrable anti-AD antibody. McGuire et al., (1971) recorded complement-fixing antibody 20 days post-infection. The titers increased linearly and at 6 weeks large quantities were observed. Although the fluorescent antibody procedure and the complement-fixation test are valuable research tools, they are not easily adapted to wide-scale field testing in control programs.

Recently Cho and his co-workers (1972, 1973) have applied CEP to AD and have detected antibody in experimental mink at 7 days after infection. High titers were attained by 2 months post-inoculation. The sensitivity and specificity of the 3 assay methods for AD antibody (immunofluorescence, complement fixation and CEP) have recently been compared (Crawford et al., 1975). All 3 systems were specific for AD antibody. Complement fixation and immunofluorescence were more sensitive than CEP. However, rapidity and simplicity made CEP the test of choice for routine diagnosis. In all specific serologic tests for AD, it is likely that there is a quantitative relationship between the dose of the virus and the onset of positivity. Greenfield et al. (1973) were the first to compare the IAT test with the CEP. About 50% of 70 mink (CHS and non-CHS) that were positive by CEP reacted to the IAT. Ingram and Cho (1975) have used their CEP procedure on farms having a high annual incidence of AD. On one particular farm, the prevalence was 46% when the mink were first tested in December 1973. After the reactors were pelted, subsequent testing in February 1974, revealed a 10% prevalence and in November, 1974, a 2.9% prevalence.

Kammer and Anderson (personal communication) devised a plan to reduce the prevalence of AD on two farms. First, as a screening procedure, mink of all genotypes were tested by the IAT by the owner. The reactors were pelted and serum from 14,000 IAT-negative mink were then tested by CEP. Although all of the results have not been compiled, Kammer and Anderson have observed that the IAT detects 50–60% of the positive mink (if one arbitrarily considers that CEP identifies 100% of the affected mink). Thus, we apparently have the necessary tools to control and perhaps eradicate the disease on individual farms.

7.7.3 Decontamination

Because of the likelihood of indirect transmission by contaminated pens and nest boxes, a regular program of pen cleaning and disinfection must also be instituted in order to control the disease. The general features of disease control have recently been outlined by Hansen (1974). Haagsma (1969b) reported that the usual methods of cage disinfection are not sufficient to kill the AD virus. He found that the virus in a filtrate was not completely inactivated by chloramine, BTC 2125, *O*-phenylphenol and 1% or lower solutions of sodium hydroxide, but the virus was inactivated by 2% sodium hydroxide (8.4). The virus withstood heating at 80°C for 10 min but not at 90°C for the same time period. Burger et al. (1965) reported that crude tissue suspensions were infectious after heating for 90–95°C for 15 min but not after 30 min. Hansen et al. (1974) described heat stability trials in which organ material containing AD virus was heated to 80°C for 20 min and remained infective. However, heating the same material at 125°C under 1.37 atmospheric pressure appeared to destroy the virus. Eklund et al. (1968) concluded on the basis of heat stability trials that the heat resistance of AD virus was not as marked as scrapie virus, but was greater than most animal viruses (see 8.4).

7.7.4 *Vaccination*

Contemporary prospects for the control of AD by vaccination are dismal. The experimental AD vaccines alluded to by Russell (1962) were not effective when used on commercial farms and were eventually abandoned. Karstad et al. (1963) vaccinated mink with 1, 2 or 3 doses of formalin-treated tissue collected from AD-infected mink. The vaccinates were not immune when subsequently challenged with diseased tissue suspensions.

The results of Porter et al. (1972a) have dampened the enthusiasm for research on vaccines. They found that formalin-inactivated virus vaccine injections followed by live virus challenge in non-CHS mink increased AD virus infectivity and markedly enhanced the severity of lesions. Although the pathogenetic mechanism is obscure, the undesirable implications for such a vaccine are obvious. No one has an attenuated strain of AD virus that could be used as live virus vaccine. In addition to the production of antibody with the possible formation of objectionable immune complexes, there is the possibility of persistence of such an attenuated virus and the emergence of virus mutations of increased virulence. The paper by Gajdusek et al. (1971) concerning the control of chronic degenerative diseases with vaccines is especially germane to the present discussion.

7.8 *Comment*

AD has been regarded as a prime target for research aimed at a host–virus interaction that permits viral persistence and produces immunologically mediated lesions. Very rapid strides have been made within the past decade. The discovery of specific serologic tests and in vitro cultivation are significant breakthroughs that will allow meaningful pathogenesis and epizootiologic research.

The central goal of the epizootiologist should be the eradication of the causative virus. Many will question whether it is possible, but most will agree that it should be our long-term objective. Until we know why some non-CHS mink can clear the virus from their tissues, if indeed they do, we will have no firm basis for vaccination. The enhancing effect of inactivated vaccines and the inherent dangers of living virus vaccines are additional considerations in the development of effective and safe prophylactic agents.

There are no 'short cuts' to the eradication or even the control of AD. The best contemporary regimen to follow would seem the intensive use of countercurrent immunoelectrophoresis to identify and remove subclinical cases from the herds at risk. Additional information is needed before the success of maintaining an AD-free herd can be assured. We need to know: (1) the infectious period and factors affecting virus shedding; (2) virus stability under field conditions; (3) decontamination procedures and (4) effective and economically feasible management and isolation procedures.

The ecology of AD deserves more study. Presently, we feel that mink are the most important reservoir and source of the virus, but the role of other mustelids such as the skunk in the perpetuation of the virus in nature should be elucidated. The use of specific serologic tests will clarify the epizootiology of AD. More information is needed on the zoonotic aspect of AD. We have only cursory information about the biology of the virus. It is highly likely that some of the isolates are a mixture of viruses of differing virulence for mink. Attempts should be made to separate possible major virus types in each isolate and apply rigid classical virologic studies so that they can be defined in terms of their biologic and antigenic characteristics.

References

BEGUEZ-CESAR, A. (1943) Bol. Soc. Pediat., 15, 900.
BURGER, D., GORHAM, J. R. and LEADER, R. W. (1965) In: (D. C. Gajdusek, C. J. Gibbs, Jr. and M. Alpers, Eds) Natl. Inst. Neurol. Dis. Blindness, Monogr. 2. Slow, Latent and Temperate Virus Infections. U.S. Government Printing Office, Washington, D.C.) p. 307.
CHO, H. J. and INGRAM, D. G. (1972) J. Immunol., 108, 555.

CHO, H. J. and INGRAM, D. G. (1973) Can. J. Comp. Med., 37, 217.

CRAWFORD, T. B. (1972) Fed. Proc. Abstr., No. 2372, 31, 635.

CRAWFORD, T. B., MCGUIRE, T. C., PORTER, D. D. and CHO, H. J. (1975) submitted for publication.

EKLUND, C. M., HADLOW, W. J., KENNEDY, R. C., BOYLE, C. C. and JACKSON, T. A. (1968) J. Infect. Dis., 118, 510.

GAJDUSEK, D. C., GIBBS, C. J. and LIM, K. A. (1971) Proc. Int. Conf. on Application of Vaccines against Viral, Rickettsial and Bacterial Diseases of Man. (Scientific Publication, Pan American Health Organization) No. 226, p. 566.

GORDON, W. S., BROWNLEE, A. and WILSON, D. R. (1939) Proc. III Int. Congr. Microbiol. New York, p. 362.

GORHAM, J. R., LEADER, R. W. and HENSON, J. B. (1964) J. Infect. Dis., 114, 341.

GORHAM, J. R., LEADER, R. W., PADGETT, G. A., BURGER, D. and HENSON, J. B. (1965) In: (D. C. Gajdusek, C. J. Gibbs, Jr, and M. Alpers, Eds) Natl. Inst. Neurol. Dis. Blindness, Monogr. 2. Slow, Latent and Temperate Virus Infections. (U.S. Government Printing Office, Washington, D.C.) p. 279.

GREENFIELD, J., WALTON, R. and MCDONALD, K. R. (1973) Res. Vet. Sci., 15, 381.

HAAGSMA, J. (1969a) Neth. J. Vet. Sci., 2, 19.

HAAGSMA, J. (1969b) Tüdschr. Diergeneesk., 94, 824.

HANSEN, M. (1974) Dansk Pelsdyravl, 37, 313.

HANSEN, M., LUND, S. and WOLLER, J. (1974) Dansk Pelsdyravl, 37, 288.

HARANG, A. K. (1966) Medlemsbl. Nor. Veterinaerforen., 5, 135.

HARTSOUGH, G. R. (1975) U.S. Fur Rancher, 55, 8.

HARTSOUGH, G. R. and GORHAM, J. R. (1956) Nat. Fur News, 28, 10.

HELGEBOSTAD, A. (1970) Nord. Veterinaermed., 22, 21.

HELMBOLDT, C. F. and JUNGHERR, E. L. (1958) Am. J. Vet. Res., 19, 212.

HEMMINGSEN, B. and HEJE, N. I. (1964) Nord. Veterinaermed., 16, 881.

HENSON, J. B., GORHAM, J. R and LEADER, R. W. (1962a) Nat. Fur News, 34, 8.

HENSON, J. B., GORHAM, J. R., LEADER, R. W. and WAGNER, B. M. (1962b) J. Exp. Med., 116, 357.

HENSON, J. B., GORHAM, J. R. and LEADER, R. W. (1963) Texas Rep. Biol. Med., 21, 37.

HENSON, J. B., GORHAM, J. R., PADGETT, G. A. and DAVIS, W. C. (1969) Arch. Pathol., 87, 21.

HENSON, J. B., LEADER, R. W. and GORHAM, J. R. (1961) Proc. Soc. Exp. Biol. Med., 107, 919.

INGRAM, D. G. and CHO, H. J. (1974) J. Rheum., 1, 74.

INGRAM, D. G. and CHO, H. J. (1975) Educ. Bull. No. 47, Ontario Veterinary College, p. 8.

JOHANSEN, A. (1974) Dansk Pelsdyravl., 37, 74.

KARSTAD, L., PRIDHAM, T. J. and GRAY, D. P. (1963) Can. J. Comp. Med. Vet. Sci., 27, 124.

KENYON, A. J. (1966) Am. J. Vet. Res., 27, 1780.

KENYON, A. J., HELMBOLDT, C. F. and NIELSEN, S. W. (1963a) Am. J. Vet. Res., 24, 1066.

KENYON, A. J., MAGNANO, T., HELMBOLDT, C. F. and BUKO, L. (1966) J. Am. Vet. Med. Assoc., 149, 920.

KENYON, A. J., TRAUTWEIN, G. and HELMBOLDT, C. F. (1963b) Am. J. Vet. Res., 24, 168.

KONO, Y., KOBAYASHI, K. and FUKUNAGA, Y. (1973) Arch. Ges. Virusforsch., 41, 1.

KRAMER, J. W., DAVIS, W. C. and PRIEUR, D. J. (1975) Fed. Proc. Abstr. No. 3631, 34, 861.

LARSEN, A. E. (1965) Nat. Fur News, 37, 18.

LARSEN, A. E. (1966) Am. Fur Breeder, 39, 12.

LARSEN, A. E. and GORHAM, J. R. (1975) Vet. Med./Small Anim. Clin., 70, 291.

LARSEN, A. E. and PORTER, D. D. (1975) Infect. Immun., 11, 92.

LODMELL, D. L., BERGMAN, R. K., HADLOW, W. J. and MUNOZ, J. J. (1971) Infect. Immun., 3, 221.

LODMELL, D. L., HADLOW, W. J., MUNOZ, J. J. and WHITFORD, H. W. (1970) J. Immun., 104, 878.

LUTZNER, M. A., LOWRIE, C. T. and JORDAN, H. W. (1967) J. Hered., 58, 299.

MALLEN, M. S., AGALDE, F. L., BALCAZAR, M. R., BOLIVAR, J. I. and MEYRAN, S. (1950) Am. J. Clin. Pathol., 20, 39.

MCGUIRE, T. C., CRAWFORD, T. B., HENSON, J. B. and GORHAM, J. R. (1971) J. Immun., 107, 1481.

PADGETT, G. A. (1969) Fed. Proc. Abstr. No. 2384, 28, 685.

PADGETT, G. A., GORHAM, J. R. and HENSON, J. B. (1967) J. Infect. Dis., 117, 35.

PADGETT, G. A., LEADER, R. W., GORHAM, J. R. and O'MARY, C. C. (1964) Genetics, 49, 505.

PADGETT, G. A., REIQUAM, C. W., HENSON, J. B. and GORHAM, J. R. (1968) J. Pathol. Bacteriol., 95, 509.

PAN, I. C., TSAI, K. S. and KARSTAD, L. (1970) J. Pathol. 101, 119.

PERRYMAN, L. F., BANKS, K. L. and MCGUIRE, T. C. (1975) J. Immunol., 115, 22.

PORTER, D. D., DIXON, F. J. and LARSEN, A. E. (1965) J. Exp. Med., 121, 889.

PORTER, D. D. and LARSEN, A. E. (1964) Am. J. Vet. Res., 25, 1226.

PORTER, D. D. and LARSEN, A. E. (1974) Progr. Med. Virol., 18, 32 (Karger, Basel).

PORTER, D. D., LARSEN, A. E., COX, N. A., PORTER, H. G. and SUFFIN, S. C. (1975) Fed. Proc. Abstr. No. 4121, 34, 947.

PORTER, D. D., LARSEN, A. E. and PORTER, H. G. (1969) J. Exp. Med., 130, 575.

PORTER, D. D., LARSEN, A. E. and PORTER, H. G. (1972a) J. Immunol., 109, 1.

PORTER, D. D., LARSEN, A. E. and PORTER, H. G. (1972b) Fed. Proc. Abstr. No. 2371, 31, 635.

RUSSELL, J. D. (1962) Nat. Fur News 34, 8.

SANDHOLM, M. and KANGAS, J. (1973) Zbl. Vet. Med. B, 20, 206.

SHACKELFORD, R. M. (1950) Pilsbury Publishers, Inc., New York.

SHEN, D. T., GORHAM, J. R., HARWOOD, R. F. and PADGETT, G. A. (1973) Arch. Ges. Virusforsch., 40, 375.

TAYLOR, R. F. and FARRELL, R. K. (1973) Fed. Proc. Abstr. No. 3403, 32, 882.

TRAUTWEIN, G., SCHNEIDER, P. and ERNST, E. (1974) Zentralbl. Veterinaermed., 21, 467.

CHAPTER 8

Purification and structure of Aleutian disease virus

Hyun J. CHO

8.1 Evidence of viral aetiology

The transmissible nature of Aleutian mink disease (AD) was first described in 1962 (Karstad and Pridham, 1962; Henson et al., 1962; Russell, 1962; Trautwein and Helmboldt, 1962). The disease was reproduced by inoculating cell-free filtrates of infected mink tissue suspensions (Karstad and Pridham, 1962; Henson et al., 1962) or the pellets obtained after ultracentrifugation (Henson et al., 1963). Based on this indirect evidence, it was assumed that the causative agent of AD was a virus. In 1973, Kenyon et al. (1973) and Cho and Ingram (1973a) independently isolated the virus employing immunological techniques, and the ultrastructure of the virus was observed (see 8.6).

8.2 Sources of virus

The virus is readily recovered from the serum, organs and urine of infected mink. Highest virus titers of 10^8 and 10^9 ID_{50} per gram of spleen, liver and lymph node were found at 10 days after experimental infection (Porter et al., 1969). At times later than 10 days the virus titers in the tissues slowly fall, so that 2 or more months after infection, titers of 10^5 ID_{50} per gram of spleen are commonly found (Porter and Larsen, 1974). Eklund et al. (1968) recovered the virus from lymph nodes, kidney, spleen, submaxillary salivary gland, liver, intestine, blood clot and serum. The infectivity titer of the serum was 10^5 ID_{50} per ml, and only spleen, mesenteric lymph nodes and kidneys

Slow virus diseases of animals and man, edited by R. H. Kimberlin
© *North-Holland Publishing Company 1976*

showed similar or higher infectivity titers than serum. Because of the large amount of infective serum present in most tissue, the infectivity could be attributed to viremia. Thus, Eklund et al. (1968) considered that only the mesenteric lymph nodes, spleen and kidneys could be the sources of virus, whereas Porter et al. (1969) did not consider kidney as the site of virus replication. Several other workers also found the AD virus (ADV) in whole blood, serum, bone marrow, spleen, faeces, urine and saliva (Kenyon et al., 1963; Gorham et al., 1964). As an experimental source of the virus, several workers commonly used a 10% suspension of spleen from infected mink (Porter et al., 1969; McGuire et al., 1971; Cho and Ingram, 1972; Kenyon et al., 1973).

8.3 Methods of virus assay (see also 7.7.2)

The virus assay is generally performed by mink inoculation followed by diagnosis of AD. The criterion of infection is either an increased level of serum gamma globulins usually determined by cellulose acetate electrophoresis of serum, by the detection of either gross and/or histopathological lesions in tissues of mink, or by detection of ADV antibody.

Histopathological lesions and hypergammaglobulinemia are usually detectable one month after experimental infection, whereas using serological tests, ADV antibody can be detected as early as 7 days after infection. Larsen and Porter (1975) have recently shown that approximately one quarter of the unselected pastel mink which were infected by ADV could clear the viremia and failed to develop hypergammaglobulinemia and lesions. Consequently, the use of serum protein changes or lesion production can underestimate the frequency of infection in pastel mink. Specific ADV antibody was detected in these non-persistently infected mink but at much lower titers than in persistently infected mink. Until the development of serological tests, the microscopic examination of the mink tissues was the most accurate method of diagnosis of the disease. The pathology of AD is described in detail in the following chapter (9.3).

Serological diagnosis is based on the detection of ADV antibody in the serum of the virus infected mink. For the detection of antibody to ADV, Porter et al. (1969) employed an indirect immunofluorescent test using labelled rabbit anti-mink IgG. Liver sections of mink infected 9 days with the virus were the source of antigen for this test. Liver sections were suitable since the distribution of infected cells was uniform and since few or no immunoglobulin-containing cells were present. It was found that a few mink infected for 9 days and all those infected 10 days or more developed serum antibody. By 60 days after infection, when hypergammaglobulinaemia was marked, the mink has exceptionally high antibody titers (Porter et al., 1969). Recently Porter and co-workers (1975) used tissue culture cells infected with ADV as the source of antigen for the indirect immunofluorescence test. Infected tissue culture cells have shown the same specificity and sensitivity as infected liver sections.

McGuire et al. (1971) employed a complement-fixation test for the detection of antibody to ADV. A fluorocarbon extract was prepared from spleens and lymph nodes from mink infected for 9 days. This extract was filtered through a 0.2 μ Millipore filter and used as antigen in this test. ADV antibody was detected by this method at 20 days post-infection but not at 14 days in mink of the Aleutian genotype. During the course of infection, the antibody titers varied from 8000 to 260,000.

Employing a fluorocarbon-extracted antibody-free ADV antigen, which was prepared by activation with acid, or heat, or with repeated fluorocarbon treatment, serological tests of counterimmunoelectrophoresis and immunodiffusion tests have been developed (Cho and Ingram, 1972, 1973b). Using these techniques, ADV antibody was detected regularly in the serum of mink 7 days after experimental infection. The antibody titers in the serum increased throughout infection, with titers of about 100 being reached 15 days after infection. Titers of 1000–5000 were reached at 30 days, and exceeded 5000 at 2 months post-infection and thereafter. The counterimmunoelectrophoresis test is considered to be the most practical technique for the diagnosis of the disease at present because it is simple, rapid, specific and sensitive and is capable of detecting infection early in the course of the disease. Only 0.02 ml of serum sample is required for this test and the reaction is clear cut. Another advantage of the test is that whole blood can be used instead of serum (Cho and Ingram, 1973b).

Another method of virus assay can be performed by the detection of ADV antigen from infected mink tissues or from infected cell cultures. Using a direct immunofluorescence test, Porter et al. (1969) found ADV antigen in the macrophages of the liver, spleen and lymph node in 34 of 59 mink infected for 8–18 days, and 8 of 14 mink 60 days after infection, but not in any of the 17 mink infected for only 1–6 days. ADV antigen was found in the cytoplasm of macrophages in these tissues. However, it was not determined whether the virus replicated in these cells or whether the virus was phagocytized as an immune complex (Porter et al., 1969).

In vitro multiplication of ADV has been difficult to achieve. Success was

reported by Basrur et al. (1963) and by Gray (1964) who employed mink
kidney and testis cell cultures. They found nuclear and cytoplasmic changes
in the infected cells. The infectious agent was present in the 8th serial passages
in the tissue culture when tested by mink inoculation. However, other
workers have been unable to reproduce this finding. Recently Yoon et al.
(1973, 1974) claimed that ADV replicates in cultures of mouse L cells (Yoon
et al., 1973) and mink kidney cells (Yoon et al., 1974). They reported cytopathic effects in the ADV-inoculated monolayers and they also observed
the incorporation of [^{3}H]uridine and ^{14}C-labelled amino acid into the
virus in mouse L cells and in mink kidney cell cultures. However, these
observations remain to be confirmed by other workers. We have been quite
unable to demonstrate a cytopathic effect induced by the virus in cultures
of mouse L cells, mink kidney cells, spleen cells, macrophages or buffy coat
cells.

   Employing immunofluorescent techniques, Porter and Larsen (1974) have
shown limited induction of ADV antigen in 5 different mink cell strains, in
primary kidney cells from the African green monkey, and in the human cell
strain WI-38. The antigen was found in the nucleus 2 days after infection
and small amounts of antigen were sometimes found in the cytoplasm 3 and
4 days after infection. Employing a feline renal cell line (CRFK), Porter
et al. (1975) could readily isolate ADV when incubated at 31.8 °C but not
at 37 °C. However, once adapted to cell culture, the virus titers were highest
at 37 °C. After 10 passages in the cell culture, the virus produced AD in
mink, and the ratio of in vivo to in vitro infectivity was 1:1. The growth of
ADV was abolished by the presence of fluorodeoxyuridine in the culture
medium.

8.4 Physico-chemical properties of the virus in crude extracts and of partially purified virus

8.4.1 Preliminary classification of AD virus

Table 8.1 summarizes the known properties of ADV. Several characterization studies of the virus have been performed on tissue suspensions prepared
from chronically infected mink, thus, the reported properties are almost
certain to be those of virus–antibody complexes. ADV is infectious despite
being complexed with antibody (Porter and Larsen, 1967). Even the virus
extracted from tissues collected as early as 9–13 days after infection is already bound to antibody (Cho and Ingram, 1972, 1974; Crawford, 1973).

TABLE 8.1

Physico-chemical properties of Aleutian mink disease virus.

Description	Properties	Reference
Size	23–25 nm	2, 9, 11
Morphology	Icosahedral	2, 7, 9, 11
Buoyant density in CsCl	1.405–1.430 g/ml	9,12
Nucleic acid	No direct test yet	
Classification	Parvovirus	3, 8, 9
Nuclear replication	Yes	8, 9
Dialysis	Not dialysable	10
Ether	Stable	1, 4, 9
Fluorocarbon	Stable	1
pH 3	Stable	6
Deoxycholate	Stable	1
Protease enzymes	Stable	1
Nuclease enzymes	Stable	1
Ultraviolet light	Inactivated	5
Heat 56°C, 30 min	Stable	4, 9
60°C, 30 min	Partially inactivated	4, 9
80°C, 30 min	Inactivated	4, 9
0.05 N NaOH	Inactivated	1
0.5 N HCl	Inactivated	1
4 N urea	Partially inactivated	1
0.5% iodine	Inactivated	1
1% chloramine (pH 9.4)	Stable	6
2% *O*-phenyl phenol (pH 12.5)	Stable	6

(1) Burger et al., 1965; (2) Cho and Ingram, 1973a; (3) Cho and Ingram, 1974; (4) Eklund et al., 1968; (5) Goudas et al., 1970; (6) Haagsma, 1969; (7) Kenyon et al., 1973; (8) Porter et al., 1974; (9) Porter et al., 1975; (10) Tabel and Ingram, 1970; (11) Yoon et al., 1974; (12) Cho, this chapter.

Therefore, it should be kept in mind that the properties of purified virus dissociated from antibody are not necessarily the same as those of the virus–antibody complexes. Until methods for purifying the virus were developed, the properties of antibody-free virus could not be studied. Techniques for purifying the virus have been developed recently (Cho and Ingram, 1972, 1974; Crawford, 1973; Kenyon et al., 1973), and therefore it is now possible to characterize the purified virus.

Goudas et al. (1970) examined the effect of UV-irradiation on the virus in the serum. The infective serum was exposed to UV light at 253.7 nm with

3 different doses: 1.17×10^4, 1.17×10^5 and 2.34×10^5 ergs/mm². The results indicated that the ADV was inactivated by all 3 doses of UV light. The sensitivity of ADV to UV irradiation differs from that of scrapie agent or transmissible mink encephalopathy agent. Similar doses of UV-irradiation produced much smaller losses of infectivity of the latter agents (Alper et al., 1967; Marsh and Hanson, 1969).

Conflicting reports have been published on the classification and nucleic acid type of the virus. Kenyon et al. (1973) claimed that the virus preparation eluted by acid buffer from an affinity chromatography column (see 8.6) contained RNA as determined by the orcinol method and suggested that ADV may be a member of the closely related cardioviruses and enteroviruses which are classified as picornaviruses. The findings of Yoon et al., (1973, 1974) give further support to the suggestion that ADV is a picornavirus, since [³H]uridine labelled the virus in mouse L cell culture (Yoon et al., 1973) and virus grown in mink kidney cells had an icosahedral structure with a diameter of 25 nm (Yoon et al., 1974). However, they did not determine whether the virus could be grown in serial passage in cell culture or if the virus grown in cell culture stimulated the production of specific anti-ADV antibody in mink. On the other hand, Cho and Ingram (1974), who have also obtained purified virus, consider ADV to be a member of the parvovirus group of single-stranded DNA viruses. In size and morphology the virus resembles a parvovirus more closely than a picornavirus. Its stability to many chemicals, to low pH, and to high temperature (Burger et al., 1965; Eklund et al., 1968; Haagsma, 1969) and its initial replication in the nucleus (Porter and Larsen, 1974) are characteristic of certain parvoviruses (Toolan, 1972; Tinsley and Longworth, 1973). Acridine orange staining (0.01 %) of the purified virus gave a brillant flame red colour when examined under UV light, indicating that the virus has a single-stranded nucleic acid (Cho and Ingram, 1974). Recently, Porter and Larsen (1974) stated that if ADV belongs to a known group of viruses, the parvovirus group seems most likely, followed by the picornavirus group. Porter et al. (1975) further demonstrated that the ADV grown in feline renal cell line has similar properties to the parvovirus group.

8.4.2 Size of the virus

Buko and Kenyon (1967) reported that the AD agent was smaller than serum albumin molecules in that the infective agent passes through dialysis tubing. However, Tabel and Ingram (1970) found that under strictly controlled conditions, with intact dialysing membranes, the diffusate of a spleen extract from ADV-infected mink did not produce the disease in mink and concluded that the virus cannot pass through an intact dialysing membrane.

In filtration experiments, several workers have found that the virus from crude tissue extracts passes a Millipore membrane of 50 nm pore size (Basrur et al., 1963; Buko and Kenyon, 1967; Eklund et al., 1968), however, the virus present in a suspension of spleen did not pass a 35–100 nm Gradocol membrane after serially passing through membranes of larger pore sizes. The Gradocol membrane does not permit the passage of type 2 polio virus (Eklund et al., 1968). Fluorocarbon-activated ADV, which was dissociated from antibody, passed a 50 nm pore size Millipore membrane but was retained in a Diaflo membrane, XM 300 (Amicon) having a pore size about 18 nm (Cho and Ingram, 1973a).

Tsai and co-workers (1969) reported aggregates of virus-like particles in the form of crystalline arrays within the cytoplasm of endothelial cells in the kidneys of ADV infected mink. These particles measured about 25 nm in diameter. However, others have failed to find similar particles in endothelial cells at any stage of the infection (Porter and Larsen, 1974).

By electron microscopy, negatively stained ADV had a diameter of 23 nm (Cho and Ingram, 1973a, 1974) whereas Epon-embedded virus was about 20 nm (Kenyon et al., 1973). Yoon et al. (1974) reported that the virus harvested from mink kidney cell cultures had an icosahedral structure, 25 nm in diameter. Porter et al. (1975) demonstrated that ADV grown in CRFK culture had a diameter of 24 nm and icosahedral symmetry.

8.4.3 Other physical properties

In a recent experiment in our laboratory, it was shown that fluorocarbon-activated ADV (see 8.5) banded at a buoyant density of 1.405 to 1.416 g/ml in cesium chloride. In this experiment, 3.5 ml of cesium chloride (14.4 g of cesium chloride in 23.4 ml of 0.05 M Tris–HCl, pH 7.4) was put into a centrifuge tube and 1 ml of a concentrated suspension of ADV (the antigen titer was 1:3000 as determined by counterimmunoelectrophoresis) was mixed with the top part of the cesium chloride solution. After centrifugation at 243,000g in a Beckman SW 50.1 Ti rotor at 25 °C for 72 h, a heavy band was visible at a density of 1.405–1.416 g/ml as determined by its refractive index. This band was collected together with two other fractions in the density regions of 1.29–1.31 g/ml and of 1.33–1.35 g/ml. To minimize the cross contamination between fractions of different densities, two more cycles of cesium chloride density gradient ultracentrifugation were performed on these three fractions separately, under the same conditions. After each density gradient ultracentrifugation, the fractions were diluted with the Tris–HCl buffer, concentrated by centrifugation at 300,000g for 1 h, and the pellets were resuspended in 1 ml of the buffer. The antigenicity and infectivity of these three fractions were determined by counterimmunoelectrophoresis and by mink inoculation, respectively. High titers of infectivity and anti- genicity were detected at a buoyant density of 1.405–1.416 g/ml as shown in Table 8.2. Numerous virus particles, both complete and empty, were observed in this fraction by an electron microscopic examination of the negative-stained preparation as shown in Fig. 8.1. Our earlier report of a lower value of the density of 1.36 ± 0.02 g/ml in cesium chloride (Cho and Ingram, 1974) may have been due to: (a) the much lower concentration of ADV which made the band undetectable in the density gradient, and (b) the large volume of each fraction (0.6–0.7 ml) made it difficult, by weighing 0.1 ml fractions, to reach an accurate value of the density of the virus. Recently, Porter et al. (1975) have shown that ADV grown in CRFK cultures has a density of 1.415–1.430 g/ml in cesium chloride. This value is quite close to our present estimate of 1.405–1.416 g/ml and is consistent with the density of a parvovirus (Tinsley and Longworth, 1973). However, Yoon et al. (1974) reported that ^{14}C-labelled ADV from mink kidney cell culture gave a sharp

TABLE 8.2

Determination of the infectivity and antigenicity of fractions of a cesium chloride density gradient of Aleutian disease virus.

Dilutions	Density of fractions (g/ml)		
	1.29–1.31	1.33–1.35	1.405–1.416
10^{-4}	0/2*	1/2	
10^{-5}	0/2	0/2	2/2
10^{-6}	0/2	0/2	3/3
10^{-7}			3/3
10^{-8}			2/3
10^{-9}			0/2
Antigen titers	1/8**	1/160	1/2560

* Number of mink infected/number of mink inoculated. Each mink received 1 ml intraperitoneally.

** Antigen titers were determined by counterimmunoelectrophoresis.

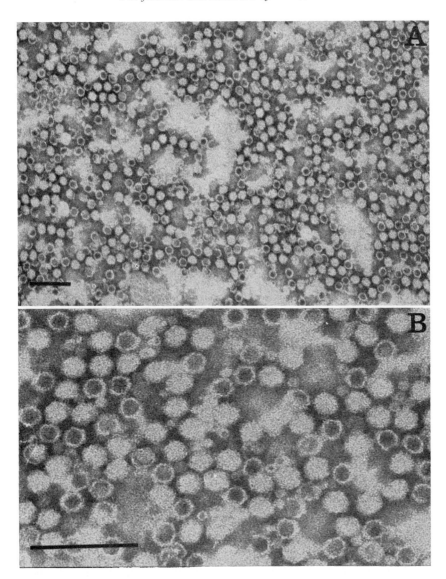

Fig. 8.1. Aleutian disease virus banded at a buoyant density of 1.405–1.416 g/cm³ in cesium chloride gradient. Negative stained with 2% phosphotungstic acid, pH 7.2. Bar: 100 nm.

peak at a density of 1.34 g/ml in a cesium chloride gradient. Employing a sucrose gradient, Yoon et al. (1973) and Crawford (1973) indicated that ADV had a buoyant density of 1.21 g/ml. The different results might be due to the use of two different gradient solutions, cesium chloride and sucrose, or to different methods and degrees of purification of the virus. It seems, however, the characteristics of the virus used by Kenyon et al. (1973) and Yoon et al. (1973, 1974) differed from those of the virus used by Porter et al. (1969, 1975) and by Cho and Ingram (1972, 1973a, 1974). Stocks of ADV were exchanged between Porter and Cho in 1973 and it was shown that both strains of the viruses have similar properties virologically and immunologically. Apparently the virus used by Kenyon's group has many characteristics similar to a picornavirus (RNA virus), whereas Porter's and Cho's virus has the properties of a parvovirus (DNA virus).

ADV is more heat resistant than most animal viruses but is far less resistant than the scrapie agent or the agent of transmissible mink encephalopathy (11.6.1; Table 15.2). The infectivity titer of a 650 nm Millipore membrane filtrate of an ADV-infected tissue suspension was not affected at 56 °C, but it was reduced by at least 1 $\log_{10}$ at 60 °C and by at least 3 $\log_{10}$ at 80 °C when the material was heated for 30 min (Eklund et al., 1968). A 10^{-1} or 10^{-2} dilution of unfiltered viral suspension was heated at 60 °C and at 80 °C for periods ranging from 30 min to 4 h. Exposure of the virus to 60 °C had no apparent effect on its infectivity, but virus was not detected after it was exposed to 80 °C for 30 min (Eklund et al., 1968). In this experiment it seemed that suspensions of unfiltered crude virus showed greater stability to heat than the filtered virus. In further studies tissue culture grown ADV was unaffected at 56 °C for 30 min, whereas heating at 60 °C for 30 min reduced the infectivity from ID_{50} $10^{6.2}$ to $10^{5.5}$ and after exposure at 80 °C for 30 min, infectivity was not demonstrated (Porter et al., 1975). In contrast ADV in crude tissue preparations from infected mink was relatively heat resistant. Suspensions of diseased tissues have remained infective after heating for 30 min at 80 °C, for 3 min at 99.5 °C (Gray, 1964) or for 15 min at 90–95 °C (Burger et al., 1965). It should be noted that the greater heat stability of ADV in the crude preparations may be explained by a stabilizing effect of protein impurities on the preparation. However, some skepticism should also remain on the findings since these workers did not use end-point titrations of the infectivity.

8.4.4 Response to chemical treatments

Eklund et al. (1968) reported that a 400 nm Millipore filtrate of ADV from an infected tissue suspension was rapidly inactivated by 0.3% formalin. The virus survived exposure to 0.3% formalin for 4 h but not for 8 h at 37°C. However, the infectivity titer of the virus exposed for 4 h was at least 2 $\log_{10}$ less than that of untreated virus. ADV present in a crude tissue suspension was rather more resistant to formalin. A crude tissue suspension remained infective when it was treated for 24 h at 37°C with 0.5% formalin (Russell et al., 1963). However, Karstad et al. (1963) found that a 15% tissue suspension from infected mink was inactivated after 2 days at 37°C and 19 days at 4°C in 0.3% formalin.

The infectivity of ADV was not affected by fluorocarbon treatment (Burger et al., 1965) or by ether (Burger et al., 1965; Eklund et al., 1968; Porter et al., 1975). These results are consistent with the fact that the virus does not have an envelope. The virus was stable at pH 3.0 (Haagsma, 1969). Fluorocarbon-extracted partially purified ADV preparation was resistant to treatment with proteolytic enzymes and nuclease separately or combined (Burger et al., 1965). The virus was not inactivated by 1% chloramine, pH 9.4, or by 2% O-phenylphenol, pH 12.5 (Haagsma, 1969). Burger et al. (1965) reported that the fluorocarbon-extracted ADV was partially susceptible to 4 M urea and completely inactivated by 0.5% iodine, 0.5 N hydrochloride and 0.5 N sodium hydroxide, whereas Haagsma (1969) claimed that 0.25 N or lower concentrations of sodium hydroxide did not inactivate the virus. It has not been determined whether a purified virus preparation free from ADV antibody would show a similar stability to these chemicals and enzymes.

8.5 Preliminary extraction and purification of the virus

To facilitate virus extraction from infected mink tissue, it was desirable to infect mink with a suspension containing as high a titer as possible. To increase infectivity of the inoculum, mink were inoculated intraperitoneally with 1.0 ml of 10% spleen homogenate of an infected mink and 4 serial passages were carried out at 10-day intervals. Spleens from mink of the 4th passage or later were removed at 10 days and used as the source of inoculum (Cho and Ingram, 1973b). At the second serial passage of the virus, Porter et al. (1969) could increase the infectivity titer by 10^3–10^4 times more than that found in chronically infected mink.

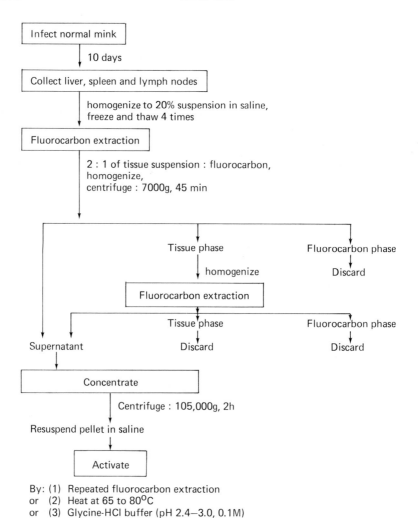

Fig. 8.2 Preparation of Aleutian mink disease viral antigen.

The procedures for extracting ADV from infected mink tissues are shown in Figure 8.2. Spleen, liver and lymph nodes were collected from mink 10 days post-infection, and these organs were homogenized and made into a 20% suspension in saline. These materials were frozen and thawed 4 times. This suspension was mixed with fluorocarbon (DuPont Chemical, Montreal) (2:1 ratio), homogenized with an omnimixer at 10,000 r.p.m. for 2 min. The homogenates were centrifuged at 7000g for 45 min. The aqueous phase was collected and the tissue phase was homogenized again with saline and the fluorocarbon extraction was repeated. Both aqueous phases were pooled and concentrated by ultracentrifugation at 105,000g for 2 h. After ultra-centrifugation, the supernatant was discarded and the pellet was collected and resuspended in 0.85 % saline. This concentrated pellet was infective but did not react serologically in the counterimmunoelectrophoresis or immuno-diffusion reactions. This non-reactive state of the antigen is believed to be due to the complexing of antibody and antigen. The non-reactive antigen preparations were activated by acid treatment of the extracted antigen (glycine–HCl buffer, 0.1 M, pH 2.4–3.0), or by heat activation at 65–80 °C for intervals up to 2 h, or by repeated fluorocarbon extraction (Cho and Ingram, 1972, 1974). These treatments cause the dissociation of virus–antibody complexes and it is believed that the treatments result in the exposure of the antibody-binding sites on the surface of virus particles. After these treatments, the dissociated antibodies were separated by further ultra-centrifugation at 105,000g for 2 h. Of these 3 procedures, repeated fluoro-carbon extraction produced the most satisfactory antigen preparation, in our experience.

Similarly, by acid elution of fluorocarbon-extracted ADV, Crawford (1973) prepared ADV antigen free from antibody and could remove the anti-complementary effect of the antigen and render it capable of migrating through the gel during electrophoresis.

8.6 *Isolation and ultrastructure of the virus*

The ADV antigen preparation was further purified by column chromatog-raphy. It was eluted with the void volume on a column of Sephadex G-200 (Pharmacia) but using Sepharose 6B (Pharmacia), antigen was eluted with an average partition coefficient (K_{av}) of 0.13. Virus particles were identified by electron microscopy in the preparations of purified ADV antigen from columns of Sepharose 6B. By mink inoculation, it was shown that both Sepharose 6B purified antigen and the immune precipitates formed in a counterimmunoelectrophoresis were infective (Cho and Ingram, 1973a). Thus it was concluded that the ADV antigen present in the precipitin line was indeed located on the surface of the virus particle. Further evidence that the antigen was located on the virus was obtained by examining, under the electron microscope, the immune precipitate formed in the liquid phase after counterimmunoelectrophoresis. In negative-stained preparations of the immune precipitates, numerous virus particles and aggregated virus–antibody

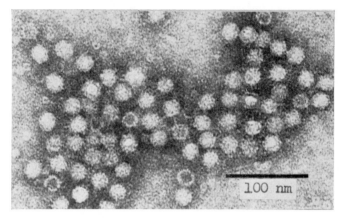

Fig. 8.3. Aleutian mink disease virus–antibody complexes. Negative stained with 2%
phosphotungstic acid, pH 7.0 (Cho and Ingram, 1974).

complexes were observed as shown in Fig. 8.3. ADV has a diameter of
23 nm, with a spherical shape and icosahedral structure. Both complete and
empty virus particles were observed, and the virus is similar in morphology
and size to certain parvoviruses or picornaviruses (Cho and Ingram, 1973a,
1974).

Kenyon et al. (1973) independently isolated the virus by affinity chromatography using a column composed of Sepharose 4B bound with ADV antibody
as an immunoadsorbent. ADV-free IgG was prepared from serum of
chronically infected mink by DEAE-cellulose chromatography. This
purified antibody was coupled with Sepharose 4B which was activated with
an aqueous solution of CNBr. The infective antigenic material was prepared
from mink tissues harvested 7 days after infection and reacted with the
immunoadsorbent. The virus containing column was washed until the
absorbance at 280 nm of the eluate was negligible.

Dissociation of the immunoadsorbed virus with 0.75 M NaCl and then
with a glycine–HCl gradient released infective virus particles. The eluates
were resolved by counterimmunoelectrophoresis with AD hypergammaglobulinemic serum. The zones between the antigen and antiserum wells
were prepared for Epon embedding and examined with an electron microscope after staining the grid with a solution containing 1% lead ion as a
mixture of acetate and citrate. Electron microscopy revealed virus particles
resembling a picornavirus with a diameter of 20 nm (Kenyon et al., 1973).

8.7 General conclusions

The transmissible nature of Aleutian mink disease was reported as early as 1962. However, the isolation of the virus has been achieved by two independent groups of investigators only recently, in 1973.

Electron microscopy has revealed that the virus has an icosahedral structure and is 23–25 nm in diameter. The nucleic acid type and classification of the virus has not been definitely determined. However, certain properties of the virus, i.e. its morphology and size, its initial replication in the nucleus, its density in cesium chloride, its stability to many chemicals, low pH and relatively high temperature, are similar to certain parvoviruses.

Many of the physico-chemical properties of the virus reported previously had been determined using crude tissue extracts from chronically infected mink; thus they represent the properties of virus–antibody complexes. Recent developments in procedures for virus purification, virus propagation in cell culture achieved by Porter et al. (1975) and in serological methods, such as immunofluorescence, complement fixation, counterimmunoelectrophoresis and immunodiffusion tests will enable rapid progress towards a better understanding of the characteristics of the purified virus in the near future.

Acknowledgements

I thank Dr. David D. Porter for his information on cell culture of AD virus before publication and his valuable comments on this manuscript. Special thanks are due to Dr. S. E. Magwood, Director of this Institute, for considerable advice and consultations on many aspects of the paper during the course of its preparation.

References

ALPER, T., CRAMP, W. A., HAIG, D. A. and CLARKE, M. C. (1967) Nature, 214, 764.

BASRUR, P. K., GRAY, D. P. and KARSTAD, L. (1963) Can. J. Comp. Med., 27, 301.

BUKO, L. and KENYON, A. J. (1967) Nature, 216, 69.

BURGER, D., GORHAM, J. R. and LEADER, R. W. (1965) In: (D. C. Gajdusek, C. J. Gibbs and M. Alpers, Eds) Natl. Inst. Neurol. Dis. Blindness, Monogr. 2. Slow, Latent and Temperate Virus Infections. (U.S. Government Printing Office, Washington, D.C.) p. 307.

CHO, H. J. and INGRAM, D. G. (1972) J. Immunol., 108, 555.

CHO, H. J. and INGRAM, D. G. (1973a) Nature, New Biol., 234, 174.

CHO, H. J. and INGRAM, D. G. (1973b) Can. J. Comp. Med., 37, 217.

CHO, H. J. and INGRAM, D. G. (1974) J. Immunol. Meth., 4, 217.

CRAWFORD, T. B. (1973) Fed. Proc., 32, 842.

EKLUND, C. M., HADLOW, W. J., KENNEDY, R. C., BOYLE, C. C. and JACKSON, T. A. (1968) J. Infect. Dis., 118, 510.

GORHAM, J. R., LEADER, R. W. and HENSON, J. B. (1964) J. Infect. Dis., 114, 341.

GOUDAS, P., KARSTAD, L. and TABEL, H. (1970) Can. J. Comp. Med., 34, 118.

GRAY, D. P. (1964) MVSc Thesis, University of Toronto.

HAAGSMA, J. A. (1969) Tijdschr. Diergeneesk., 94, 824.

HENSON, J. B., GORHAM, J. R., LEADER, R. W. and WAGNER, B. M. (1962) J. Exp. Med., 116, 357.

HENSON, J. B., GORHAM, J. R. and LEADER, R. W. (1963) Nature, 197, 206.

KARSTAD, L. and PRIDHAM, T. J. (1962) Can. J. Comp. Med., 26, 97.

KARSTAD, L., PRIDHAM, T. J. and GRAY, D. P. (1963) Can. J. Comp. Med., 27, 124.

KENYON, A. J., HELMBOLDT, C. F. and NIELSEN, S. W. (1963) Am. J. Vet. Res., 24, 1066.

KENYON, A. J., GANDER, J. E., LOPEZ, C. and GOOD, R. A. (1973) Science, 179, 187.

LARSEN, A. E. and PORTER, D. D. (1975) Infect. Immun., 11, 92.

MARSH, R. F. and HANSON, R. P. (1969) J. Virol., 3, 176.

MCGUIRE, T. C., CRAWFORD, T. B., HENSON, J. B. and GORHAM, J. R. (1971) J. Immunol., 107, 1481.

PORTER, D. D. and LARSEN, A. E. (1967) Proc. Soc. Exp. Biol. Med., 126, 680.

PORTER, D. D., LARSEN, A. E. and PORTER, H. G. (1969) J. Exp. Med., 130, 575.

PORTER, D. D. and LARSEN, A. E. (1974) Progr. Med. Virol., 18, 32.

PORTER, D. D., LARSEN, A. E., COX, N. A., PORTER, H. G. and SUFFIN, A. C. (1975) Fed. Proc., 34, 947.

RUSSELL, J. D. (1962) Nat. Fur News, 34, 8.

RUSSELL, J. C., BENNETT, J. M. and MARTY, E. W. (1963) Proc. U.S. Livestk. Sanit. Ass. 462.

TABEL, H. and INGRAM, D. G. (1970) Arch. Ges. Virusforsch., 32, 53.

TRAUTWEIN, G. W. and HELMBOLDT, C. F. (1962) Am. J. Vet. Res., 23, 1280.

TINSLEY, T. W. and LONGWORTH, J. F. (1973) J. Gen. Virol., 20, 7.

TOOLAN, H. W. (1972) Progr. Exp. Tumor Res., 16, 410.

TSAI, K. S., GRINYER, I., PAN, I. C. and KARSTAD, L. (1969) Can. J. Microbiol., 15, 138.

YOON, J. W., KENYON, A. J. and GOOD, R. A. (1973) Nature, New Biol., 245, 205.

YOON, J. W., DUNKER, A. K. and KENYON, A. J. (1974) Fed. Proc., 33, 605.

CHAPTER 9

Pathology and pathogenesis of Aleutian disease

James B. HENSON, John R. GORHAM, Travis C. McGUIRE and Timothy B. CRAWFORD

9.1 Introduction

Aleutian disease (AD) was first described in 1956 by Hartsough and Gorham, approximately 15 years after the discovery of the Aleutian mutation. The latter occurred spontaneously on a commercial mink ranch in Northern Oregon, USA (see 7.3). After discovery of the Aleutian mutation, animals with this beautiful blue pelt were commercially raised in large numbers and were called Aleutian due to the resemblance of the pelt colour to that of the Aleutian fox. It was recognised by mink ranchers when large numbers of Aleutian-type mink were first being raised that animals of this genotype were less hardy than others. There was no indication of the reason for this lack of hardiness until it was recognised by Leader et al. (1963) that Aleutian type mink have the Chediak–Higashi syndrome which makes them more susceptible to bacterial infections than normal. Early experience on commercial mink ranches suggested that AD occurred primarily, if not exclusively, in Aleutian-type mink, so the disease was called Aleutian Disease. Later observations have proven, however, that the disease may occur in any genotype of mink and also affects ferrets.

The early morphologic descriptions of the disease stressed the occurrence of the marked plasmacytosis, glomerulonephritis and arteritis. Obel (1959) suggested that the disease might be a plasma cell myeloma. Other observers

Supported in part by NIH grant AI 06477.

noted the similarity between the lesions occurring in affected mink and collagen diseases in man. Similarities were pointed out between the lesions in AD and systemic lupus erythematosus (SLE), polyarteritis nodosa and others. These early observations suggested that there might be similar pathogenetic mechanisms involved in AD and these poorly understood human diseases and that AD might be a model disease or an 'experiment in nature' that could be used to delineate basic disease processes in these human diseases. With this in mind, research initiated on AD in a number of laboratories has resulted in the accumulation of the data to be described below.

It was also recognised early in research on AD that the agent persisted throughout the course of the disease. At about that time, attention was being focused on the occurrence of persistent viral infections as the aetiology of human and other animal diseases. The mechanism(s) of viral persistence in AD became another aspect of the disease that has been emphasised in research. Present information would suggest, however, that AD mechanistically is dissimilar to a number of other persistent or 'slow virus' infections, such as some of those described in detail in this publication. It seems likely that there is a spectrum of persistent infections which involve different mechanisms and which produce a spectrum of organ-system dysfunctions.

The purpose of this paper is to describe AD, define its pathogenesis and relate the pathogenetic mechanisms to other diseases described in this publication and to other human and animal diseases.

9.2 Clinical disease

The clinical disease and course observed in experimentally and spontaneously infected mink are similar. Aleutian type mink *(aa)* develop a more rapid clinical course and die within a shorter time span after experimental inoculation than do non-Aleutian type mink *(Aa* or *AA)*. Similarly, ferrets inoculated with AD virus develop some of the lesions, but rarely succumb to the disease. Thus, there is a difference in the clinical course observed in different genotypes of mink. Aleutian mink almost invariably die 2–5 months after inoculation. On the other hand, non-Aleutian type mink may live for months (8–12) or years and some non-Aleutian type mink appear to recover from the disease. We have shown that approximately 25% of non-Aleutian type mink infected with AD virus recover from the disease, while Larsen and Porter (1975) have reported that similar percentages of pastel mink clear the virus. Both experimentally and spontaneously infected ferrets appear to suffer few ill effects from the disease (Ohshima et al., unpublished). Examination of the tissues of such ferrets will demonstrate both gross and microscopic lesions as well as hypergammaglobulinemia.

The clinical disease in mink is characterized principally by gradual loss of weight resulting in severe emaciation with maintenance of a voracious appetite. Severely affected animals frequently develop ulcers in oral mucosa and may have malina. Electrophoretic separation of serum proteins reveals hypergammaglobulinemia, which is an important feature of the disease, (Henson et al., 1961, 1962; Ingram and Cho, 1974; Porter and Larsen, 1974; Tables 9.1 and 2). Infected animals become anaemic and have other alterations detectable by clinical laboratory tests. Inoculation of AD virus into

TABLE 9.1

The mean serum protein values of 14 normal and 17 spontaneously affected mink, expressed as g/100 ml serum.

	Total serum protein	Albumin	α-globulin	β-globulin	γ-globulin	Albumin/ globulin
Normal	7.26	4.39	0.66	0.99	1.22	1.3
Diseased	8.22*	3.19*	0.71	1.10	3.22*	0.6*

*   Statistically significant ($P < 0.01$).

TABLE 9.2.

Gamma globulin levels by months for 6 spontaneously diseased and 1 normal mink, expressed as % total serum protein.

	γ-globulin level and date serum collected					
	July 1	Aug. 9	Sept. 9	Oct. 11	Nov. 3	Nov. 29
Affected	7.3	14.3	37.2	50.0*	—	—
Affected	5.2	20.9	35.3	39.4	45.0*	—
Affected	6.7	16.3	10.3	—	43.4	—
Affected	4.5	—	31.5	36.1	45.5	—
Affected	7.0	24.2	30.4	—	50.6*	54.8*
Affected	5.6	8.1	31.8	38.7	44.6	53.2*
Normal	5.8	15.3	14.7	10.8	10.3	12.2

*   Clinical signs present.

178   *J. B. Henson et al.*

ferrets causes little or no clinical disease with affected ferrets becoming
hypergammaglobulinemic and developing plasmocytosis in a number of
tissues (Kenyon et al., 1967; Ohshima et al., unpublished).

9.3 Lesions

9.3.1 Gross lesions

The gross lesions observed in animals dying of AD, regardless of genotype,
are similar. They include emaciation, loss of all body fat, lymphadenopathy,
splenomegaly and hepatomegaly (Helmboldt and Jungherr, 1958; Henson,
1964; Henson et al., 1966; Pan et al., 1970; Trautwein and Seidler, 1972).
Severely affected livers contain many pin-head size, pale tan to gray foci
of cellular infiltrates scattered throughout the parenchyma and are more
difficult to cut than normal due to fibrosis. The kidneys early in the disease
have petechial hemorrhages scattered over their surfaces and become
enlarged and turgid with a bulging, wet, cut surface. Later the kidneys
become pale, shrunken and irregular, are more difficult to cut and have

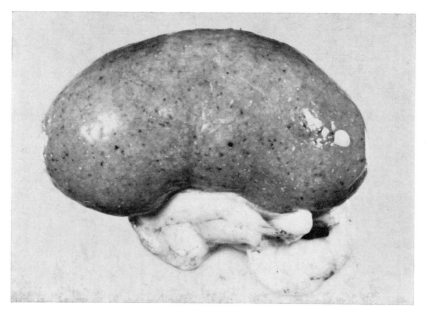

Fig. 9.1. Kidney from mink with subacute AD showing scattered petechial hemorrhages,
paleness and a slightly granular and irregular surface. × 5.

small inflammatory cell infiltrations in the cortical areas, sometimes accompanied by small cortical cysts. Kidneys intermediate between these two extremes are pale and slightly irregular on their surfaces and have scattered petechial hemorrhages (Fig. 9.1).

The lymphoid system is severely affected with lymphadenopathy and splenomegaly detectable early after experimental infection. Both the nodes and spleen become progressively enlarged and may be 4–10 times normal. In severely affected mink the nodes become very prominent, emphasised by lack of body fat, and both the nodes and enlarged spleen may be palpated in the live animal. Affected spleens are greatly enlarged and fairly firm with little blood flowing from the bulging, freshly cut surface. Affected lymph nodes have no discernable demarcation between the cortex and the medulla and are grey to light tan in colour.

Gastric ulcers occur in some animals as do ulcers in the oral mucosa. Hemorrhage and infarction of the brain have been noted in a small percentage of mink with coning of the cerebellum due to the resultant pressure.

Gross lesions in spontaneously and experimentally infected ferrets include lymphadenopathy and splenomegaly (Kenyon, 1966; Kenyon et al., 1967; Ohshima et al., unpublished). In some instance, pin-head size foci of cellular infiltrates may be observed in the liver; renal changes are very unusual. In addition, the enlargement of the lymphoid system does not reach the proportions observed in mink. Ferrets infected with AD virus are usually in good body condition and rarely die.

9.3.2 Microscopic lesions

The microscopic lesions in both AD-affected mink and ferrets are dominated by proliferation of plasma cells (Helmboldt and Jungherr, 1958; Obel, 1959; Henson et al., 1966). The plasma cell proliferation precedes the development of other lesions to be described below and develops in a predictable time sequence. After initial inoculation of AD virus, small numbers of lymphocytes, macrophages and plasmablasts can be observed in the periportal areas of the liver. It appears likely that similar changes are concurrently underway in the lymphoid tissues, but are much more difficult to define due to the normal cellularity of these organs. Similar collections occur around the small muscular arteries at the cortico-medullary junction of the kidney and can also be seen perivascularly around other small muscular arteries in other tissues. The number of cells in the various locations increase with the rapid appearance of mature plasma cells. The proliferation continues until

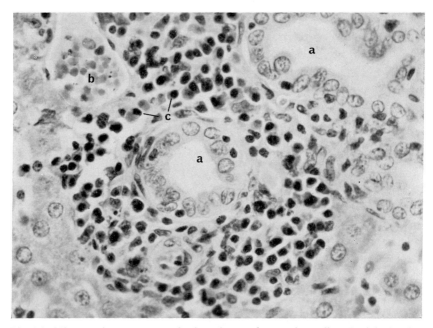

Fig. 9.2. Microscopic appearance of a hepatic portal area of an affected mink showing proliferated bile ducts (a), vein (b) and infiltration of mature-appearing plasma cells (c). Hematoxylin–Eosin. × 485.

TABLE 9.3.

Distribution of lesions in 17 spontaneously affected mink.

	Arteritis	Peri-arteritis	Portal infiltration	Bile duct proliferation	Interstitial infiltration	Nonpurulent meningitis
Kidney	7	13			17	
Bladder	4	4				
Liver	2		17	13		
Stomach	7	8				
Intestine	2	4				
Brain	4	6				7
Heart	4	5				

the population is dominated by mature plasma cells, which form sheets or large isolated masses of cells with such large collections occurring throughout the lymphoid organs, kidney and liver (Fig. 9.2). In the latter two organs these collections reach such large sizes so as to be seen grossly and are prominent in prepared histological slides observed with the naked eye. The distribution of non-lymphoid lesions in selected tissues of 17 spontaneously infected mink is given in Table 9.3.

The lymphoid changes are characterized, as indicated above, by a progressive increase in the number of plasmablasts and mature plasma cells until late in the course of the disease. The initial changes observed histologically are increased folliclar activity with numerous mitotic figures and cellular debris. These changes are suggestive of active stimulation of the lymph nodes. As the disease progresses, increased numbers of plasmablasts and mature plasma cells are observed in the medullary cords, with plasma cells becoming so numerous as to appear to completely fill the sinuses. The number of these cells in the cortex also gradually increases. Later in the disease, the follicles change from very cellular and active to relatively inactive and sparsely populated with a paucity of lymphocytes. In the advanced disease, the follicles as well as the rest of the lymphoid tissues, although still quite enlarged, appear to be depleted of lymphoid elements and are predominantly less cellular, with the cells that are present consisting of large irregular, macrophage-appearing cells and plasma cells.

Changes similar to those described for the lymphoid follicles occur in the spleen follicles with the enlarged spleen dominated cellularly in the late stages by large macrophage-appearing cells with a lack of mature-appearing lymphoid cells.

The microscopic changes in the liver are detectable very early, probably due to the lack of cellularity (lymphoid elements) in the portal areas in the normal liver. The initial changes are the occurrence of a small number of lymphocytes, plasmablasts and a few macrophage-appearing cells in the periportal areas. As the disease progresses, the population of cells within the periportal areas increases, with plasmablasts and mature plasma cells dominating (Fig. 9.2). The number of cells increases until these areas become distended by large collections of mature-appearing plasma cells. At the time that the periportal changes are occurring, the macrophage elements in the sinusoids become active and proliferate. Some lymphoid cells occur here also, but the plasma cellular series does not dominate in this anatomic location as in the periportal areas. Late in the course of the disease the hepatic cords are disrupted, with the overall appearance of the liver being hypercellularity with the dominance of large mononuclear cells and some lymphocytes. There is an increase in the bile ducts associated with the periportal cellular infiltrates. The bile ducts appear morphologically normal and do not contain any increased bile or other material (Fig. 9.2). There is dilation as well as proliferation of the bile ducts until these changes can be detected grossly when liver sections are observed by the naked eye. There is proliferation of fibroblasts associated with the increased size and number of bile ducts. All of these changes result in the hepatomegaly, grossly apparent cell infiltrates, and increased toughness of the liver. Degenerative arteritis, to be described later, frequently occurs in the small muscular arteries of the liver.

The renal lesions in AD are a dominant character of the disease and result in uremia, which is the ultimate cause of death. The initially detectable lesions are perivascular collections of small numbers of mononuclear cells around the small muscular arteries at the cortico-medullary junction. These are similar in their make-up to those described above for the liver. These early cells are mononuclear and appear to be lymphocytes, a small number of larger, irregular cells which resemble macrophages and a few plasma-

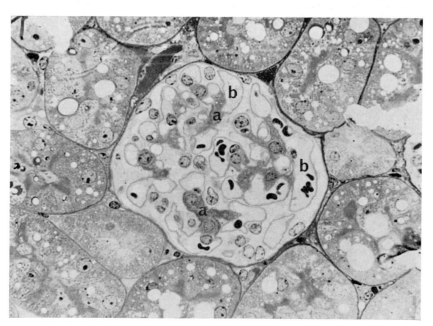

Fig. 9.3. Plastic embedded thin section of a normal mink glomerulus with normal-appearing mesangial areas (a) and many patent capillaries (b). Toluidine blue. × 300.

blasts. The total number of cells increases rapidly until large masses or sheets of cells infiltrate throughout the renal parenchyma, especially in the cortical areas. This results in a separation of the tubules and a break-down or reduplication of the tubular basement membranes. With the advanced disease, mature plasma cells form large sheets and collections in the inter- stitial areas. The early kidney lesions (perivascular accumulation) and the early periportal infiltration in the liver represent the earliest detectable diagnostic lesions by light microscopy.

Other renal lesions include the frequent occurrence of degenerative arteritis in the small muscular arteries of this organ and the occurrence of mononuclear infiltrates in the pelvis of the kidney.

Glomerular alterations are a characteristic of the disease in mink and have been characterized by light, fluorescent and electron microscopy, (Helmboldt and Jungherr, 1958; Obel, 1959; Henson et al., 1966, 1967, 1968, 1969; Pan et al., 1970; Trautwein and Seidler, 1972; Johnson and Henson, unpublished; see Figs 9.3, 4 and 5). The initial detectable glomerular

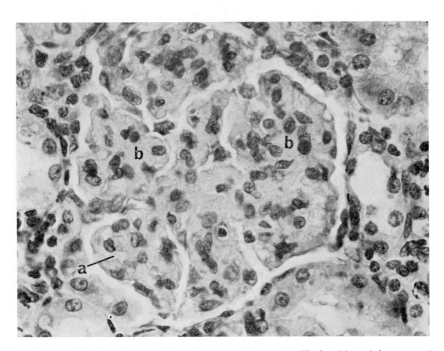

Fig. 9.4. Severely affected glomerulus with few patent capillaries (a) and increase of eosinophilic granular material (b) replacing the normal anatomic structures. Hematoxylin– Eosin. × 500.

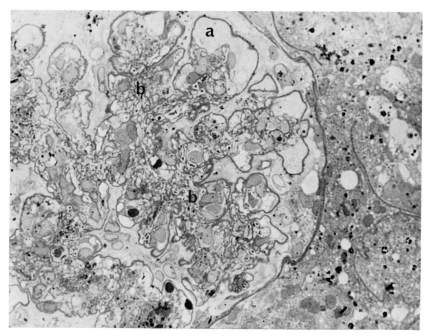

Fig. 9.5. Plastic-embedded thin section of a severely diseased glomerulus showing more detail than Fig. 9.4 with decreased number of patent capillaries (a) and increased mesangial areas (b). The latter is composed primarily of increased mesangial cells and mesangial matrix. Toluidine blue. × 500.

lesions by light microscopy include increased cellularity of the glomeruli which is most notable in the mesangial areas. Accompanying the increased cellularity is an increased amount of faintly eosinophilic, slightly granular material in haematoxylin and eosin stained sections (Fig. 9.4). Also present are a few neutrophils and, later in the disease, obvious degenerative changes with hemorrhage in some glomeruli. The mesangial areas continue to increase in size with leakage of plasma proteins into Bowman's space and the development of proteinaceous tubular casts. Late in the disease, the glomeruli become avascular, are not more cellular than normal, and are characterized by the occurrence of large amounts of the slightly granular, eosinophilic material which has been found to represent mesangial matrix and mesangial cell cytoplasm (Figs 9.4 and 5). Also observed in the disease is 'wire-loop' thickening of the peripheral capillaries, but this is not a dominant feature of the disease.

When diseased glomeruli are observed by electron microscopy, the deposition of electron-dense, granular material has been observed sub-endothelially, sub-epithelially and within the glomerular basement membrane. Similar material has also been observed in mesangial cells and entering the mesangial angle through pores between endothelial cells. Fluorescent anti-body staining of the same diseased kidneys indicates the deposition of mink immunoglobulin and C3 in a 'lumpy-bumpy' pattern. In a small number of glomeruli, AD viral antigen can also be seen after antibody elution.

There is increased amount of mesangial matrix and a proliferation of mesangial cell cytoplasm early in the course of the disease. Fusion of foot processes and other non-specific changes also occur. The mesangial increase continues to dominate the ultrastructural appearance of diseased glomeruli until the increased amount of cells and cellular material impinges on the glomerular capillaries sufficiently to block the circulation and cause uremia. Thus, the glomerulonephritis in AD is caused by the deposition of antigen–antibody complexes which result in changes in the glomerular basement membrane and stimulate the proliferation of mesangial cells. The latter proliferate and also produce increased amounts of mesangial matrix. The mesangial cytoplasm and matrix is the material observed by light microscopy as pink, slightly granular material in haematoxylin and eosin-stained sections. Increased Periodic Acid–Schiff (PAS) positive material can also be seen in the glomeruli. This represents increased amounts of mesangial matrix. Therefore, the glomerulonephritis in AD can be classified as a proliferative glomerulonephritis in which the stimulation of the mesangial cells and basement membrane damage by the deposition of immune complexes results in damage and cellular proliferation that proceeds to renal failure. Degener-ative arterial lesions occurs frequently in diseased kidneys, but this lesion and its development will be described later.

The arterial lesions in AD (Helmboldt and Jungherr, 1958; Henson, 1964; Henson et al., 1966; Porter et al., 1973; Henson and Crawford, 1974) were recognized as one of the more interesting features of the disease and were one of the first morphologic alterations that suggested a similarity in the pathogenetic mechanisms between AD and polyarteritis nodosa and SLE. The initial arterial lesions are increased size and irregularity of the endothelial cells with some evidence of increased numbers of cells in the intima. These changes are minimal, however, in the early disease. In observing the arterial lesions in spontaneously diseased animals, one of the early events appears to be the occurrence of myolysis in small muscular arteries in various locations in the body and breakdown of the internal elastic lamina (Fig. 9.6). Small numbers of neutrophils are observed in these areas also. There is an

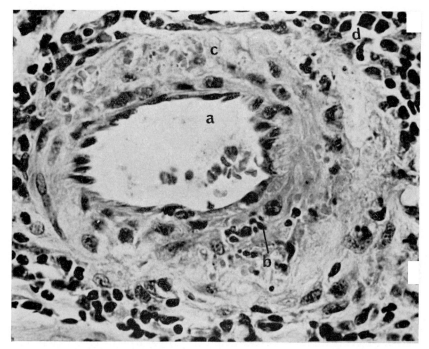

Fig. 9.6. Small muscular artery with patent lumen (a), polymorphonuclear leucocytes in media (b), early medial necrosis (c) and perivascular mononuclear infiltrates (d). Hematoxylin–Eosin. × 500.

infiltration of mononuclear cells around affected portions of diseased arteries, hemorrhage into the affected portions with erythrocytes and plasma, formation of fibrin and a few inflammatory cells which result in the morphologic appearance of fibrinoid necrosis (Fig. 9.6). Some vessels may appear to be large masses of fibrinoid with increased cellularity in the intimal areas and the entire vessel surrounded by a thick cuff of mononuclear cells, primarily lymphoid, but with some plasma cells (Figs 9.7 and 8). Late in the course of the disease, connective tissue is laid down around the periphery of the diseased vessel and there is frequently narrowing of the lumen. Generally, however, the arterial alterations are not characterized by the formation of intravascular clots and/or infarction.

Studies of the arterial lesions by fluorescent antibody techniques have indicated the early deposition of immunoglobulin, complement and AD viral antigen. These findings suggest that the arterial lesions in AD is mediated by the deposition of antigen–antibody complexes and is similar in its pathogenic mechanisms to those described for serum sickness (Cockrane and Hawkins, 1968).

The extent of the involvement of a given vessel appears to vary, similar to that described for serum sickness, with some vessels having their entire circumference affected while others are segmentally involved. The organs most frequently having arterial lesions included the liver, gastric submucosa, kidney and urinary bladder.

Microscopic involvement of other organs are primarily related to the vascular system. In practically any tissue, arterial lesions and/or perivascular accumulation of mononuclear cells, can be observed. In some brains, however, there is infarction with hemorrhage. In these brains, the basalar and other arteries are usually affected. Gastric involvement includes mucosal ulcerations and arteritis. It has not been possible, however, to establish a definite causal relationship between the affected vessels and the ulceration of the gastric mucosa. In some instances, hemorrhage and occasionally hemorraghic anaemia and death resulted from these gastric ulcerations.

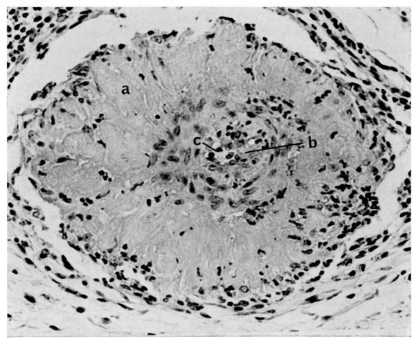

Fig. 9.7. Severely diseased artery in the gastric submucosa with extensive fibrinoid (a), subintimal proliferation (b) and patent lumen (c). Hematoxylin–Eosin. × 600.

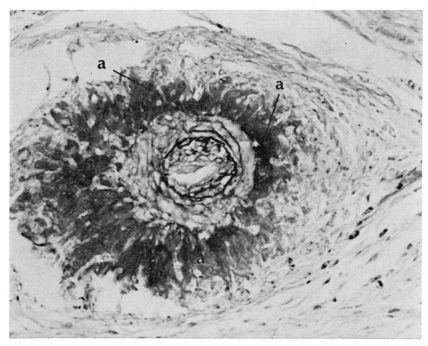

Fig. 9.8. Same artery as shown in Fig. 9.7, showing Periodic Acid–Schiff positive staining of fibrinoid. × 600.

The lung is affected in the vascular system without any indication of other parenchymal pulmonary involvement. Likewise, the heart is primarily affected in the vascular tree. Other organs are also involved, but these are similar to the vascular changes described previously.

Experimentally and spontaneously diseased ferrets have some of the lesions observed in AD-infected mink, but rarely died of the disease (Kenyon, 1967; Kenyon et al., 1967; Ohshima et al., unpublished). In fact, infected ferrets, whether with spontaneous or experimentally induced disease, usually appear in good body condition and health. Observation of the tissues, however, reveal similar changes to those described in the mink, but the ferret lesions appear to be less active.

Affected ferret livers have periportal collections of plasma cells with occasional dilation and/or proliferation of the bile ducts. Likewise, the lympho-reticular system is increased in size grossly and microscopically has increased numbers of mature plasma cells. In the ferrets, the disease does not progress to the degree described above for the terminal stages of AD in mink. The lymphoid tissues are active and have large accumulations of plasma cells. The interstitial infiltrations observed in the mink kidneys occur in ferrets, but to a lesser degree. In the ferret, peri-arterial collections of plasma cells as well as some interstitial infiltration of these cells occur. The gross distortion of the renal parenchyma by large accumulations of cells as seen in the mink, however, does not usually occur in the ferret. Glomerular lesions are mild or do not occur in infected ferrets, so that uremia does not occur. Likewise degenerative activities are unusual in the ferret.

Thymic alterations have been reported in diseased ferrets with the occurrence of follicles in diseased thymuses. Our observations of a large number of ferrets, however, have failed to show follicles and have shown infiltration of plasma cells. In our material, the thymic lesions have been minimal. Generally, the disease in ferrets appears to progress slower and does not appear to be as active. Importantly in the ferret disease, glomerular alterations are minimal or absent.

9.4 Virus replication and distribution

Mink have been experimentally infected with AD virus and blood collected periodically thereafter and titrated in mink to determine the infectivity titre. It was possible to detect infectivity soon after inoculation with the amount of virus in the blood reaching peak titres of 10^5–10^7 mink infectious doses (ID_{50}) 7–10 days after inoculation. The titre in the blood then decreased 3–4 weeks after inoculation and was maintained at a level of 10^4–10^5 ID_{50} for the duration of the 4-month observation period. Porter et al. (1969) reported maximum virus titres of 10^8–10^9 per gram of spleen, liver and lymph node 10 days after experimental infection. Eklund et al. (1968) have demonstrated early AD virus replication with decline of titre 2 months or more after infection to spleen titres of 10^5 ID_{50}/g and serum titres of 10^4 ID_{50}/ml. These studies indicate that the virus in experimental AD replicates rapidly in all genotypes of mink, reaches peak titres approximately 1–2 weeks after inoculation, then declines approximately 3–4 weeks after inoculation and is maintained at a relatively stable level in the circulation in Aleutian-type mink. In some non-Aleutian mink, the virus is eliminated 1–2 months after inoculation with elimination noted in about 25% of our non-Aleutian mink. Other studies have demonstrated the persistence of infectious virus in AD-infected mink for months or years after inoculation.

We have studied the distribution of AD viral antigen using fluorescent antibody techniques (Crawford, 1972; Henson and Crawford, unpublished).

These studies utilized serum from chronically infected mink as a source of antibody for demonstrating antigen in infected mink tissues. Specificity was demonstrated by the occurrence of fluorescence in infected and not in non-infected tissues, blocking of fluorescence by non-fluorescenated antisera, absorbtion of antibody by appropriate antigen-containing tissues and by the use of antisera collected from a number of ranches in the United States which had no possibility of contact between them. The technique was demonstrated to be specific. Porter et al. (1969) had previously demonstrated a similar specificity. Fluorescence was noted in the cytoplasm of cells as numerous green dots or stippling. In some cells, a small amount of fluorescence as green strands or dots were seen in the nucleus.

The tissues with the greatest number of antigen-containing cells in our studies have been the lymphoid organs and liver. In intraperitoneally inoculated mink, the mesenteric node contained the largest number of fluorescent cells, mostly in the cortical areas. Peripheral nodes in similarly inoculated mink contained scattered single or small groups of fluorescent cells, usually in the cortex. When groups of fluorescent cells occurred, the number varied from 3 to 12. Fluorescent cells were rarely noted in the medulla of lymph nodes.

Single or multiple fluorescent cells were seen in the livers of experimentally infected mink with the cells usually located peri-sinusoidally. Few fluorescent cells were seen in the portal areas. The cell type containing antigen appeared to be Kupffer cells or cells of similar appearance.

In all other tissues, fluorescent cells were less numerous and occurred as isolated individual cells or small groups of 2–6 cells in the stromal tissues. In no instance was fluorescence noted in epithelial tissues. In the stomach and intestines, for example, fluorescent cells were seen in the submucosa, substantia propria and rarely in the muscularis, but not in the muscle cells themselves or the epithelium. The other organs in which the greatest number of fluorescent cells occurred was in the mammary gland and placenta of pregnant mink.

Antigen-containing cells were seen 2–4 days after infection and persisted for up to 4 months. The largest number of cells were observed at 7–10 days after experimental infection. The number of fluorescent cells probably does not represent a valid estimate of the total number of antigen containing cells, however. It has been shown in this laboratory that infected mink begin producing anti-AD antibody approximately 3–4 days after inoculation and that virus obtained this early after inoculation may be complexed with antibody.

The virus strain used to inoculate mink also influences the number of fluorescent cells observed in experimentally infected mink. When sapphire mink were inoculated with 'Pullman' strain and with a strain obtained from Dr. D. Porter ('Utah' strain; see 7.4.2), the latter resulted in greater numbers of fluorescent cells with more seen in infected sapphire mink than in non-Aleutian type mink. Eklund et al. (1968) have reported higher virus titres in Aleutian as compared to non-Aleutian mink.

Recent studies by Porter et al. (1975) indicate that AD virus can be propagated in vitro in cat cells. In these tissues the fluorescence occurs predominantly in the nuclei of infected cells. The discrepancy between the location of fluorescence in vitro seen by Porter and that in the tissues of the diseased mink and ferrets is unknown at this time. The microscopic appearance of antigen-containing cells was green, brightly fluorescing cytoplasmic droplets with only a small amount of fluorescent material in the nuclei seen as a few small fluorescent droplets and/or strands. Whether this material in the cytoplasm represents phagocytosed antigen or antigen being produced within these cells, remains to be proven.

We have also studied the distribution of AD viral antigen in pregnant female mink and unborn kits. The females were inoculated intraperitoneally with 10^4 ID_{50} of 'Pullman' strain virus in the form of a spleen homogenate in the last third of pregnancy. The females were killed 14 days later and their tissues and tissues of the unborn kits were collected, snap frozen and examined by fluorescent antibody technique for the presence of AD viral antigen. Viral antigen was present in the tissues of the female mink as described above. The mammary gland also had large numbers of fluorescent cells in the form of individual or small groups of 2–5 cells in the supportive tissues, but not the epithelium. These cells had brightly fluorescing cytoplasm with the apple green fluorescence occurring as small droplets. The ovaries of these pregnant mink also contained numerous fluorescent cells, some of which were adjacent to, but never present within, ova. Viral antigen was also demonstrated in the supportive elements of the placenta, both maternal and foetal, but never in the epithelial elements.

Examination of the tissues of unborn foetuses revealed viral antigen in all the lymphoid organs, liver, lung, intestines, pancreas, heart and muscle. The antigen, in all cases, appeared to be in cells in the supportive tissues and in macrophages. As an example, what were thought to be Kuppfer cells in the liver contained antigen but hepatocytes did not.

These studies indicate that virus readily crosses the placenta and infects kits in utero. The presence of viral antigen in the placenta is another indication of this process. The presence of large amounts of antigen in the mammary gland was unexpected, although we had previously demonstrated that mink milk was infectious (Gorham et al., 1965). It appears, therefore, that AD virus replicates rapidly and is widespread in the tissues of the infected mink. The agent readily crosses the placenta and infects unborn kits. Padgett et al. (1967) previously demonstrated the presence of infectious virus in the tissues of unborn kits from infected, but not from uninfected females. Eklund et al. (1968) studied the distribution of AD virus in a number of organs in experimentally infected mink using mink infectivity as the indicator system. High titres were found in the serum, kidney, spleen and liver. Lower titres of virus were detected in other organs examined and included submaxillary salivary gland, brain, heart, intestine, lung, mesenteric lymph node and blood clot. These authors indicated that the mesenteric lymph node, spleen and kidneys could be considered as sources of virus. Our studies using fluorescence have never demonstrated significant amounts of viral antigen in the kidney. The urine, however, contains infectious virus and may well be one of the primary sources of virus for horizontal transmission (see 7.5.1).

9.5 Immune response

In 1961 AD-infected mink were reported to have a pronounced hypergamma-globulinaemia (Henson et al., 1961). The genesis of this hyper-response was not clear and has still not been completely elucidated, although much more is known about the immune response and the role it plays in the pathogenesis of AD. The occurrence of widespread plasmacytosis accompanied by hypergammaglobulinaemia suggested to early workers that AD might be a plasma cell myeloma. The electrophoretic pattern of serum from diseased mink indicated a polyclonal, rather than a monoclonal response, (Henson et al., 1961, 1962). Later, however, Porter et al. (1965), Eklund et al. (1968) and Tabel and Ingram (1970) reported the occurrence of a monclonal gamma response in AD-infected mink.

Studies have been conducted by a number of workers indicating the sequential increase in the level of gamma globulin after experimental inoculation. Increased amounts of gamma globulin become detectable from 3–4 weeks after experimental inoculation and continue to rise rather steadily until death occurs in Aleutian-type mink. We investigated such a response in spontaneously diseased mink maintained under field conditions and found similar results (Table 9.2). In this study, mink were bled and some were killed at monthly intervals from the time of weaning (July) through to January. The degree of hypergammaglobulinaemia was determined for each serum collected and the presence of lesions compared to the occurrence of hypergammaglobulinaemia. The results indicated that the gamma globulin response in spontaneously affected mink was similar to the experimental disease. There was an increasing gamma globulin level in mink from the initial time of detection until death occurred (Table 9.2). The level of hyper-gammaglobulinaemia appeared to parallel the severity of lesions (Tables 9.4 and 9.5).

Somewhat different findings have been demonstrated in non-Aleutian mink, (Eklund et al., 1968; Lodmell et al., 1973; Larsen and Porter, 1975). In a small percentage of such mink, there is an initial elevation of gamma

TABLE 9.4

Relationship between the degree of hypergammaglobulinemia and the occurrence and severity of the arterial, renal and hepatic lesions in experimentally infected mink.

	Level of γ-globulin (%)					
	10–19	20–29	30–39	40–49	50–59	Total
Mink with Aleutian disease	0	4	9	7	4	24
Arteritis	0	0	0	1	1	2
Renal interstitial infiltrations:						
Mild	0	3	3	1	0	7
Moderate	0	0	5	4	2	11
Severe	0	0	0	2	2	4
Total						22
Renal glomerular lesions:						
Mild	0	0	3	2	2	7
Moderate	0	0	0	4	2	6
Severe	0	0	0	0	0	0
Total						13
Hepatic portal infiltrations:						
Mild	0	4	8	2	1	15
Moderate	0	0	1	5	1	7
Severe	0	0	0	0	2	2
Total						24

J. B. Henson et al.

TABLE 9.5

Relationship between the degree of hypergammaglobulinemia and the occurrence and severity of arterial, renal and hepatic lesions in spontaneously infected mink.

	Level of γ-globulin (%)				
	10–29	30–39	40–49	50–59	Total
Mink with Aleutian disease	0	4	4	3	11
Arteritis	0	0	0	3	3
Renal interstitial infiltrations:					
Mild	0	4	2	0	6
Moderate	0	0	2	0	2
Severe	0	0	0	3	3
Total					11
Renal glomerular lesions:					
Mild	0	2	1	0	3
Moderate	0	0	2	1	3
Severe	0	0	0	2	2
Total					8
Hepatic portal infiltrations:					
Mild	0	0	1	0	1
Moderate	0	4	2	1	7
Severe	0	0	1	2	3
Total					11

globulin levels followed by a decrease to normal levels in animals that recover from the disease. In addition, the degree of gamma globulin elevation, or the rapidity of gamma globulin elevation, appears to be less or slower in developing in some non-Aleutian mink when compared to sapphire mink. This slow progression is similar to the slower progression of the clinical disease and the extended mean death times in most non-Aleutian mink when compared to Aleutian-type mink.

Early investigations on the possibility that the increased gamma-globulin represented specific antibody suggested that this immunoglobulin increase was not specific antibody (Porter et al., 1965a). Various considerations were given, including the possibility that AD represented a myeloma as well as the other possibility that the gamma globulin increase was a non-specific one triggered by some unknown action of the AD agent.

In 1967, Porter and Larsen demonstrated the occurrence of circulating virus–antibody complexes by treatment of infected mink serum with anti-mink gamma globulin. The latter treatment decreased infectivity by approximately 2 logs, suggesting to these workers that much of the circulating virus was complexed with immunoglobulin. In 1969, Porter et al. used an indirect immunofluorescent technique to demonstrate the occurrence of antibody in the serum of AD-infected mink. Since that time, antibody activity has also been demonstrated by complement fixation (McGuire et al., 1971) and precipitation techniques (Cho and Ingram, 1972). All of these techniques have suggested that the antibody level in diseased mink that are hyper-gammaglobulinaemic reaches extraordinary high levels: geometric mean titres of 1 : 100,000–260,000 have been reported. It has appeared that there is a general correlation between antibody titres and increased levels of gamma globulin, but the total amount of the immunoglobulin that is specific anti-AD antibody remains to be elucidated.

These studies indicate that AD-infected mink mount an extremely vigorous anti-viral antibody response that reaches extremely high levels. The mechanism by which this antibody response occurs is unclear, but it seems likely that AD virus is causing some defect in the normal homeostatic antibody control mechanism. Although chronic antigenic stimulation occurs, it seems unlikely that the gamma globulin response in infected mink simply represents chronic stimulation. The possibility that AD viral antigen or the virus itself serves as a 'super antigen' or by some unknown mechanism causes a circumvention of the normal T–B cell interaction resulting in a hyper-B cell response seems worthy of consideration.

It appears that AD-infected mink are somewhat immunodepressed in that they do not respond to a number of antigens as well as normal mink, (Porter et al., 1965a; Kenyon, 1966; Tabel et al., 1970; Lodmell et al., 1970, 1971; Trautwein et al., 1974; Henson, unpublished). The immune response to protein, cellular and bacterial antigens have been investigated and compared in AD-infected and non-infected mink, with the overall response in diseased mink generally less than in the normal ones. If mink are immunized with antigens and then infected with AD virus, the antibody level to the non-AD antigens usually decreases even though the animals experience greatly increased gamma globulin production after infection.

Available data indicates that there are fewer antibody-forming cells in the lymphoid tissues of AD-infected mink, when compared to normal mink, 47 days after infection (Lodmel et al., 1971). Our own data has suggested a similar depression as early as 21 days. Trautwein et al. (1974) indicated that a depression of antibody formation reflecting a reduction in the number of antibody-producing cells occurred at the time of peak hypergamma-globulinemia.

Attempts have been made to produce resistance in mink by vaccination with killed virus vaccines (Karstad et al., 1963; Porter et al., 1972). The results have been uniformly unsuccessful. In fact, Porter et al. (1972) have indicated that the disease in such sensitized mink was more severe than was the disease in non-immunized mink given similar challenge. This finding and others indicates the potential role of the immune response in the pathogenesis of the lesions in AD-infected mink.

Little is known about the cellular immune response in AD-infected mink, but Perryman et al. (1975) have suggested that such responses are suppressed. The humoral responses, are extraordinarily high in this disease, with the production of large amounts of what appears to be specific antibody. Regardless of the magnitude of the response, attempts to demonstrate virus neutralization by the immunoglobulin produced in diseased mink has failed to suggest any neutralization (Porter et al., 1969; Gorham et al., 1963). This finding is consistent with the demonstration of circulating virus–antibody complexes in the serum of infected mink (Porter and Larsen, 1967). Instead of being protected, it appears that the immune response in AD-infected mink may be one of the factors responsible for the development of lesions. This will be discussed in more detail under 'pathogenesis' (9.7).

9.6 Genetic factors

Since the early recognition of AD, genetics has appeared to influence the prevalence of the disease in populations of mink (Henson et al., 1963) and the rapidity and severity of the disease in individual animals. In addition, some non-Aleutian genotypes recover from the disease while Aleutian-type mink do not. The specific role of genetic factors in AD in unclear.

Epizootiologic data has indicated that the disease is more likely to occur and be more severe in herds of mink that are solely or predominantly Aleutian genotypes, rather than non-Aleutian (see Ch. 7). There are reported instances, however, in which explosive outbreaks of disease have occurred in pastel or dark herds (non-Aleutian mink). It is clear, however, that the disease can occur in any genotype of mink and that the tissue and functional alterations resulting in death are similar regardless of the genotype. It is also very clear that the disease progresses much more slowly in most, if not all, non-Aleutian mink when compared to Aleutian-type genotypes. This will be discussed in more detail below (9.7).

In efforts to develop genetically resistant groups of mink, we have defined non-infected animals in groups of mink in which a significant proportion of the population were spontaneously diseased. The non-affected animals were bred and their offspring examined for the occurrence of spontaneous AD. Non-diseased animals from the F-1 generation were bred and the occurrence of the disease determined in the F-2 generation. Similar trials have been conducted by Larsen and Porter (1975). In all instances, it has not been possible to determine any decrease in the prevalence of AD in either the F-1 or F-2 generations.

Regardless of these findings, it appears that practically all Aleutian-type mink are highly susceptible and practically all die from experimental inoculation of AD virus. Non-Aleutian mink, however, appear to represent a spectrum of susceptibility to the agent and also represent a spectrum of variability in the recovery from the disease after experimental infection, (Lodmell et al., 1971). Some non-Aleutian mink apparently do not develop the disease or perhaps do not replicate the agent after experimental inoculation, while others are able to eliminate the virus from their circulation and tissues even though it has reached relatively high titres and hypergamma-globulinaemia has occurred.

We conducted trials to compare the development of disease in Aleutian and non-Aleutian mink utilizing the occurrence of lesions, hypergamma-globulinaemia and other parameters (Johnson and Henson, unpublished). In these trials, a group of Aleutian and non-Aleutian mink were inoculated with AD virus. Animals were killed periodically and examined for the occurrence of gross and microscopic lesions. Glomeruli were examined by light and fluorescent microscopy for the occurrence of lesions and the deposition of immunoglobulins and C3.

The results indicated a definite difference in the progression of the disease. Aleutian mink developed lesions in a shorter time than the non-Aleutian mink and died more quickly from the disease. There was earlier deposition of immunoglobulin and C3 in the glomeruli, more frequent and rapid occurrence of necrotizing arteritis and other lesions, and more rapid elevation of globulin levels. When the tissues of the animals that had died from AD regardless of genotypes were compared, it was not possible to determine any difference by the parameters used, except that one could recognise the Aleutian-type mink because of the extremely large lysosomal or other cell-bound structures in the tissues (Padgett et al., 1964, 1968). Any animal that died from the disease had similar severity of lesions, regardless of genotype.

In a trial conducted in this laboratory, 25% of the non-Aleutian mink inoculated with 10^4 ID_{50} of 'Pullman' strain virus recovered from the disease after initially being viremic. In addition, we have followed similar animals for as long as 5 years and they have continued to maintain antibody, but have not developed clinical disease. Similar findings have been shown in this laboratory in spontaneously and experimentally infected ferrets. The latter develop hypergammaglobulinemia and marked plasma cell proliferation, but appear not to succumb to the disease. Whether all these ferrets are in fact carrying infectious virus in their tissues has not been shown. At any rate, it appears that the ability of certain genotypes of mink and of ferrets to cope with the agent varies considerably with the Aleutian-type mink almost always dying, whereas some of the non-Aleutian mink and most ferrets recover. Eklund et al. (1968) and Porter and Larsen (1974) have indicated that recovery or resistance could not be directly attributed to the Aleutian gene, however. Since mink are extremely heterogenous genetically, it may well be that the increased susceptibility to AD virus has been genetically established in some non-Aleutian mink.

Lodmell et al. (1970) have indicated a difference in the immune response of Aleutian and non-Aleutian mink with the suggestion that the Aleutian mink were somewhat impaired in their antibody-producing capabilities. Preliminary evidence from this laboratory has suggested that there may be a difference in the ability of Aleutian and non-Aleutian mink to clear injected complexes from the glomeruli, with the Aleutian-type mink being less capable of such clearance (Johnson and Henson, unpublished). Thus, genetic factors do play an important and yet to be clearly elucidated role in AD. It seems possible that the ability of the agent to replicate, the immune responses and the ability of the animal to catabolize and/or eliminate circulating complexes may singly or in concert influence the occurrence and progression of AD infection in different genotypes of mink.

9.7 Pathogenesis

Sufficient data has accumulated to give some insight into the transmission, epizootiology and pathogenesis of AD. Virus is excreted in the urine, and probably the feces, saliva and milk of infected mink (Gorham et al., 1965). Epidemiologic studies have demonstrated that the infected female is very important in the disease transmission and that vertical as well as horizontal transmission occurs (7.5).

After experimental inoculation of the agent, there is rapid viral replication with peak titers occurring 8–10 days after inoculation. Approximately 3–4 weeks after infection the virus titer decreases and remains present in the serum of Aleutian-type mink and some non-Aleutian mink. Infectious virus can be detected in many organs and viral antigen has been demonstrated in most tissues including the mammary gland, the placenta, and the tissues of unborn feti. Approximately 20–25% of non-Aleutian mink eliminate virus. The virus does persist, however, in many infected mink and in ferrets.

The immune response directed against viral antigens is an early and vigorous one, with antibody detected by fluorescent techniques, complement fixation, and precipitation. Attempts in this laboratory to obtain non-antibody-coated virus by immunosuppression of experimentally inoculated mink and/or early collection of tissues has indicated that there is early production of antibody. Similarly, studies to detect antibody by complement fixation in animals infected for only a few days indicate early antibody production.

Porter and Larsen (1967) studied the occurrence of circulating infectious virus–antibody complexes in the serum. These workers were able to show that approximately 2 logs of infectivity could be removed from the infected serum by treating it with anti-mink immunoglobulin, whereas treatment of similar serum with anti-albumin did not cause such a significant loss of infectivity. This indicates that circulating infectious virus–antibody complexes occur in AD.

The early and continuous plasma cell proliferation noted in this disease and the suggestion that the large amounts of gammaglobulin produced is specific antibody, indicates that there is either a continued antigenic stimulation in infected mink and/or there is a dysfunction in the control mechanisms maintaining antibody homeostasis. The exact mechanism(s) causing the hypergammaglobulinaemia are yet to be shown. It appears, however, that these types of mechanisms result in the large numbers of plasma cells noted in the disease as well as the great amount of antibody produced.

A number of lesions result from the immunologic response directed against viral antigens. The glomerulonephritis results from deposition of antigen–antibody complexes. Examination of diseased glomeruli by fluorescent antibody techniques have shown immuno-reactants and viral antigen, and these and ultrastructural studies have suggested that the initial insult in the glomerulus is the deposition of complexes. These complexes cause alterations in the basement membranes and stimulation of the mesangial cells and mesangial matrix, probably in an effort to catabolize the phlogogenic complexes deposited therein. The result is a proliferative glomerulitis with ever increasing size of the mesangium due to increased numbers of cells and increased mesangial matrix which finally results in compression of capillaries and uremia. The arterial lesions also appear to be the result of deposition of antigen–antibody complexes and probably have a pathogenesis similar to that reported in serum sickness (Cockrane and Hawkins, 1968). In AD, it appears that there is a continued deposition of these complexes with arterial lesions of varying longevity present in the individual animal.

The pathogenesis of the hepatitis in AD has not been elucidated. Work in this laboratory has suggested that the proliferation of virus in the liver coupled with the occurrence of antibody and possibly sensitized cells cause the lesions. Viral antigen occurs in cells that resemble Kupffer cells. Whether this represents phagocytosed antigen or viral proliferation in these cells remains to be proven. There is an accumulation of other mononuclear cells in association with the antigen containing cells in these peri-sinusoidal locations and it seems likely that the cells are occurring as a result of immunologic processes.

The plasma cell proliferation in the portal areas would appear to be an expression of the generalized plasma cell proliferation occurring throughout the animal. We have observed few cells containing viral antigen in the portal areas. The bile duct proliferation appears to be a secondary phenomenon associated with physical aspects of the cell masses in the portal areas. There is no suggestion of bile stasis, of cellular alteration of bile duct epithelium or of occurrence or accumulation of proteinaceous or other material in the bile ducts.

It has also been suggested that the anemia in AD is immunologically mediated with possible involvement of AD viral antigen, anti-AD antibody and complement on the erythrocyte surfaces damaging the red blood cells (Cho and Ingram, 1973).

Cellular immune responses in AD have been incompletely studied as have such processes in most persistent virus infections. Perryman et al. (1975) have shown a depression in cellular immune responses in AD in preliminary studies. Such findings could be extremely important in the persistence of the agent. The possible role of antibody and/or antibody class interaction with the formation of protected virus particles or aggregates has likewise not been investigated.

We have studied the influence of cyclophosphamide therapy on the development and treatment of AD (Cheema et al., 1972). Initiation of cyclophosphamide therapy at the time of AD virus inoculation prevents the development of lesions with continued presence of the agent. These findings suggest that the agent per se does not cause significant tissue alterations and further incriminate the immune responses in the pathogenesis of the disease. The effect of the drug is probably 2-fold, influencing the degree of antibody response as well as the cellular responses. It seems unlikely that the level of drug used completely prevented the development of anti-AD antibody.

Maintenance of AD virus inoculated mink on cyclophosphamide therapy for as long as 8 weeks prevents the development of lesions. Stoppage of the therapeutic regime at that time, however, results in the development of typical AD lesions and clinical disease. Treatment of mink with fairly advanced AD with cyclophosphamide appears to stop the progression of the disease with some minor lesion resolution during the observation period. The disease again progresses in the usual fashion after cessation of therapy.

It seems, therefore, that the lesions of AD are immunologically mediated with the agent persisting without causing major tissue or functional alterations in cyclophosphamide-treated mink. The drug is incapable of permanently stopping the progression of the disease without continuous therapy and does not result in lesion resolution.

Comparison of lesion development and severity of clinical disease in AD-infected mink and ferrets shows some distinct differences. The lesions in ferrets are dominated by plasma cell proliferation in the same locations as in infected mink. The morphologic appearance of the lympho-reticular elements and other tissues in ferrets suggests that the lesions are 'less active'. There is less active lymphoid stimulation with less follicular activity, mitotic figures and cellular debris in ferrets when compared to mink. Hepatic and other lesions also appear less active. In addition, we have observed spontaneously diseased ferrets with hypergammaglobulinemia for over a year and found that they remained hypergammaglobulinemic throughout the observation period. The gamma globulin levels remained relatively stable and the animals appeared healthy. Glomerular lesions and uremia either do not occur in ferrets or are very mild. Arteritis is also not a feature of the ferret disease.

The replication rates of AD virus in ferrets is unknown. We have demonstrated AD viral antigens in small amounts in ferret lymphoid tissues and livers. These observations suggest that ferrets become infected with AD virus and that the agent persists. The agent may proliferate at a slower rate than in the mink or the control mechanisms are more effective in keeping the agent in check. The lack of occurrence of arteritis and the unusual occurrence of glomerulonephritis suggests that circulating immune complexes are not being deposited in these indicator tissues. Whether this is a reflection of the agent or the immune responses of the ferret, or both, remains to be proven. That the hypergammaglobulin response in ferrets represents antibody is clear.

It is also evident that different populations of mink differ in their response to AD viral infection. There seems to be an association with coat colour, but this probably represents a marker with disease or lack of it, determined by genetic aspects to be defined. The possible role of lysosomal dysfunction playing some role in the development of the glomerular lesions has been described earlier (7.3.2). Replication of the agent, immune responses, lysosomal function and other factors may independently or together influence the course of the disease.

The recovery of some non-Aleutian mink indicates that some mink are able to effectively eliminate the virus, or at least effectively suppress it. Ferrets likewise are able to live with the infection. It seems that there is a spectrum of virus susceptibility and disease resistance within the mink population. Aleutian-type mink represent one extreme with high susceptibility and rapid disease course with certain non-Aleutian mink representing the other extreme. Ferrets would seem to be most like the latter. The mechanisms of disease resistance, the host–parasite relationships, the lack of or ability of the host defence mechanisms to cope with the virus, the mechanisms whereby the hypergammaglobulinemia develop and the characterization of the virus all represent important areas remaining to be investigated in AD. Elucidation of these areas will contribute to our understanding of one type of persistent viral infection and of disease processes in man and other animals. AD is, and will for some time continue to be, an important model for disease investigation.

9.8 Comparison to human and other animal diseases

Persistent or slow virus infections have been proven to be or have been suggested as the aetiologic agents in a number of human diseases such as SLE (Gyorkey et al., 1969), polyarteritis nodosa (Gocke et al., 1971), rheumatoid arthritis (Phillips, 1971), viral hepatitis (Blumberg et al., 1971), kuru (Gajdusek and Gibbs, 1973), Creutzfeldt-Jakob disease (Gajdusek and Gibbs, 1973) and others. Similarly, these types of infections occur in animals as exemplified by equine infectious anemia (Henson and McGuire, 1971), scrapie (Eklund et al., 1964), hog cholera (Cheville and Mengeline, 1969), African swine fever (De Tray, 1957), lymphocytic choriomeningitis (Oldstone and Dixon, 1969) and others. In some of these diseases the morphologic lesions and possible pathogenetic mechanisms appear to be similar to AD. In others, such as kuru and scrapie, there are considerable differences between them and AD. It appears that there is a spectrum of types of persistent infection with the agent replicating slowly in some and rapidly in others. Likewise, the immunologic processes play an important role in the pathogenesis of some and not others.

The inclusion of AD under a general heading of 'slow virus' infections is a misnomer. In this context, the diseases described in this publication are an example of the spectrum of host–parasite situations generally represented under the general heading of slow or persistent viral infections. As an example, one can contrast scrapie to AD in a number of ways. In AD, the agent replicates rapidly with a requirement for the immune response in order for the disease to develop. In contrast, no role for the immune processes has been demonstrated in scrapie (Ch. 14). The scrapie agent appears to replicate slowly with clinical disease developing years after the initial exposure to the agent. Also, the organ systems involved or at least the organ system which results in the predominant clinical signs of the disease differ. In scrapie, the chronic involvement of the nervous system with slowly progressive degeneration (Ch. 12) is in contradistinction to the development of lesions in the organ systems involved in AD. In both diseases, it seems that the agent may replicate in the lymphoid system, but the lympho-reticular responses in AD are directed towards a hyperactivity, whereas such is not the case in scrapie (Ch. 14).

Such chronic host–parasite relationships are not unique for virus infection. Similar observations have been made in such conditions as hemoparasitic diseases. In trypanosomiasis, for example, the organism persists for months or years in the individual animal and causes a number of lesions and functional disorders such as anemia and immunosuppression. There are possible genetic factors that modulate the severity of disease in different breeds of cattle. It appears in trypanosomiasis that the organism undergoes antigenic modulation in the host by unknown mechanisms. Similar findings have been reported in equine infectious anemia virus in fection in horses (Kono et al., 1973). The virus undergoes continuous antigenic modulation similar to the antigenic changes in trypanosomiasis described above. There are a number of other examples also.

It seems likely that persistent infection by a variety of types of organisms is a relatively common occurrence. Within the framework of persistent infections, there are probably a spectrum of mechanisms that determine the chronic host–parasite relationship. One such example would be 'slow virus infections' in which the agents replicate slowly. Many other factors can be operational and influence the duration of the disease. Agent replication rates, immune response, type of agent, replication site of the agent, host genetic factors and ability of host cells to degrade deposited complexes and other injurious products are but a few of these factors.

These types of factors should be taken into consideration in attempts to investigate and define the pathogenetic mechanisms in persistent viral infections and in the elucidation of aetiological agents and host–parasite relationships in many of the poorly understood diseases encountered in human and veterinary medicine today.

References

BLUMBERG, B. S., MILLMAN, I., SUTRICK, A. I. and LONDON, W. T. (1971) J. Exp. Med., 134, No. 3, Part 2, 320.

CHEEMA, A., HENSON, J. B. and GORHAM, J. R. (1972) Am. J. Pathol., 66, 543.

CHEVILLE, N. F. and MENGELINE, W. L. (1969) Lab. Invest., 20, 261.

CHO, H. J. and INGRAM, D. G. (1972) J. Immunol., 108, 555.

CHO, H. J. and INGRAM, D. G. (1973) Can. J. Comp. Med., 37, 217.

COCKRANE, C. G. and HAWKINS, D. C. (1968) J. Exp. Med., 127, 137.

CRAWFORD, T. B. (1972) Fed. Proc. 31, (Abstract 2372) 635.

DE TRAY, D. E. (1957) Am. J. Vet. Res., 18, 811.

EKLUND, C. M., HADLOW, W. J., KENNEDY, R. C., BOYLE, C. C. and JACKSON, T. A. (1968) J. Infect. Dis., 118, 510.

EKLUND, C. M., KENNEDY, R. C. and HADLOW, W. J. (1964) In: (D. C. Gajdusak, C. J. Gibbs, Jr, and M. Alpers, Eds) Natl. Inst. Neurol. Dis. Blindness, Monogr. 2. Slow Latent, and Temperate Virus Infections. (U.S. Government Printing Office, Washington, D.C.) p. 207.

GAJDUSEK, D. C. and GIBBS, C. J. (1973) Prosp. Virol., VIII, 279.

GOCKE, D. J., HSU, K., MORGAN, C., BOMBARDIERI, S., LOCKSIN, M. and CHRISTIAN, C. L. (1971) J. Exp. Med., 134, No. 3 Part 2, 330.

GORHAM, J. R., LEADER, R. W. and HENSON, J. B. (1963) Fed. Proc., 22, (Abstract) 265.

GORHAM, J. R., LEADER, R. W. and HENSON, J. B. (1965) J. Infect. Dis., 114, 341.

GYORKEY, F., MIN, K. W., SINKOVICS, J. G. and GYORKEY, P. (1969) N. Engl. J. Med., 280, 333.

HARTSOUGH, G. R. and GORHAM, J. R. (1956) Natl. Fur News, 28, 10.

HELMBOLDT, C. F. and JUNGHERR, E. L. (1958) Am. J. Vet. Res., 19, 212.

HENSON, J. B. (1964) Ph.D. thesis, Washington State University.

HENSON, J. B. and CRAWFORD, T. B. (1974) Adv. Cardiol., 13, 183.

HENSON, J. B., GORHAM, J. R., LEADER, R. W. and WAGNER, B. M. (1962) J. Exp. Med., 116, 357.

HENSON, J. B., GORHAM, J. R. and LEADER, R. W. (1963) Texas Rep. Exp. Biol. Med., 21, 37.

HENSON, J. B., GORHAM, J. R., PADGETT, G. A. and DAVIS, W. C. (1969) Arch. Pathol., 87, 21.

HENSON, J. B., GORHAM, J. R. and TANAKA, Y. (1967) Lab. Invest., 17, 123.

HENSON, J. B., GORHAM, J. R., TANAKA, Y. and PADGETT, G. A. (1968) Lab. Invest., 19, 153.

HENSON, J. B., LEADER, R. W. and GORHAM, J. R. (1961) Proc. Soc. Exp. Biol. Med., 107, 919.

HENSON, J. B., LEADER, R. W. and GORHAM ,J. R. (1966) Path. Vet., 3, 289.

HENSON, J. B. and MCGUIRE, T. C. (1971) Am. J. Clin. Pathol., 56, 306.

INGRAM, D. G. and CHO, H. J. (1974) J. Rheum., 1, 74.

KARSTAD, L., PRIDHAM, T. J. and GRAY, D. P. (1963) Can. J. Comp. Med. Vet. Sci., 27, 124.

KENYON, A. J. (1966) Am. J. Vet. Res., 27, 1780.

KENYON, A. J., HOWARD, E. and BUKO, L. (1967) Am. J. Vet. Res., 28, 1167.

KONO, Y., KOBAYASHI, K. and FUKUNAGA, Y. (1973) Arch. Ges. Virusforsch., 41, 1.

LEADER, R. W., PADGETT, G. A. and GORHAM, J. R. (1963) Blood, 22, 477.

LARSEN, A. E. and PORTER, D. D. (1975) Infect. Immun., 11, 92.

LODMELL, D. L., BERGMAN, R. K., HADLOW, W. J. and MUNOZ, J. J. (1971) Infect. Immun., 3, 221.

LODMELL, D. L., HADLOW, W. J., MUNOZ, J. J. and WHITFORD, H. W. (1970) J. Immunol., 104, 878.

MCGUIRE, T. C., CRAWFORD, T. B., HENSON, J. B. and GORHAM, J. R. (1971) J. Immunol., 107, 1481.

OBEL, A. L. (1959) Am. J. Vet. Res., 20, 384.

OLDSTONE, M. B. A. and DIXON, F. J. (1969) J. Exp. Med., 129, 483.

PADGETT, G. A., GORHAM, J. R. and HENSON, J. B. (1967) J. Infect. Dis., 117, 35.

PADGETT, G. A., LEADER, R. W., GORHAM, J. R. and O'MARY, C. C. (1964) Genetics, 49, 505.

PADGETT, G. A., REIQUAM, C. W., HENSON, J. B. and GORHAM, J. R. (1968) J. Pathol. Bacteriol., 95, 509.

PAN, I. C., TSAI, K. S. and KARSTAD, L. (1970) J. Pathol., 101, 119.

PERRYMAN, L. E., BANKS, K. L. and MCGUIRE, T. C. (1975) Fed. Proc., 34 (Abstract 3680) 870.

PHILLIPS, P. E. (1971) J. Exp. Med., 134, No. 3, part 2, 313.

PORTER, D. D., DIXON, F. J. and LARSEN, A. E. (1965a) J. Exp. Med., 121, 889.

PORTER, D. D., DIXON, F. J. and LARSEN, A. E. (1965b) Blood, 25, 736.

PORTER, D. D. and LARSEN, A. E. (1967) Proc. Soc. Exp. Biol. Med., 126, 680.

PORTER, D. D. and LARSEN, A. E. (1974) Progr. Med. Virol., Vol. 18 (Karger, Basel) p. 32.

PORTER, D. D., LARSEN, A. E., COX, N. A., PORTER, H. G. and SUFFIN, S. C. (1975) Fed. Proc. 34 (Abstract 4121) 947.

PORTER, D. D., LARSEN, A. E. and PORTER, H. G. (1969) J. Exp. Med., 130, 575.

PORTER, D. D., LARSEN, A. E. and PORTER, H. G. (1972) J. Immunol., 109, 1.

PORTER, D. D., LARSEN, A. E. and PORTER, H. G. (1973) Am. J. Pathol., 71, 331.

TABEL, H. and INGRAM, D. G. (1970) Can. J. Comp. Med., 34, 329.

TABEL, H., INGRAM, D. D. and FLETSH, S. M. (1970) Com. J. Comp. Med. Vet. Sci., 34, 320.

TRAUTWEIN, G. W. and HELMBOLDT, C. F. (1962) Am. J. Vet. Res., 23, 1280.

TRAUTWEIN, G. and SEIDLER, D. (1972) Zentralbl. Vet. Med., 17, 144.

TRAUTWEIN, G., SCHNEIDER, P. and ERNST, E. (1974) Zentralbl. Vet. Med., 21, 467.

PART IV

Scrapie

The disease

The virus

Virus–host interactions

The relationship of scrapie with other diseases

CHAPTER 10

Scrapie in sheep and goats

A. G. DICKINSON

10.1 General introduction to scrapie

Scrapie is a fatal, progressive degenerative disorder of the CNS which occurs as a natural infection in sheep and goats. The wide variety of strains of the infectious causal agent share a number of physico-chemical properties which indicate that their molecular structure is probably outside the present range for conventional viruses. There are no in vitro methods of detecting scrapie agents, therefore all work on them has to be related to tests of infectivity by injection of appropriate animals. These tests are essentially of two types: those which attempt to determine how much agent is present by titration and those which detect differences in the biological properties of the agents.

During the 1950s, many investigators still doubted whether scrapie was caused by an independent organism with the general features, but not necessarily the chemical structure, of a virus. Two frequently raised alternatives, both of which avoided the possibility that there was replication of a virus-like component of the inoculum, were (1) that scrapie was already present in the animals as a latent infection which was activated by injection with particular types of inocula or, alternatively, (2) that scrapie tissue homogenate acted in a broadly similar manner to the antigen responsible for allergic encephalitis. Mackay, Smith and Stamp (1960) presented a variety of results of sheep experiments from which they concluded that neither of these explanations was acceptable.

Even though critical experiments were lacking, hypotheses have continued to appear, encouraged by the agents' unusual properties. These included a membrane hypothesis (Gibbons and Hunter, 1967; Hunter, 1972), a linkage-substance hypothesis (Adams and Field, 1968) and a membrane colicin-like

Slow virus diseases of animals and man, edited by *R. H. Kimberlin*
© *North-Holland Publishing Company 1976*

agent (Hunter et al., 1968). What is certain is that scrapie agents include some replicating informational molecule but these more exotic theories could not accommodate certain findings, namely (a) there are different, accurately copied, strains of agent with stable properties under standard passage conditions, (b) these strain differences are independent of the host's genotype and (c) there can be infection with more than one strain of agent, which can compete for some scarce, limiting component in the host (Dickinson and Meikle, 1971; Dickinson et al., 1972, 1975a; Dickinson, 1975; see 10.6.8 and 11; Ch. 14). The group in this laboratory chose to study the agents' properties on as wide a basis as possible, avoiding theorising about their structure; we regard it as too premature to formulate a hypothesis in terms, so clear-cut, that experiments can be devised which might refute it.

The composition of the agent is still unknown and the only agreed remnant of the hypotheses is that most of the infectivity detectable by present methods in the brains of mice with advanced disease, is associated with cell membranes; the form of this association, and any significance which it may have, are unknown. These aspects are under investigation (Ch. 11) but a preliminary comment is needed on the widely held belief that nucleic acids have been excluded as the informational molecules. This view arose from the agents' extreme resistance to inactivation by large doses of 254 nm UV-irradiation (Wilson, 1954; Alper et al., 1967) but this can equally well indicate that present views on the UV stability of nucleic acids are too limited or that such molecules can be chemically protected or repaired in ways not yet understood.

This chapter is intended as a description of published and unpublished work, in its contemporary context, which has been significant for the development of the subject and also as a strategic guide to those attempting new isolations of scrapie and related agents and their preliminary investigation. It is hoped that the reader who proceeds to examine the literature directly, will be able to do so with an awareness of its limitations.

In order to construct a balanced picture of the current position with scrapie in its natural hosts, a preliminary description will be given of what might be termed 'classical' scrapie, using Suffolk sheep for the purpose. Against this background it will be easier to set out the difficulties of working with this group of diseases (Ch. 15), of which scrapie is the best understood. There are diagnostic and basic problems, and operational difficulties due to the time scale of experimental work. Also, it is rarely known which strains of agent are involved, it is difficult to assemble suitable control groups in the natural host species and, in many respects, it is inappropriate to select research techniques and priorities by analogy with conventional acute virus infections.

10.2 'Classical' scrapie in sheep

Natural scrapie has been studied most extensively in Suffolk sheep. Death due to scrapie occurs most frequently in sheep between 2 and 5 years old, with rare cases in slightly younger sheep and a minority spread over the later half of the 10–15 year potential lifespan (Fig. 10.1) (Dickinson et al., 1964, 1965). These later cases are often not seen because husbandry practices limit the average lifetime to about 5 years, which further concentrates the losses among ewes to their most fertile phase. Reasons will be given why the 2–5 year-old cases, or most of them, have been infected throughout life, with the infective agent replicating and progressively spreading asymptomatically to various organs (10.9.1). Thus, in such cases, age at death and incubation period are roughly the same, the issue being whether infection occurred before, at or soon after birth.

Clinical symptoms are not seen for most of this incubation period and start insidiously, often as impaired social behaviour with an affected sheep either leading or trailing a flock which is being driven; these contrasting excitable or lethargic symptoms usually lead in Suffolks to locomotor in-coordination, particularly of the hind legs, and this ataxia can be very

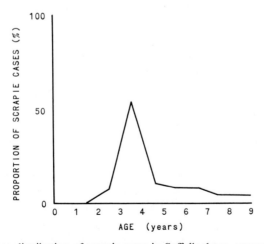

Fig. 10.1. The age distribution of scrapie cases in Suffolk sheep, corrected to allow for the progressive commercial culling of older sheep.

pronounced in such a heavily built breed. Fine trembling is often present and pruritus accompanies the ataxia – the animal can become preoccupied with biting its extremities and rubbing certain parts of its body against fencing until raw wounds develop, which can extend to large areas of the body. The initial focus of this 'scraping' in Suffolks is usually at the base of the tail and the animal can lose condition and start wasting even before incessant rubbing displaces normal feeding. A 'nibbling reflex' involving lips, tongue and jaws is often elicited when the animal rubs itself against fencing or when scratched by hand. The duration of clear signs of illness is usually 1–2 months but can range from 2 weeks to 6 months, though there is difficulty in stating an upper limit because of equivocal early signs. One of the first practical problems is that pruritus and debilitation are not always seen and, with the natural disease, clinical signs cannot in all cases be regarded as diagnostic by themselves, because the limits of their variety are not understood.

Confirmation by brain histopathology appears to present no great difficulty in Suffolk sheep but an open mind must be kept even on this, because we know that Cheviot sheep can present great difficulties in this respect (12.7 and 10.7.2).

Different breeds (or species) can vary in the details of their typical clinical syndrome, particularly in the relative preponderances of incoordination and pruritus (see review by Palmer, 1959). For example, in a recent French outbreak with a 20% incidence, a proportion of the cases occurred at 10–14 months old as 'paralytic forms' with only a 2-week course (Joubert et al., 1972). Variation in the course of the disease presumably reflects not only genetic differences between hosts but also strain differences in the agent, because it is possible to produce radically different clinical syndromes within a single genotype, breed or species by injecting them with different strains of agent and these effects can include the incubation period and the rate of progression of clinical symptoms (Dickinson, unpublished).

10.3 Economic significance

Scrapie occurs, or has occurred, in most areas of the world but European countries have known it as a continuing problem for over two centuries. It accounts for a fairly high proportion of adult deaths in certain breeds and flocks – a 10% incidence is not uncommon and it is occasionally much higher. The incidence of scrapie is likely to be higher than the limited data indicate. Sporadic sheep deaths following unidentified illness are too readily accepted as inevitable and, because scrapie is so widely feared, there is a tendency to slaughter individual sheep on the slightest unconfirmed suspicion, to protect the breeder's reputation, thus confounding any statistics. Fortunately scrapie does not now appear to be as serious in Britain as it was in the 18th century when it provoked the following utterance:

'This disorder has been known to be fatal to the greatest part of a flock and is considered as the most calamitous circumstance the sheep owners have to dread' (Claridge, 1795). If indeed it is less frequent in Britain now than it was 200 years ago, more work is needed on the nature of the disease and its epidemiology to understand the declining incidence because an epidemic could recur with serious consequences, given the re-emphasis on sheep which world food shortage is likely to encourage.

Given the emphasis on fat-lamb production a common attitude is that the economic losses can be largely avoided because the animals are killed too young for scrapie to appear. However, this attitude could be too complacent. Scrapie is maternally transmitted in sheep, so that lambs, though showing no symptoms, can be infected and harbouring progressively increasing quantities of agent. Circumstances have been found in which scrapie can occur from 7 months old (10.9). We should not assume, at this stage of our knowledge, that scrapie agents are never transmissible to man from infected meat, particularly as we know that some types of cooking would not inactivate the infectivity.

Financial losses arise from several sources: direct loss due to death from scrapie or from unnecessary culling, impaired rearing of lambs which have affected mothers, large indirect effects due to loss of reputation by individual breeders or the wider repercussions in the disruption of the world sheep trade and, last, the damage to genetic improvement schemes where 'avoidance of scrapie' can be a precondition limiting the choice and therefore the selection potential.

10.4 The development of scrapie research

10.4.1 Scrapie in sheep

In 1936 Cuillé and Chelle reported the experimental transmission of scrapie by injecting healthy sheep with brain and spinal cord from an affected sheep. This work received rapid, but inadvertent, confirmation from the Moredun Research Institute in the famous 'Louping Ill vaccine' event which Gordon, Brownlee and Wilson reported in 1939. One of the three 1935 batches of vaccine also contained scrapie, with the result that 7% of the recipients of the 18,000 doses in this batch developed scrapie. As the L.I. vaccine was made from formalin-inactivated sheep brain, this highlighted from the start that scrapie agent could withstand concentrations of formalin (0.35% for at least 3 months) which inactivated conventional viruses. Experimental transmission work on scrapie, mainly in Cheviot and Cheviot-cross sheep, has continued at Moredun since 1938, some early parts of which were published by Gordon (1946, 1957) and Wilson, Anderson and Smith (1950).

The latter workers established an isolate of scrapie infectivity ('SSBP/1'; see 10.6) which had been serially transmitted by intracerebral injection for 9 passages without noticeable changes of clinical signs, incubation periods or in the relatively mild brain lesions. They also confirmed the observations of Cuillé and Chelle that the agent was filterable through gradocol membranes like a virus.

The early work by Gordon from 1938–42 had indicated that the infectivity could be diluted out and very crude titrations were possible by subcutaneous injection of sheep. But this work also drew attention to some of the difficulties. Observations were continued for over 4 years and most of the cases occurred during the last 2 years, though controls showed that about a fifth of the cases were due to natural scrapie: but even with highly infective inocula more than half the sheep survived – almost certainly, we now realise, because they were genetically not susceptible to the strains of agent involved (see 10.7.2). A small subgroup in these experiments indicated that intracerebral instead of subcutaneous injection, gave cases more quickly. Subsequent work – initially by Wilson – therefore concentrated on intracerebral injections, which resulted in 30–40% shorter incubations and was aimed to avoid the problem of natural scrapie by injecting lambs and terminating the observations a year later, i.e., before the age at which scrapie had been seen in Cheviots. In retrospect it now seems likely that the repercussions of natural scrapie cannot be entirely avoided by this means, because of competition between scrapie agents (14.5). Also, it had not been possible to establish whether titrations of infectivity could be satisfactorily completed within 12 months, because the lower doses usually have much longer incubation periods.

Wilson's careful description of some of the early work and its limitations (Wilson et al., 1950) was accompanied by a reluctance to publish too hastily the results of work in such a radically new area of veterinary medicine – it was at the end of an era dominated by the development of new vaccines by relatively simple techniques. Much of the pioneering work of Wilson has remained unpublished, but its main features were generally known by the small number of workers in this field by 1955. For example, he searched extensively for antibodies to scrapie in infected sheep but failed to find any. He also initiated studies of the physico-chemical stability of scrapie agent which showed that it was not totally inactivated by the following treatments; boiling for 30 min; temperatures of 80 °C (1 h), -20 °C or -70 °C (2 months); incubation with 3% formalin, 1% formalin + 1% ether, 5% chloroform or 4% phenol (each at 37 °C, for at least 2 weeks); methanol precipitation or 254 nm UV-irradiation (410 nm gradocol filtrate given at least 10^6 ergs/ mm^2) (Wilson, unpublished).

The existence of an agent as stable as scrapie raised questions of whether it could be excreted and produce long-term infection of pastures or be infectious by natural routes such as orally or by skin abrasions.

Greig (1940, 1950) had been convinced that scrapie could be transferred to healthy sheep by grazing a pasture alternately with affected sheep. His suggestive evidence must be regarded as insufficient and this issue has still not been settled although we now know that scrapie can be transmitted infectiously (10.10), but it should not be assumed that all ages of sheep are equally receptive to all strains of agent. Another uncertainty was whether indoor sheep pens could be safely reused and how pens or buildings could be decontaminated. Some facilities at Moredun and also at Compton (A.R.C. Institute for Research on Animal Diseases) have been exposed to many strains of scrapie during the past 30 years, but particularly to the SSBP/1 sheep-passaged source or an offshoot from it (10.6). In 1959, Stamp and colleagues were able to write regarding SSBP/1 work in Moredun sheep facilities that 'as the experimental sheep were discarded 12 months after injection, the method of keeping the experimental animals seemed acceptable for there is considerable evidence that uninoculated controls living alongside numerous cases of [SSBP/1] scrapie for longer periods of time (up to $3\frac{1}{2}$ years) do not develop the disease' [due to contagion]. This is still regarded as the position for work with SSBP/1 but there has been little subsequent refinement of criteria which could lead to an altered conclusion.

It was found in early work that agent was present not only in CNS tissue but also in lymph nodes, spleen and cerebrospinal fluid, but it appeared to be absent from serum and skin (Stamp et al., 1959). These authors also reported Wilson's failure to infect 5 Cheviots orally with SSBP/1 brain homogenate, though he had found intravenous and intradermal injections and skin scarifications to be effective routes. However, Pattison and Millson (1961) subsequently showed that oral infection was possible in sheep and goats using SSBP/1-derived brain suspension. The oral route appeared to be an efficient one for infecting potentially susceptible sheep because sub-cutaneous injection of the survivors with SSBP/1, 19 months after oral dosing with it, produced no further cases, whereas there were subsequent cases in the controls which had received normal brain orally on the first occasion (Gordon, 1966).

From 1955 attention tended to move to work in goats, which answered many of the points raised by the sheep work. Besides the cost and the uncertainty whether some were already incubating scrapie, the main obstacle to using sheep for injection experiments was the low incidence, even with high intracerebral doses: with SSBP/1 in Cheviots the 'take' averaged only about a third. What was needed, at least for preliminary investigations of pathogenesis, microbiology and biochemistry, was an experimental host in which all could develop scrapie after a reasonably short incubation. There were four approaches to achieve this: testing various breeds of sheep to find a better one than Cheviots for SSBP/1 experiments (10.7.1), genetical selection to try to produce groups with 100% incidence (10.7.2 and 3), testing the response of goats (because scrapie had been reported by Chelle (1942) to occur in them) and attempting to transmit the disease to laboratory animals, particularly mice (10.8 and Ch. 14).

10.4.2 Scrapie in goats

The transmission of SSBP/1-derived scrapie to all of a group of crossbred goats at Compton by intracerebral injection (Pattison, 1957) was followed by its passage from goats to goats, again with 100% incidence, and also back to Cheviot sheep (Gordon and Pattison, 1957). The clinical symptoms and brain lesions in these Cheviots were essentially the same as in the original sheep, which validated continuing the work in goats. Agent was found to be present in the following tissues of affected goats: brain, cerebrospinal fluid, pituitary, sciatic nerve, spleen, salivary gland, adrenal, pancreas, liver and, as traces late in incubation, in muscle, but none was detected in thyroid, blood, urine, saliva, milk or faeces (Pattison and Millson, 1960, 1961, 1962; Pattison, 1964). These workers also mentioned the probability that lower doses of infectivity produced longer incubation periods and this feature has been extensively exploited in work with mice (14.4.4).

One point should be stressed before considering some complications which arose in the work with goats. The crossbred goats used at Compton were an uncontrolled genetic mixture, with Anglo–Nubian, Toggenburg, Saanen and British Alpine ancestry. They were 100% susceptible to the small range of strains of agent used and, although the incubation periods for a single inoculum could range from 7 to 30 months, the incubation was often much more uniform than this. However, it would be unwarranted to make the sweeping extrapolation that all goats are 100% susceptible to scrapie, without specifying the genotypes, ages and routes of injection or doses and strains of agent, and the maximum observation period needed. Reasons for this caution will become clear from the work with mice (Ch. 14) and from further details for sheep given in this chapter.

It has become progressively clear that scrapie is a naturally occurring disease of goats (Chelle, 1942; Mackay and Smith, 1961; Hourrigan et al., 1969 and personal communication; Harcourt and Anderson, 1974) and, though the majority of the reported cases were in direct contact with affected sheep, this was not so in all cases. One supposed advantage in the original use of goats was to avoid complications of natural scrapie but even though it was not known to have occurred in the Compton goat flock over the period 1950–1964 (Pattison, 1964), some reservations must be held because, for one reason, it was not a closed flock.

Scrapie was produced in 6.5% of goats injected with brain homogenates from several apparently normal goats (Pattison and Millson, 1961; Pattison, 1964, 1965). There could be a number of explanations for this. Unrecognised natural scrapie might have been involved, although other experimental details given make this unlikely as an origin for all of the cases. The practical difficulties of working with scrapie, including the possibility of cross-contamination and the much more demanding decontamination standards than for conventional microbiology, are the simplest explanations for such results. This also seems the most likely explanation for the claim that goat-passaged scrapie agent was dialysable (Pattison and Sansom, 1964) and later that scrapie could be 'unmasked' by in vitro treatments from normal mouse tissue (Pattison and Jones, 1968). A more likely alternative would be to conclude that the greater the number of laboratory procedures, the greater the risk of contaminating the material with scrapie (Mackay, 1968).

Work described so far has tended to emphasise the problematic findings, experimental difficulties and obstacles to interpretation in scrapie work. What then, are the prerequisites for critical work with scrapie and related agents?

10.5 *Laboratory standards for work with scrapie-like agents*

The phrase 'scrapie-free environment' is a recurring shorthand in sheep and goat papers and should properly be understood to imply that the environment was not known to be harbouring any scrapie agent (i.e. buildings, fields, vehicles, sheep dogs, personnel, benches, instruments, glassware, etc.). Of these, apart from disposable equipment, one might presume that the greatest assurance could be held regarding instruments and glassware which could receive standard microbiological decontamination by autoclaving or dry heating. This would be a false assurance because some, at least, of the scrapie agents can survive conventional decontamination procedures.

For example, ME7 agent is not fully inactivated in tissue smears by 24 h dry heating at 160°C or by autoclaving at 20 p.s.i. (126°C) for 20 min (Taylor and Dickinson, unpublished) and another agent source, as a saline homogenate, was reported to survive autoclaving at 20 p.s.i. (126°C) for 30 min (Field and Joyce, 1970). Detailed specification of the conditions for complete inactivation are needed in terms of the degree of dispersion of the tissue, because agent is more protected within fragments of tissue. Irradiation (UV or γ) is practically useless for decontamination and chemical sterilisation is generally quite unsatisfactory (e.g. with formalin, alcohol, urea (8 M), peracetic acid (2%), ethylene oxide (16 h, 55°C) although hypochlorite solutions (10% 'Chloros') and 90% phenol may prove to be adequate scrapicides. We have found that no scrapie agent survives four separate 1-h autoclaving cycles at 26 p.s.i. (131°C) and this, therefore, is used in these laboratories as a standard sterilisation regime.

Scrapie has such unconventional properties that there has been a widespread willingness to accept uncritically the most tenuous evidence and use it to support extravagant theories. It is, unfortunately, necessary to emphasize that the basic criteria of Pasteur and Koch are still valid and should be applied with even greater rigour. Technical lapses are likely even in the best-run busy, microbiology laboratory. With most conventional microorganisms, unexpected – possibly accidental – results can be relatively quickly repeated and source materials crosschecked using in vitro tests. But with scrapie-like agents the time scale is so great that such detective work is an extremely formidable task and, in addition, only one laboratory in this field is developing strain-typing techniques. These techniques have already been able to prevent a false literature developing in one area of the subject: extensive work was starting in mice based on a claim that TME agent had been transmitted to mice. The incident was, however, shown to be almost certainly due to contamination with mouse-passaged ME7 scrapie agent and was not TME (Barlow, 1972; Dickinson, unpublished). The stringency with which special working standards need to be used with scrapie-like agents is clearly much greater where serial passages are intended than with, say, biochemical work based on a single large pool of infected tissue.

Although it may be necessary, on rare occasions, to attempt to repeat important unexpected results in entirely new laboratories with new personnel (see Mackay, 1968) this is usually impracticable and can have disadvantages. Rigorous aseptic techniques, combined with highly developed strain-typing procedures and, where appropriate, competent independent repetition of the work, are the only satisfactory bases for accepting the more important results or unprecedented interpretations.

10.6 Lineage and ramifications of the SSBP/1 scrapie source

The SSBP/1 (Scrapie Sheep Brain Pool) passage line has played a large part in scrapie research but comprehensive details of it have not been published before. The line was started in 1945 by D. R. Wilson from a pool of three natural cases (one Cheviot and two Cheviot crosses) and has been passaged largely, but not entirely, through Cheviot sheep (Fig. 10.2). This source was used for most of the early work on pathogenesis and on the agent's physico-chemical properties, for a test of the response of 24 breeds (10.7.1) for two major selection experiments (10.7.2 and 3) and for transmission to goats and mice.

If SSBP/1 contained only one strain of agent it would be simple to present and interpret this large body of work in an integrated form but it is now known to contain at least three agent strains which have very different biological properties in terms of their replication kinetics (Table 14.2) and the lesions which they produce in mice (see 12.2.3 and 5). SSBP/1 is not exceptional in this respect because, as will be shown later, multiple infection is very common in natural scrapie (10.8). Characterisation of strains of agent is done entirely by transmission to and passage in specific strains of mice but it must be emphasised that there is no reason to assume that all strains of scrapie are transmissible to mice (Table 10.1).

Before proceeding to consider the constituent strains in SSBP/1 in detail, it is necessary to digress about the two major passage lines of scrapie in goats, the 'drowsy' and 'scratching' lines shown in Fig. 10.2. The first indication that there might be different strains of agent came from the observation in these two passage lines in crossbred goats that their clinical

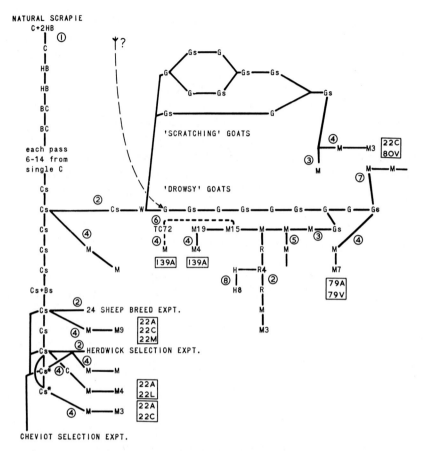

Fig. 10.2. The passage-history of the SSBP/1 source of scrapie infectivity and of agents derived from it (or attributed to it – see text). Sheep: C = Cheviot, HB = Cheviot × Border Leicester cross, B = Scottish Blackface, BC = Scottish Blackface × Cheviot cross, W = Welsh Mountain. 's' after letter indicates pool of more than 1 brain used. Goats: G. Rats: R. Mice: M (refer to original papers for strains used, e.g., M4 = 4 passages in mice). Tissue culture: TC (72 passages). * subcutaneous injection (all others intracerebral), ø LIP Cheviot, 1760 day incubation period, †? suggested origin of 'drowsy' component strains (see text). Points at which material was transferred to different workers: ① W. Smith ② I. H. Pattison ③ R. L. Chandler ④ A. G. Dickinson ⑤ W. J. Hadlow ⑥ D. A. Haig and M. C. Clarke ⑦ C. J. Gibbs ⑧ R. H. Kimberlin. □ Strains of agent identified.

syndromes differed, as indicated by the names, and that this was more clear-cut after several passages (Pattison and Millson, 1961). This observation alone does not, of course, establish that the difference was caused essentially by different strains of agent. The differences between the two lines were sufficiently consistent to rule out genetic diversity in the goats as the cause but a feasible alternative could have been that there was only a single type of scrapie agent but accompanied in one of the lines by another organism, of a more conventional type, which was being passaged along with scrapie and modifying the syndrome. This type of explanation remains possible as a source of differences between some scrapie isolates. However, it has been excluded as the explanation in the case of a number of strains of agent in inbred mice where it can be shown that an agent's characteristics can remain unaffected by such procedures as boiling, formol treatment, serial passage from limiting dilutions (cloning) and recovery after mixture with a different scrapie strain (Dickinson, 1975 and unpublished).

From what follows it will become clear that the difference between the 'drowsy' and 'scratching' lines is due to agent-strain differences (though each line appears to be a mixture of two or more strains). The problem is to try to decide the most likely origin of these line differences, because the alternative explanations of the evidence raise questions of fundamental and practical significance for this group of diseases. The most important issue is whether transmission of a single (i.e., homogenous) strain of agent to a new species depends on 'adaptation' of the agent – 'adaptation' implying some change – but discussion of this will have to wait till a later section (10.11).

Work on some components of SSBP/1 is still incomplete but two of them, 22A and 22C, are now well characterised and have played important roles in probing the molecular events involved in agent replication (14.5). Much less is yet known in detail about two other strains, 22M and 22L, isolated from the 21st and 24th sheep-passages, respectively; they certainly differ from 22A and 22C. A strain of agent indistinguishable from 22C has also been isolated from goats injected with the 'scratching' type of scrapie. The agent 22C has therefore been isolated from sheep and goats at least 12 passage-steps apart (SSBP/1/25 and goat 'scratching' line pass 5, Fig. 10.2). It is now becoming clear from work in progress that the goat 'scratching' line contains another component (80V) in addition to 22C, but it is too soon to check its identity as one of the other components already isolated directly from SSBP/1.

The goat 'drowsy' line has been more extensively analysed than the 'scratching' line but no components of SSBP/1 have been detected in the 8th passage. This negative evidence is the most convincing in the case of 22A which would be expected to be at a selective advantage, therefore more easily found, with the technique used (namely, serial passage in VM, *sinc*[p7] mice; see 14.4.4). Instead there are at least three components in the 'drowsy' line, 79V and two others which produce intense white matter vacuolation in C57BL mice – 139A (synonym 'Chandler') and 79A. None of the four agents isolated from the SSBP/1 Cheviot line (i.e., 22A, 22C, 22L and 22M) produce intense vacuolation of white matter in C57BL mouse brain and neither does direct injection of C57BL mice with SSBP/1 sheep brain at the 16th, 21st, 24th or 25th passages (Fraser et al., 1974 and unpublished). The same holds for 80V and 22C isolated from the goat 'scratching' line.

What, then, is the origin of these three strains of agent in the 'drowsy' line? The possibilities are that (a) the components of the 'drowsy' line originated quite separately from the known injection sequence recorded in Fig. 10.2, or (b) 79A and 139A are present in SSBP/1 at very low titre compared with 22A or 22C, with the result that severe white matter vacuolation does not have a chance to develop in the first passage C57BL mice, or (c) the components of the 'drowsy' line were produced by some obligatory change on 'adapting' to certain goats in the 'drowsy' lineage – obligatory because new strains appear to have replaced previous ones.

The first alternative is the easiest one to accept – the observed facts could simply result from a goat at an early stage of the passages, perhaps the first goat, already incubating natural scrapie due to agents 79A, 79V and 139A when it was injected with SSBP/1-derived inoculum; the only other condition needed to satisfy this possibility would be that 79A, 79V and 139A replicated more rapidly in subsequent goats than SSBP/1 components, which would be likely if they had a higher titre in the original goat brain and also accords with the observation that the incubation period of the 'drowsy' line tends to be shorter than that of the 'scratching' line. Reluctance to accept the second alternative is based on consistent failure to detect agents resembling 79A and 139A in either SSBP/1 from sheep or in 'scratching' agent from goats: both these agents would be at a competitive advantage over all the other strains detected, because of their quicker incubation in C57BL. As for the third category – obligatory 'adaptation' – this seems the least satisfactory alternative to accept because the progress and techniques available in scrapie work are still insufficient for critical analysis to recognise 'mutational' or host-specified changes in agent properties given that SSBP/1 is a mixture of strains.

Agents isolated from 21 other sources have now been examined in varying detail by passage in mice and a variety of strains of agent have been found, but none is identical with any of the strains mentioned in connection with SSBP/1 (10.8). From these other sources, the agent ME7 has been recovered on several occasions and multiple infection with several agent-strains has been common. The question which must therefore be posed is whether the conclusions, which will be drawn from the use of SSBP/1 in the account which follows, are of general validity or even whether they are applicable to the majority of strains. The broad picture emerging is that the conclusions about 'susceptibility' are likely to apply to many but not all strains of agent.

10.7 Some genetical aspects of response to experimental infection

10.7.1 The 24-breed experiment

In a search for a more suitable breed than Cheviots, W. S. Gordon injected SSBP/1, either intracerebrally or subcutaneously, into small groups of most British sheep breeds, in the hope that one would have a 100% incidence (Gordon, 1959 and personal communication). None did. The samples of the breeds were not necessarily representative, however, as they came from a very limited number of flocks. For example, cases were confined to one of the two sub-groups of Suffolks. The incidence in the different breeds was broadly the same for the two injection routes. The published conclusions were based on the number of cases which developed within 24 months and the average incidence ranged from 78% in Herdwicks and 72% in Dalesbreds to 0% in Dorset Downs. Avoidance of natural scrapie was again a reason for limiting the observation period but this obscured the fact that the Dorset Downs were not 'fully resistant' because a small proportion of cases developed after between 25 and 47 months for which the SSBP/1 injection was almost certainly responsible (Gordon, personal communication). The unsatisfactory sampling of the breeds used was further emphasised by the subsequent finding that a very large group of Herdwicks injected subcutaneously with SSBP/1 only had a 35% incidence (Pattison, 1966).

Sheep from more breeds have now been tested with SSBP/1: intracerebral injection produced one case after 26 months in Finnish Landrace sheep but five others are still healthy at 40 months, as are all six of the primitive breed, Soay (Dickinson, unpublished).

10.7.2 Genetical selection in Cheviot sheep

Since 1961 Cheviot sheep have been selected as two lines from a representative foundation group of 303 ewes and 15 rams all chosen to be free from natural scrapie. Half are being selected for an increased incidence of scrapie following

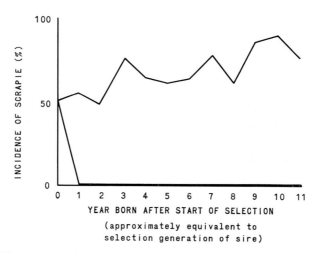

Fig. 10.3. The response of Cheviot sheep to genetical selection for or against the production of scrapie by subcutaneous injection of SSBP/1. After the first generation selection was restricted to the rams.

subcutaneous injection with SSBP/1, the other half for a decreased incidence (interim report by Dickinson et al., 1968). The two lines now differ in incidence by about 90% (Fig. 10.3). There was a deliberate attempt to minimise inbreeding and this policy has retarded selection progress, but it was being kept to a minimum so that the two eventual divergent populations would not have fertility problems and would, therefore, be more easily available for scrapie experiments.

It is now almost certain from recent line-cross and back-cross segregation results, that a single gene controls recessive resistance to subcutaneously injected SSBP/1. Whether these homozygous resistant individuals permit agent to replicate extraneurally is unknown but it seems quite possible that they do, though this is likely to be slow or only after a long delay if inferences from the mouse scrapie-control system are valid (Dickinson, 1975). The basis for this speculation is that at least half, and possibly all, of these genetically 'resistant' sheep can develop scrapie if injected intracerebrally with SSBP/1. It may therefore be more appropriate to distinguish the two Cheviot phenotypes as 'short incubation period' (SIP) and 'long incubation period' (LIP) rather than 'susceptible' and 'resistant'. Using 10% SSBP/1 homogenate intracerebrally (1 ml), SIP incubations average 7 months (range 3–11 months) and for the subcutaneous route (5 ml) they average 10 months (range 4–25 months). In contrast, LIP cases only occur following intracerebral injection and range from 18 months through to old age.

It is unknown which strain or strains of agent in SSBP/1 are basically responsible for killing the Cheviots but the fact that they persist in a long-passaged line suggests that they can all take an active part in the brain of SIP sheep. However, the following results indicate that some of the SSBP/1 components may not replicate in LIP sheep. On each occasion when SSBP/1 from SIP Cheviots has been analysed (using mice) both 22A and 22C have been found but on the only occasion that SSBP/1 from an LIP Cheviot was used, 22A was present but accompanied by 22L, not 22C (Fig. 10.2): this LIP Cheviot had an incubation period of 59 months. It should be added that 22L is presumably also present in the SIP sheep but, given the mouse passage techniques used, its presence would be obscured by 22C and any 22L would tend to be lost in these mouse passages.

In an earlier section (10.2) reference was made to diagnostic problems with natural scrapie; as the range of clinical symptoms is unknown, natural scrapie has to be defined as a lethal disease and the presence of certain types of brain vacuolation is regarded as diagnostic. The rather unsatisfactory nature of these criteria is highlighted by the results of the Cheviot selection experiment where there are the following independent criteria for deciding whether the animal has scrapie: routine injection of sheep of specified genotype with a standard inoculum by a constant route, with the knowledge that the results will be lethal after a fairly predictable time and preceded by progressive ataxia which is often accompanied by pruritus. Even before it was possible to control all these variables, Zlotnik (1960) was aware that scrapie produced by SSBP/1 was sometimes accompanied by hardly any diagnostic lesions. As the selection experiment has progressed, it has become clear that 13% of the unequivocal clinical cases do not have clear-cut diagnostic lesions and would have to be regarded as negative if vacuolation were the only criterion or the overriding one (12.6 and 7).

A problem has developed with this experiment which has made many of the sheep useless for the intended purposes in that there has been a severe outbreak of scrapie (Dickinson, 1974). Briefly, the foundation flock was held and bred in geographical isolation from known sources of scrapie and remained free from any signs of the disease for a number of years. When sheep were to be tested with SSBP/1 inoculum they were transported to another farm for injection and observation. After 7 years a case occurred in the isolated breeding flock. This and all but one of the subsequent cases have been in the positive selection line and over 30% of this line are now affected in each generation. The brain lesions in these cases are severe, which effectively rules out any component of SSBP/1 as the cause.

Subcutaneous injection of eleven LIP Cheviots with brain from one of these natural cases (source 141, Table 10.1) has produced scrapie in five of them with incubations of 17, 19, 27, 44 and 46 months. In addition, intracerebral injection of thirteen LIP Cheviots gave five cases with 13, 14, 15, 15 and 22-month incubations. These findings show that the results using SSBP/1 should not be generalised to all strains of agent. On the basis of present information, especially the detailed knowledge of scrapie in mice, there is no reason to assume that agents do not exist which will produce cases more readily in the 'negative' line than in the 'positive' line.

10.7.3 *Genetical selection in Herdwick sheep*

Work similar to that in the Cheviots was initiated in 1961 in Herdwick sheep by W. S. Gordon, again using SSBP/1 (Nussbaum et al., 1975). The results so far are similar in both experiments. In the Herdwicks inbreeding was not deliberately kept to a minimum in the 'positive' line and a very small proportion of sheep fully susceptible to subcutaneous injection with SSBP/1 have been produced having incubation periods averaging $6\frac{1}{2}$ months. There was a small proportion of instances (3%) where a mating involving two negative line Herdwicks produced a progeny which developed scrapie much later than for the SSBP/1 susceptible groups. This category was not observed following subcutaneous injection in the selected Cheviots but they could be analogous to the Cheviot LIP category (which only appeared following intracerebral injection) because incubation was generally much shorter in the Herdwicks than in the Cheviots. If this is the explanation, it suggests that some LIP Cheviots injected subcutaneously (and other undetected 'negative' Herdwicks) could be permitting delayed extraneural replication but not reaching clinical disease before senility. This situation is known to exist in mice with incubation being longer than lifespan (Dickinson et al., 1975b).

In the Herdwicks also there has been an outbreak of scrapie in the positive line (Pattison, 1974) although it is not known what proportion will eventually be affected as the incidence is rising rapidly (Kimberlin, personal communication).

10.8 *Isolation of agents from scrapie in sheep and goats*

Two successful attempts to transmit scrapie to mice started independently in 1960 (Chandler, 1961; Dickinson and Smith, 1966) but Zlotnik and Rennie (1962) were the first to succeed using tissue from a natural case. A number of authors have since transmitted scrapie to mice from natural cases (Gibbs et al., 1965; Hanson et al., 1971; Gustafson et al., 1972), although it is difficult to accept one claimed transmission to mice using sheep serum.

The two most extensive studies using tissues from natural cases have not yet been reported. These are by Hadlow (personal communication) using, separately, many areas of sheep brain and various extraneural tissues from 12 Suffolks and a Cheviot case, and by Dickinson and Fraser, details of which follow. Attempts to improve and standardise isolation and agent identification techniques over the last decade have led to certain changes in methods, but it is now becoming clear that the region of sheep brain used has to be taken into account in assessing the results and comparing those of different workers. Until recently we have used only mid-line cerebral cortex, for three reasons (1) it was the simplest area to remove using rigorous aseptic techniques, (2) it avoided encroaching on areas important for histopathological diagnosis, (3) Gordon's early titrations in sheep (1960) had not indicated gross differences in infectivity between medulla and the rest of the brain. Homogeneity of titre throughout the brain is not supported by Hadlow's results (personal communication) and a similar conclusion is emerging from work with some mouse-passaged agents in mice (Dickinson, unpublished); the indications are that infectivity is higher in the more vacuolated areas. Cerebral cortex tends to be a relatively undamaged area in natural scrapie.

In the following results, therefore, one reason why some sources failed to infect mice could be that a relatively low titre brain region was used, the other major reason being the strains of agent involved; the issue is discussed more fully in 10.11.

Table 10.1 shows the current results of the first attempt to consider the epidemiology of scrapie in terms of the strains of agent involved. In the light of such information it will eventually be possible to decide whether the experiments described in previous sections are relevant to the disease in the field. What has been attempted is necessarily crude and much more work is required before it can be discerned whether the somewhat arbitrary choice of a small range of unrelated inbred mouse strains for primary isolation (Table 10.1, caption) has imposed restrictions on the array of agent strains which can be detected from sheep. The 26 source cases investigated so far were deliberately chosen to represent different breeds and areas of Britain

A. G. Dickinson

TABLE 10.1

Isolation of scrapie agents from naturally infected cases. Unless otherwise noted, injections were intracerebral with 0.02 ml 10% saline homogenate of midline cerebral cortex into a panel of C57BL, VM, C57BL × VM, BALB/c and RIII and/or BSVS mice. VM are $sinc^{p7}$, others are $sinc^{s7}$.

Breed of affected sheep	Source number	Primary passage		Mouse passages				Comments on agents isolated
		Scrapie incidence (%)	Incubation period range (days)	Passages completed		Agents isolated		
				C57BL passages[b]	VM passages	In C57BL	In VM	
Suffolk	43	2[a]	325	(1+)2	(1+)2	ME7	ME7	
	51	100	382–699	4	1	51C	?	Many plaques
	58	100	251–503	4	4	ME7	ME7	
	113H	100[a]	337–538	(1+)1	(1+)1	ME7?	ME7?	
	131	2[a]	535	0	2	—	131A[e]	May be same as 104A
	138	100	366–586	3	2	138A	138V[e]	Many plaques
	ME7	100[a]	>400	(2+)12	(2+)8	ME7	ME7	Original ME7 isolate
	FMS	100[a]	380–420	(4+)1	(4+)1	ME7	ME7	
Suffolk × Colbred	111	98	531–719	1	0	ME7?	—	
Cheviot	141	0	(to 826)	—	—	—	—	See text
	157	18	479–783	1	0	?	—	
	158	0	(to 859)	—	—	?	—	
	159	100	187–551	1	1	?	?	

Cheviot × Border Leicester	87	100	407–652	3	3	87A	87V[e]
Border Leicester	148	39	445–788	1	0	?	?
Border Leicester cross[c]	149	18	489–782	1	1	?	?
Scottish Blackface	104[d]	100	530–733	4	4	ME7	104A[e]
	132[d]	0	(to 778)	—	—	—	—
	153[d]	100	398–620	1	2	?	153A
Dorset Horn	143	0	(to 979)	—	—	—	—
Dorset Down	31	90[a]	420–556	(1+)4	(1+)2	ME7	31V[e]
Welsh Mountain	133	17	536–755	3	(1+)1	ME7	?
Swaledale	106	10	593–824	(1+)1	0	106A	—
Southdown	125	100	330–370	(1+)3	(1+)3	125A	125V
Finnish Landrace	164	0	(to 760)	—	—	—	—
Goat	124	95	315–658	3	3	ME7?	124V[e]

[a] Details not standard: 31, only injected into random-bred *sinc*^{s7} mice; 43, only injected into LM *sinc*^{s7} mice; 113H, inoculum was spleen removed surgically from 15-month-old pre-clinical sheep and only injected into VL *sinc*^{s7} mice; 131, inoculum was 500g 10 min supernatant; ME7, for details see Zlotnik and Rennie (1963); FMS, supplied at 4th mouse (*sinc*^{s7}) passage by Dr. D. A. Haig; 125, supplied at 1st mouse passage by Mr. J. C. Rennie.

[b] Earlier passages not in C57BL or VM shown in parenthesis.

[c] Sheep 149, typical clinical case but brain had insufficient diagnostic lesions.

[d] Contact cases (Table 10.2; Dickinson et al., 1974).

[e] 22A-group agents (see 14.4.4d, 14.5 and Dickinson and Meikle, 1971).

(15 flocks) and also to concentrate on one flock (Moredun flock: Suffolks 43, 51, 58, 113H, 131, ME7; Scottish Blackfaces 104, 132, 153).

The dominant feature of the results is that ME7 agent (or one at present indistinguishable from it) occurs in over half the cases where analysis has been sufficient and that it is present in many breeds. It will be recalled that ME7 is unfortunately not present in the SSBP/1 source used in most experiments (10.6) so that, for example, an early question to answer is how sheep in the selection lines of Cheviots and Herdwicks (10.7.2 and 3) respond to injection with ME7. It is useful that Zlotnik and Rennie (1965) have given results which start to cover the gap by using ME7 in goats, unselected Cheviot sheep, rats and hamsters; further ME7 work is in progress in hamsters, mink and squirrel monkeys (Marsh, Hanson and Dickinson, unpublished) and in selected Cheviots.

The second conclusion from Table 10.1 is that infection of a natural case with more than one strain of agent is not unusual and, of particular importance, that such multiple infection can occur by contagion (Scottish Blackface 104, see 10.10). Although agents may differ in the ease with which they can spread laterally these results establish that ME7 is naturally infectious to flock mates. The third important feature in the Table is that agents are often present which produce a high frequency of amyloid plaques in the brain. Experimental use of such agents has only started recently and their study is important because several human dementias and senile humans characteristically show this type of lesion (12.9).

A fourth point to emerge is that no agents have been detected which always produce in mice the very intense vacuolation of white matter which is so characteristic of two of the strains isolated from goat 'drowsy' scrapie Some of the agents shown in Table 10.1 can produce moderately severe white matter vacuolation but only in certain genotypes of mice and a comprehensive understanding is still lacking of the factors which lead to variation in white matter damage (12.5).

A limited amount of work has been undertaken using some of the Table 10.1 source sheep for inoculation of goats, other sheep, hamsters, mink and squirrel monkeys. This work is intended to explore in more detail why strains of scrapie and related agents differ in their host ranges (Ch. 15). It has already been reported that scrapie from two different naturally affected sheep differed in their attempted transmission to mink and randombred mice: the Suffolk source produced scrapie after 12–14 months in mink and after 15 months in mice, in contrast, the Cheviot source produced no cases in mink but the mice were affected after 16 months (Hanson et al., 1971;

15.5). There is a similar example from Table 10.1: the Cheviot source 141 failed to produce scrapie in any of the standard panel of mice (see Table 10.1, caption) and has also failed when used in 9 other mouse strains (BRVR, SM, VL, DH, A2G, BSC, MM, LM). However, intracerebral injection of this 141 material produced scrapie in hamsters after 12 months, but not in mink after 23 months (Marsh, Hanson and Dickinson, unpublished) and it produced 5/13 cases in LIP Cheviots after 13–22 months, which is very early (though it has not been tested in SIP Cheviots). Given this array of results the failure to produce disease in mice is more likely to be due to the particular agent strain(s) than to low titre of infectivity in the area of brain used – whether the mice are infected, but fail to be affected during their lifetime, is an entirely different question.

10.9 Correlations between relatives

When one considers the good start which Gregor Mendel gave to genetics with his study of gene segregation, it is unfortunate that there have since been many examples where the observation that related individuals had certain features in common led to the hasty conclusion that these similarities must be caused by genes. Inadequate experimentation and analysis have then often been propped up with the suggestion that 'the gene had incomplete penetrance', given as an excuse rather than a careful finding. Especially with diseases, there are various reasons other than their chromosomal genes, why relatives can show similar traits (see discussions by Dickinson, 1972; Reanney, 1975).

Poor sampling and inadequate analysis resulted in the confounding of different diseases and to the wrong conclusion by Bosanquet, Daniel and Parry (1956) that scrapie was a primary myopathy, followed by the vigorously pursued claim that scrapie was entirely caused by a recessive gene and not by an independent pathogenic microorganism (Parry, 1962, 1973). This conclusion was refuted in a critical re-analysis of these and additional data (Dickinson et al., 1965) and it was established that maternal transmission of an independent causal agent accounted for many of the observed cases of similarity in scrapie incidence in related sheep.

10.9.1 Maternal transmission

There are several reasons for concluding that maternal transmission occurs in natural and experimental scrapie in sheep. These are (1) a significant

TABLE 10.2

Natural scrapie incidence in Suffolk (S), Scottish Blackface (B) sheep and their crosses.

Purebred or F1 cross			Backcross			Backcross			F2 cross		
Ewe × Ram	Incidence (%)	(nᵇ)	Ewe × Ram	Incidence (%)	(n)	Ewe × Ram	Incidence (%)	(n)	Ewe × Ram	Incidence (%)	(n)
S+ S+	94	(197)									
S+ S−	81	(31)									
S− S+	18	(11)									
S− S−	18	(27)									
[a]S+ B−	63	(27)	S+ F1+	100	(4)	F1+ S+	100	(4)	F1+ F1+	77	(13)
			S+ F1−	100	(3)	F1− S+	100	(3)	F1+ F1−	59	(44)
S− B−	37	(83)	S− F1+	25	(4)						
			S− F1−	50	(28)						
B− S+	33	(27)	B− F1+	11	(9)	F1+ B−	0	(2)	F1− F1+	45	(22)
			B− F1−	12	(8)	F1− B−	17	(6)	F1− F1−	25	(28)
B− B−	28	(21)									
B− B−	0	(>20,000)									

7–8-year observation period → Contact group

15-year observation period → Isolated flock

[a] Breed and eventual scrapie staties.

[b] Total number of sheep alive at 18 months, which immediately precedes occurrence of earliest clinical scrapie. Minimum age of survivors is 5 years.

concordance rate in twins – that is a preponderance of pairs with both affected or both unaffected and a deficiency of pairs where only one is affected, (2) in reciprocal crosses between infected and uninfected sheep, the incidence in the progeny tends to follow the maternal incidence (Table 10.2), (3) when ewes are injected subcutaneously with SSBP/1 scrapie shortly before or soon after conception (but the progeny are not injected after birth), a proportion of the lambs develop scrapie unexpectedly early, starting from 7 months old (Gordon, 1966; Dickinson et al., 1966). Why natural maternal transmission does not normally produce such early cases, is unexplained.

Naturally affected ewes often have more than one pregnancy before they die and the present evidence indicates that lambs are equally liable to be infected by maternal transmission whether they are born a year before or contemporary with their dam being clinically affected. Data are insufficient to extend this conclusion to even earlier lambings but the indications are the same. There is no direct information that maternal transmission depends on infection of the embryo, but this may simply reflect lack of work to resolve the point, because efficient sheep assay systems have only recently been available. The maternal transmission of SSBP/1, mentioned above, is the most convincing indication that agent is passed to the embryo.

This possibility is supported by two other findings. The first is that agent is present in sheep placenta (Pattison et al., 1974) and the second is that scrapie, quite unlike conventional viruses, depends on the recipient being immunologically mature for infection to be successful by some extra-neural routes (Outram et al., 1973). Sheep are known to be immuno-competent long before birth (Schinckel and Ferguson, 1953), whereas mice do not have such competence until some days after birth – which would explain the absence of maternal transmission in mice (Dickinson, 1967; Field and Joyce, 1970; Clarke and Haig, 1971) despite an early claim that it occurred (Gibbs et al., 1965). One would expect that goats would resemble sheep in these respects but the only reported attempt failed to detect any maternal transmission in goats when both parents had been injected intra-cerebrally, before mating, with an unspecified scrapie agent; the same scrapie source was shown to be capable of infecting goats prenatally when injected directly into the foetal brain two months before birth, but infection only occurred in one twin goat when only one embryo had been injected (Pattison, 1964). Whether maternal transmission occurs naturally in goats or in those injected by extraneural routes or with other scrapie agents remains to be determined, but if this route is absent or less efficient than in sheep, scrapie is more likely to be epidemiologically self-limiting in goats than sheep.

10.9.2 The familial problem of natural scrapie

As scrapie incidence can follow a familial pattern for other than genetical reasons (Table 10.2), can any precise or useful genetical conclusions be drawn? We are not yet in a position to estimate or accurately control a number of crucial factors, which precludes simple genetic analysis of the data: estimation of 'heritability' would be meaningless in the circumstances, for reasons given elsewhere (see Dickinson, 1975). The uncontrolled factors are (1) the agent strains present and whether there is any secular change in their representation during a long experiment (the Suffolk × Blackface experiment in Table 10.2 lasted 14 years), (2) the extent of maternal trans-mission, which appears to vary widely, (3) the degree and age of exposure to lateral transmission of the infection (10.10) and (4) the possibility that effectively non-pathogenic strains of scrapie might exist and, if present, would only be apparent by the phenomenon of 'blocking' or hindering the pathogenesis of pathogenic strains encountered later (Dickinson et al., 1972, 1974, 1975a). Where some of these factors can be controlled, as the selection experiments show (10.7), simpler interpretation is possible.

Experimental flocks can be produced where all the animals develop scrapie naturally. For example, our Suffolk scrapie-line was founded on the progeny of affected ewes and now continues using only rams and ewes which will become cases. In the years 1968–71, 114 lambs were born in this line which survived to 2 years old: of these, 4 died from unrelated causes before $2\frac{1}{2}$ years old, 109 developed scrapie and one died at 5 years old without signs of scrapie. In these circumstances scrapie is virtually self-limiting, because ewes are hardly able to leave a fertile female lamb before they die. Here again, we are unable to assign values to the relative contributions of maternal transmission and 'genetic susceptibility' – that all were susceptible, given affected mothers, is obvious but whether all would be equally sus-ceptible to the same agents encountered first at, say, weaning age is unknown.

10.10 Scrapie as an infectious disease

Ten years ago there was only indirect evidence that scrapie must be in-fectious (Dickinson et al., 1965), later it was shown to be infectious for sheep and goats housed in continuous contact with affected sheep (Brotherston et al., 1968; Haralambiev et al., 1973) and more recently to be infectious to, on average, at least a quarter of the exposed sheep sharing pastures with affected sheep (Table 10.2; Dickinson et al., 1974; Hourrigan, personal communication). Probably the most intriguing earlier evidence concerned scrapie ('rida') in Iceland (Pálsson and Sigurdsson, 1959) where eradication of various diseases (2.10), including scrapie, had involved destroying all sheep in large areas and leaving them free from sheep for 1–3 years: restocking was from areas where scrapie had never been recorded but during the subsequent decade it reappeared on 30 of the farms where it had previously been occurring. This pointed to long-term persistence of the agent and these authors suggested that an intermediate host might be involved.

As scrapie is infectious to flockmates, two issues need resolving: how agent leaves an infected sheep and how it enters others. As described previously (10.9.1; 10.4.1 and 2) the placenta is one source of infection but agent has not been detected in normal excretions, though more work is needed because some of the reports of lateral transmission do not involve affected parturient ewes (e.g., Brotherston et al., 1968) and the comment is often heard that scrapie was imported into a flock with a purchased ram. It has been found that orally administered ME7 scrapie passes through the mouse alimentary canal with little, if any, loss of activity (Taylor and Dickinson, unpublished). The same could be true with sheep, which could disperse placental or other ingested agent over the pastures via the faeces. The question therefore becomes one of the degree of persistence of infectivity in pastures, but no critical work on this has been reported.

Several natural routes of entry have been found, which appear to be fairly efficient means of producing infection. These include oral dosing (Pattison and Millson, 1961; Gordon, 1966; Pattison et al., 1974), scarification (Stamp et al., 1959) and via the conjunctiva (using boiled inoculum, Haralambiev et al., 1973). Experiments are needed to check possible routes, in addition to the prenatal one, which might give preferential infection of lambs with infected mothers: these could include gulping of foetal fluids at birth (if these fluids were infective), muzzling and licking behaviour of ewe and newborn lamb could provide oral, conjunctival or umbilical-wound routes to the lamb for agent from recently ingested placenta and the possibility needs considering whether the latter could lead to transient excretion of agent in milk.

As a commercial disease-control measure for scrapie, rigorous culling of affected maternal lines is the simplest useful one to advocate because maternal transmission amplifies the result of a single infectious event, but this would not have prevented the outbreaks described above in Cheviots and Herdwicks (10.7.2 and 3). Extensive slaughter of flockmates and the tracing of 'source flocks' has been a costly, but effective, method of control in the U.S.A. (Hourrigan et al., 1969). Early diagnosis, which may be possible by cytopherometry, needs developing (Field and Shenton, 1974). Whether there are reservoirs of infection in other species needs further investigation but the possibility that some sheep, such as the LIP Cheviots, could act as carriers cannot yet be discounted. One possibility which has been considered is the large scale application of the genetical selection techniques already described. This could produce short-term alleviation, but whether it would give long-term benefits is conjectural. Plant breeders (in the case of rusts and other diseases) and poultry breeders (in the case of leukosis) have been disillusioned to find that genetical uniformity of resistance to the current strains of the disease has become a disadvantage when new strains appeared or the stock was moved to a different area. If the disease in mice is a basis for judgement, scrapie is no exception to this situation.

Now that we know that scrapie is infectious, is it possible to understand why infection with more than one strain of agent seems to be common? In mice we know that agent replication proceeds relatively rapidly in lympho-reticular organs and that there can be a very long titre plateau in, for example, the spleen before agent replication in the brain leads to death (14.2 and 3). As events seem to be similar in sheep (Eklund and Hadlow, 1969) this long extraneural phase could well account for natural infection by more than one strain of agent, because maternal or lateral transmission of infection from an extraneural source could include agents too slow to have killed the sheep but able to 'catch up' extraneurally during the plateau phase of the quicker strains. If this is so, a wider spectrum of agents may be available for isolation from extraneural organs than from brain.

10.11 What is the 'species barrier'?

From the information already presented in section 10.8 it is clear that the ease with which agents can be isolated from sheep by passage to mice varies from source to source and some of the reasons for this have already been discussed. The group of slow encephalopathies to which scrapie belongs show great differences in their species ranges as experimental infections (Table 15.3). But the variety in this respect between different scrapie sources – essentially, but not only, because of their diversity of strains of agent – is greater than has been generally realised. Already mentioned

(10.8) are the Suffolk and Cheviot sources which differed in their relative transmissibility to mink and randombred mice (Hanson et al., 1971) and the Cheviot source 141 which failed to affect a large range of inbred mice but produced scrapie readily in LIP Cheviots and hamsters. At a genetically much finer level than differences between species, it is likely that a single gene difference accounts for the failure to produce scrapie from Suffolk 131 in *sinc*^{s7} mice but success in *sinc*^{p7} mice and for the converse from Suffolk-cross 111 (Table 10.1).

There are at least three conceivable types of reasons for these differences in apparent transmissibility, apart from gross variation in agent titre. These are: (a) the strain of agent is unable to reach replication sites in the new host either because of its essential structure or because of some phenomenon involving donor tissues with which it happens to be associated; (b) the agent in question fails to replicate even though it can reach the sites in the recipient at which other agent-strains replicate, or (c) replication is so slow, or the initiation of replication is so delayed, that disease does not occur during the new host's lifespan – the recipient is infected but not affected. In the first two categories agent from the inoculum may persist on a long-term basis in the recipient or be passively excreted or be actively degraded; which of these occur is uncertain. The question of agent turnover has hardly been raised in the literature but there is some evidence for degradation processes (Outram et al., 1973 and unpublished) and it is possible that the third category above is the outcome of a balance in which replication just exceeds degradation.

The mice injected with Cheviot 141 may be a category (c) case of extremely delayed replication, because a few amyloid plaques (probably the earliest scrapie lesion, where they occur) have been found in some of them and similarly with TME injected mice (Bruce, Fraser and Dickinson, unpublished).

One other perspective on scrapie 'species barriers' must be emphasised. The divergent system, in mice, of agent isolation and identification which has been developed in these laboratories (i.e., passage in C57BL *sinc*^{s7} or VM *sinc*^{p7} mice) yields 0, 1 or 2 strains of agent from a single source but an appropriate trivergent (or higher) system might show that greater agent-strain diversity is present in the sheep. This interpretation depends on one premise, namely, that the different agent-strains are not all arising during the mouse passages – originating, that is, by either broadly random mutational events or by some process of host-directed modification analogous with methylation of phage bases. The available data (Table 10.1 and unpublished) clearly supports this premise because (a) passage of ME7 either

12 times in C57BL mice or 8 times in VM mice did not alter its characteristics, (b) ME7 has been re-isolated either alone or accompanied by other agents, (c) when 22A was isolated from SSBP/1 at various passages and from different selected genotypes of Cheviot sheep, there was one occasion when it was not accompanied by 22C (10.8), (d) if a 22A-group agent (see Dickinson and Meikle, 1971) is present in a sheep, it is detected and isolated by passage (and eventually cloning) in VM mice, but such agents have not always been present, and (e) experimental multiple infection with different strains of agent can be produced in mice in which the agents preserve their identities, and in some cases they can be shown to compete for scarce replication sites (Dickinson et al., 1972, 1975a and unpublished). Competition has recently been detected in VM mice using ME7 and 22A, both having been passaged many times in VM mice.

The evidence is therefore sufficient to conclude that mixtures of agents occur in natural infections. If scrapie, or a related disease, was found to be transmissible from the natural host to two other species but not between the latter, the most likely explanation (discounting experimental error) would be that the original infection was heterogenous and that different component strains replicated in the two recipient species. This type of observation should not be regarded as evidence for host-modification of 'the' agent because it would not be justified to assume that the infection was homogenous. The present evidence does not, however, preclude some host-directed modification of agent properties either in intra- or inter-specific passages. Something analogous to host-specified 'coating' would be the easiest type of possibility to detect and this was one of the possibilities considered by Dickinson and Outram (1973) for uncloned ME7 in two strains of $sinc^{s7}$ mice. However, any host-directed event of this type will be difficult to prove unless it can be established that only a single strain of agent is involved.

The term 'species barrier' has frequently been used to denote a different effect from those considered above. Many workers have noticed that incubation is often prolonged on passage to a new species, compared with later passages. Whether it takes only the first passage to the new species or several passages in it before the incubation period becomes stable, depends on more than one factor. A prolongation restricted to the interspecific passage, unaccompanied by any other changes, could simply be the result of infectivity being more easily degraded, if associated with very alien tissues, and thus having a lower effective titre, but data on this are lacking. Also, apart from genetic variability in the new host (e.g., *sinc* segregation in mice), the rate at which incubation stabilises with passage will depend on the variety of agent strains present in the source. The possibility is open that interspecific passage may sometimes involve an abnormal sequence of pathogenesis, analogous with that suggested for heat-treated agent (Dickinson and Fraser, 1969).

Several authors have referred to scrapie as being 'adapted' to a new species rather than using the operational term, passaged. Use of the term adapted has often loosely implied that the agent is different from that in the former species. However, there are two entirely separate ways in which this could occur, depending on whether, in the former species, the infectivity was homogenous (a single strain) or heterogenous (a mixture of strains). If homogenous, then adaptation must imply a changed agent, in the sense of one not present in the former host – this change could either be in the agent's informational code or as a new host/agent complex, if part of the structure is host specified. However, if heterogenous in the former host, any change seen in the new one can arise from selection of one or more strains previously present, which may or may not be accompanied by the appearance of new agent structures. In all the reports these alternatives are confounded because known single-strain (i.e., cloned) agent has not been the starting point. However, in transmission of uncloned ME7 from sheep–mice–goats–mice, Zlotnik (1965) found that a 'species barrier' was not necessarily present at the second and third interspecific steps and this observation has now been repeated with uncloned ME7 using the sequence mice–sheep–mice (Dickinson, unpublished).

References

ADAMS, D. H. and FIELD, E. J. (1968) Lancet, ii, 714.

ALPER, T., CRAMP, W. A., HAIG, D. A. and CLARKE, M. C. (1967) Nature, 214, 764.

BARLOW, R. M. (1972) J. Clin. Pathol., 25, (suppl. R. Coll. Pathol.), 6, 102.

BOSANQUET, F. D., DANIEL, P. M. and PARRY, H. B. (1956) Lancet, ii, 737.

BROTHERSTON, J. G., RENWICK, C. C., STAMP, J. T., ZLOTNIK, I. and PATTISON, I. H. (1968) J. Comp. Pathol., 78, 9.

CHANDLER, R. L. (1961) Lancet, i, 1378.

CHELLE, P. L. (1942) Bull. Acad. Vet. France, 15, 294.

CLARIDGE (1795) In: Letters and Papers on Agriculture to the Society Instituted at Bath, 7, 72.

CLARKE, M. C. and HAIG, D. A. (1971) Br. Vet. J., 127, 31.

CUILLÉ, J. and CHELLE, P. L. (1936) C. R. Acad. Sci., Paris, 203, 1552.

DICKINSON, A. G. (1967) Lancet, i, 1166.

DICKINSON, A. G. (1972) (Zurkowski, Ed.) Proc. Int. Symp. Anim. Genet., Warsaw, 50.

DICKINSON, A. G. (1974) Nature, 252, 179.

240 *A. G. Dickinson*

DICKINSON, A. G. (1975) Genetics, 79 (suppl.), 387.

DICKINSON, A. G. and FRASER, H. (1969) Nature, 222, 892.

DICKINSON, A. G., FRASER, H., MEIKLE, V. M. H. and OUTRAM, G. W. (1972) Nature, New Biol., 237, 244.

DICKINSON, A. G., FRASER, H., MCCONNELL, I., OUTRAM, G. W., SALES, D. and TAYLOR, D. M. (1975a) Nature, 253, 556.

DICKINSON, A. G., FRASER, H. and OUTRAM, G. W. (1975b) Nature, 256, 732.

DICKINSON, A. G. and MEIKLE, V. M. H. (1971) Molec. Gen. Genet., 112, 73.

DICKINSON, A. G. and OUTRAM, G. W. (1973) J. Comp. Pathol., 83, 13.

DICKINSON, A. G. and SMITH, W. (1966) Report of Scrapie Seminar, 1964, ARS 91-53, U.S. Dep. Agric., 251.

DICKINSON, A. G., STAMP, J. T., RENWICK, C. C. and RENNIE, J. C. (1968) J. Comp. Pathol., 78, 313.

DICKINSON, A. G., STAMP, J. T. and RENWICK, C. C. (1974) J. Comp. Pathol., 84, 19.

DICKINSON, A. G., YOUNG, G. B. and RENWICK, C. C. (1966) Report of Scrapie Seminar, 1964, ARS 91-53, U.S. Dep. Agric., 244.

DICKINSON, A. G., YOUNG, G. B., STAMP, J. T. and RENWICK, C. C. (1964) Anim. Prod., 6, 375.

DICKINSON, A. G., YOUNG, G. B., STAMP, J. T. and RENWICK, C. C. (1965) Heredity, 20, 485.

EKLUND, C. M. and HADLOW, W. J. (1969) J. Am. Vet. Med. Assoc., 155, 2094.

FIELD, E. J. and JOYCE, G. (1970) Nature, 226, 971.

FIELD, E. J. and SHENTON, B. K. (1974) Am. J. Vet. Res., 35, 393.

FRASER, H., BRUCE, M. and DICKINSON, A. G. (1974) Proc. VII Int. Congr. Neuropathol., in press.

GIBBONS, R. A. and HUNTER, G. D. (1967) Nature, 215, 1041.

GIBBS, C. J., GAJDUSEK, D. C. and MORRIS, J. A. (1965) Natl. Inst. Neurol. Dis. Blindness, Monogr. 2, Slow, Latent and Temperate Virus Infection. (U.S. Government Printing Office, Washington, D.C.) p. 195.

GORDON, W. S. (1946) Vet. Rec., 58, 516.

GORDON, W. S. (1957) Vet. Rec., 69, 1324.

GORDON, W. S. (1959) Proc. 63rd Annu. Meeting U.S. Livestock Sanit Assoc., 286.

GORDON, W. S. (1960) Report of Special Meetings on Scrapie, ARS 91-22, U.S. Dep. Agric., 1.

GORDON, W. S. (1966) Report of Scrapie Seminar, 1964, ARS 91-53, U.S. Dep. Agric., 8 and 19.

GORDON, W. S., BROWNLEE, A. and WILSON, D. R. (1939) Proc. III Int. Congr. Microbiol., New York, 362.

GORDON, W. S. and PATTISON, I. H. (1957) Vet. Rec., 69, 1444.

GREIG, J. R. (1940) Vet. J., 96, 203.

GREIG, J. R. (1950) J. Comp. Pathol., 60, 263.

GUSTAFSON, D. P., MARSH, R. F. and HANSON, R. P. (1972) Am. Soc. Microbiol., Annu. Meeting (Abstract).

HANSON, R. P., ECKROADE, R. J., MARSH, R. F., ZURHEIN, G. M., KANITZ, C. L. and GUSTAFSON, D. P. (1971) Science, 172, 859.

HARALAMBIEV, H., IVANOV, I., VESSELINOVA, A. and MERMERSKI, K. (1973) Zentralbl. Vet. Med. B., 20, 701.

HARCOURT, R. A. and ANDERSON, M. A. (1974) Vet. Rec., 94, 504.

HOURRIGAN, J. L., KLINGSPORN, A. L., MCDANIEL, H. A. and RIEMENSCHNEIDER, M. N. (1969) J. Am. Vet. Med. Assoc., 154, 538.

HUNTER, G. D. (1972) J. Infect. Dis., 125, 427.

HUNTER, G. D., KIMBERLIN, R. H. and GIBBONS, R. A. (1968) J. Theor. Biol., 20, 355.

JOUBERT, L., LAPRAS, M., GASTELLU, J., PRAVE, M. and LAURENT, D. (1972) Sci. Vet., 74, 165.

MACKAY, J. M. K. (1968) Nature, 219, 182.

MACKAY, J. M. K. and SMITH, W. (1961) Vet. Rec., 73, 394.

MACKAY, J. M. K., SMITH, W. and STAMP, J. T. (1960) Vet. Rec., 72, 1002.

NUSSBAUM, R. E., HENDERSON, W. M., PATTISON, I. H., ELCOCK, N. V. and DAVIES, D. C. (1975) Res. Vet. Sci., 18, 49.

OUTRAM, G. W., DICKINSON, A. G. and FRASER, H. (1973) Nature, 241, 536.

PALMER, A. C. (1959) Vet. Rev. Annot., 5, 1.

PÁLSSON, P. A. and SIGURDSSON, B. (1959) Proc. VII Nord. Vet. Congr., Helsinki, 179.

PARRY, H. B. (1962) Heredity, 17, 75.

PARRY, H. B. (1973) Nature, 242, 63.

PATTISON, I. H. (1957) Lancet, i, 104.

PATTISON, I. H. (1964) Vet. Rec., 76, 333.

PATTISON, I. H. (1965) In: (D. C. Gajdusek, C. J. Gibbs, Jr, and M. Alpers, Eds) Natl. Inst. Neurol. Dis. Blindness, Monogr. 2. Slow, Latent and Temperate Virus Infections. (U.S. Government Printing Office, Washington, D.C.) p. 249.

PATTISON, I. H. (1966) Res. Vet. Sci., 7, 207.

PATTISON, I. H. (1974) Nature, 248, 594.

PATTISON, I. H. and JONES, K. M. (1968) Nature, 218, 102.

PATTISON, I. H., HOARE, M. N., JEBBETT, J. N. and WATSON, W. A. (1974) Br. Vet. J. 130, 65.

PATTISON, I. H. and MILLSON, G. C. (1960) J. Comp. Pathol., 70, 182.

PATTISON, I. H. and MILLSON, G. C. (1961) J. Comp. Pathol., 71, 171.

PATTISON, I. H. and MILLSON, G. C. (1962) J. Comp. Pathol., 72, 233.

PATTISON, I. H. and SANSOM, B. F. (1964) Res. Vet. Sci., 5, 340.

REANNEY, D. C. (1975) J. Theor. Biol., 49, 461.

SCHINCKEL, P. G. and FERGUSON, K. A. (1953) Aust. J. Biol. Sci., 6, 533.

STAMP, J. T., BROTHERSTON, J. G., ZLOTNIK, I., MACKAY, J. M. K. and SMITH, W. (1959) J. Comp. Pathol., 69, 268.

WILSON, D. R., ANDERSON, R. D. and SMITH, W. (1950) J. Comp. Pathol., 60, 267.

ZLOTNIK, I. (1960) Nature, 185, 785.

ZLOTNIK, I. (1965) In (D. C. Gajdusek, C. J. Gibbs, Jr, and M. Alpers, Eds) Natl. Inst. Neurol. Dis. Blindness, Monogr. 2. Slow, Latent and Temperate Virus Infections, (U.S. Government Printing Office, Washington, D.C.) p. 237.

ZLOTNIK, I. and RENNIE, J. C. (1962) J. Comp. Pathol., 72, 360.

ZLOTNIK, I. and RENNIE, J. C. (1963) J. Comp. Pathol., 73, 150.

ZLOTNIK, I. and RENNIE, J. C. (1965) J. Comp. Pathol., 75, 147.

CHAPTER 11

The physico-chemical nature
of the scrapie agent

G. C. MILLSON, G. D. HUNTER and R. H. KIMBERLIN

11.1 Introduction

The experimental transmission and serial passage of scrapie in sheep clearly demonstrated that the disease is induced by a replicating agent, and the further ability of infectivity to pass through a gradocol membrane of APD 410 nm led early workers to conclude that the disease was caused by a virus-like agent. However, some workers were led to question the viral aetiology of the disease as a result of studies made at the Moredun Institute, Edinburgh, by D. R. Wilson who was one of the first to discover some of the unusual physico-chemical properties of the agent such as its resistance to heat, formalin, chloroform and UV light (10.4.1).

This early work was hampered by the expense of conducting experiments on large animals, by the limited and variable susceptibility of sheep to experimental infection, and by the unpredictable length of incubation periods in sheep, and also in goats which were later used as an experimental host (10.4.2). A valuable contribution to scrapie research was therefore made by Chandler in 1961 when he successfully transmitted the disease to mice, and subsequently showed that they were 100% susceptible to the disease with a predictable incubation period after experimental inoculation. Dickinson and co-workers have subsequently shown that the length of the incubation period depends on many factors such as dose and route of inoculation, but particularly on the genotype of mouse with respect to the *sinc* gene and on the strain of the agent (see 14.4).

Slow virus diseases of animals and man, edited by R. H. Kimberlin
© *North-Holland Publishing Company 1976*

11.2 Experimental source of agent

The agent occurs in a wide variety of tissues in mice. Substantial titres are found in lymphoid tissues such as spleen quite early in the incubation period but, as the disease develops, the titre in the CNS increases and becomes maximal when the mice exhibit clinical signs; the concentration of agent being at least 10-fold higher than that found in spleen (Chandler, 1963; Hunter and Millson, 1964a; Eklund et al., 1967). For this reason most workers have chosen to use infected brain in studies on the purification, intracellular location and physico-chemical properties of the agent. In some instances, use has been made of scrapie-affected spleen and also of a cell line isolated originally from affected mouse brain and which is known to support agent replication in vitro (Clarke and Haig, 1970). There is as yet no in vitro test for the detection of the scrapie agent: no specific immunological response to infection has been observed, agents have not been recognised by electron microscopy, and cells have not so far been infected in vitro. Consequently, workers have been obliged to rely solely on biological methods for determining the presence and amount of agent.

11.3 Methods of assaying infectivity

Most studies of the nature of the scrapie agent have been carried out using the Chandler strain of agent in mice except where otherwise stated (10.6). In our laboratory we have studied the Chandler agent in two strains of mice (Compton Whites and LAC/G). Titrations of infectivity are carried out by injecting serial 10-fold dilutions of the material to be assayed into groups of 6–10 mice. Each mouse receives 0.025–0.05 ml of inoculum by the intracerebral route which gives the highest operational estimates of titre and the shortest incubation time (14.4.4). Following an intracerebral inoculation of a 10^{-3} dilution of whole brain from an animal in the terminal stage of scrapie, the earliest clinical signs of disease are usually seen after about 125 days. The incubation periods for higher dilutions do not often extend beyond 210 days and experiments are normally terminated after 250 days. In most experiments diagnosis of scrapie is based on the appearance of clinical signs of the disease. However, with mice where development of the disease appears to be atypical, histological examination of brains is made to establish the diagnosis. Titres are normally expressed as the dose required to produce 50% cases ($-\log_{10}ID_{50}$). The titre of scrapie agent in the brain of clinically affected mice is usually between $10^{7.0}$ and $10^{7.5}$ per 0.025 g wet wt: equivalent to about 5.0×10^8 infective units per g of brain.

Contact transfer of the agent from affected to healthy animals can take place as a result of fighting between litter mates when scrapie may be spread to healthy mice by ingestion of tissue fragments containing the agent (Pattison, 1964; Morris et al., 1965; Zlotnik, 1968). In our experience this is more likely to occur with males than when female mice are used. Scrapie-affected male mice tend to be more aggressive towards their litter mates when the first clinical signs of scrapie become evident; however, this aggression is only observed in some strains of mice.

Our experience with Chandler scrapie after many passages in mice, indicates that the titre of an inoculum can be calculated 210 days after inoculation: cases of scrapie that develop after this time rarely affect the final titre by more than 0.15–0.2 $\log_{10}$ units. This method of assaying scrapie activity is very precise and replicate titrations of a single inoculum usually vary by no more than $\pm 0.15 - \log_{10}$ units.

It has already been emphasised that bioassay in mice is the only available method for estimating the amount of infective scrapie agent. Hence one has to assume that the operational estimates of titre are directly proportional to the number of infectious units present. There is increasing evidence that this assumption may not be valid in all cases (see 14.3.3 for further discussion of this point). In particular it is now known that a number of pharmacological agents such as prednisone acetate and arachis oil can considerably modify the pathogenesis of scrapie in mice when administered close to the time of infection by an intraperitoneal route (14.4.4l; Outram et al., 1974, 1975). It will be apparent from later sections (11.5) that scrapie infectivity is closely associated with membranes and that only a very limited degree of agent purification has been achieved. Consequently it is not known whether the application of a chemical treatment to preparations containing scrapie activity is having a direct effect on the structure of the agent per se, which may change the number of infectious units, or on the associated membrane components which may affect the pathogenesis of the disease and the operational estimate of titre in the host. Most of the drugs which are known to affect the pathogenesis of scrapie have been studied using an intraperitoneal route of infection; very little is known about modified pathogenesis after an intracerebral route which is the most commonly used in titration studies. However, there is one published report which suggests that boiling the ME7 strain of scrapie not only reduces infectivity but also alters the pathogenesis of the disease in the inoculated animals after intracerebral infection (Dickinson and Fraser, 1969).

Considerable caution must therefore be excercised when interpreting titrations of scrapie infectivity. Studies of the intracellular location of scrapie activity (11.4) may be easier to interpret if the sum of the scrapie titres in individual subcellular fractions equals the titre of the unfractionated material. However, studies of the physico-chemical properties of scrapie aimed at defining the conditions under which infectivity is lost, are harder to interpret. In studies of this kind it is particularly important to incorporate a full range of experimental controls, and to investigate the possibility of modified pathogenesis by examining incubation periods in relation to operational titres and by comparing relative titres obtained by more than one route of infection.

11.4 Intracellular location

Early studies on the isolation of the agent from infected mouse brain established that over 99% of the scrapie infectivity is found in association with membranous components of the cell (Hunter and Millson, 1967; Mould et al., 1965). Brain contains many different cell types and thus subcellular fractionation of this tissue presents certain technical problems. In addition to the morphologically discrete organelles such as the nuclei, mitochondria, ribosomes and lysosomes, there are membrane fragments derived from endoplasmic reticulum and plasma membrane; also specialised forms of plasma membrane such as the myelin sheath and synaptosomal bodies formed from 'pinched off' nerve endings.

Separation of these fractions by differential and density gradient centrifugation in sucrose revealed that the agent was associated with all cell components. However, as nuclei, mitochondria, myelin and to some extent lysosomes were increasingly purified, there was a corresponding loss of scrapie activity which indicated that activity associated with these organelles probably resulted from contamination with other membrane fragments. Resolution of plasma membrane, endoplasmic reticulum and synaptosomes was not sufficiently precise to permit the assignation of scrapie activity to any one specific fraction (Hunter et al., 1964; Millson, et al., 1971). The subcellular fractionation of scrapie-affected spleens yielded similar results and it was still not possible to achieve an adequate separation of the plasma membrane and endoplasmic reticulum, as indicated by the use of enzyme and chemical markers (Millson et al., 1971).

More definitive results were obtained using a cell line which originated from scrapie-affected mouse brain. This line has been maintained in culture for many years (Clarke and Haig, 1970) and the cells support agent replication in vitro. A crude preparation of microsomes from these cells yielded two main fractions on a dextran discontinuous gradient. One fraction was essentially plasma membrane and the other endoplasmic reticulum which contained approximately 10% of the plasma membrane; 5'-nucleotidase and NADPH–cytochrome c reductase being used as enzyme markers of the plasma membrane and endoplasmic reticulum, respectively. There was a good correlation between the distribution of 5'-nucleotidase and scrapie activity suggesting that in the cell line at least, the bulk of the scrapie agent appears to be associated with the plasma membrane (Millson and Clarke, unpublished). Since this cell line was derived from scrapie-affected brain, it seems possible that the infective scrapie agent may also be mainly associated with plasma membrane in vivo.

11.5 Attempts to purify the scrapie agent

Early purification attempts were based on the assumption that the agent was a virus which could be isolated and concentrated from tissue components by virtue of its size, density, surface charge and specific affinity for certain macromolecules. Scrapie-infected brain was disrupted and centrifuged at low speeds to sediment essentially nuclei, unbroken cells and tissue fragments. The supernatant suspension containing the partially purified agent was then subjected to various treatments designed to concentrate and purify the agent, on the assumption that infectivity was associated with discrete particles with a virus-like morphology.

11.5.1 Centrifugation studies

Gibbs (1967) centrifuged extracts of scrapie brain in caesium chloride and observed that the highest concentration of activity was in the density region of about 1.32 g/ml. In contrast, Mould et al. (1965) using mouse brain affected with the ME7 strain of scrapie and a similar gradient system found that scrapie activity was distributed throughout the gradient with a partial concentration of activity in the density region 1.10–1.22 g/ml. Ultrasonicated preparations yielded essentially the same results with a slightly increased concentration of activity in the less-dense fractions. However these authors found that exposure of the agent to caesium chloride at this concentration resulted in an overall loss of activity of nearly 99%. They further showed that most scrapie activity was in the 1.0 M region after centrifugation in a preformed sucrose density gradient: a finding confirmed by Millson (unpublished).

Kimberlin et al. (1971) showed that 20% of the scrapie activity in washed, sonicated scrapie membranes from which nuclei and myelin had been removed passed through a 50 nm membrane filter. Ninety-five percent of the infectivity in these filtrates banded at a density of between 1.12 and 1.17 g/ml in tartrate in association with a single membrane band: there was no evidence that the agent was dissociated from membrane fragments.

Adams et al. (1969) considered that the residual scrapie activity left in the soluble fraction of a brain homogenate after repeated centrifugation at 144,000 *g* for times varying from 75 to 120 min may represent unbound agent and as such may be more suitable for determining the density of agent. This 'soluble' activity was sedimented by centrifugation at 144,000 *g* for 16 h and the pelleted material was banded on caesium chloride gradients to give a sharp band possessing some infectivity at a density of 1.34 g/ml. Analytical ultracentrifugation of this material showed that particles had an *S* value of about 5.0. The scrapie activity of this fraction was very low and the authors point out that loss of infectivity in caesium chloride may reflect dissociation of the agent from some other components which are necessary for the expression of full scrapie infectivity; this point will be discussed at greater length later.

Pattison et al. (1969) demonstrated that the scrapie activity of a freeze-dried powder obtained from repeated water washings of defatted scrapie brain possessed some scrapie activity that was not sedimented after centrifugation at 100,000 *g* for 1 h, but was completely sedimented if the time of centrifugation was increased to 15 h. These authors suggested that part of the scrapie activity in the freeze-dried powder is associated with a very small particle.

Separation of the agent from tissue membranes by centrifugation has not been encouraging. The variable densities reported by different groups of workers may reflect the heterogeneity of membrane fragments with which the agent is associated, but doubtless other effects such as vesicle formation and the binding of heavy ions may contribute to the variation.

11.5.2 Chromatography and electrophoresis

Attempts have been made to purify scrapie infective particles by chromatography on calcium phosphate (Mould et al., 1964; Hunter and Millson, 1964a) on DEAE–cellulose (Hunter and Millson, 1964b) and by electrophoresis on Bio-Gel columns (Hunter et al., 1971). In none of these studies was it possible to obtain a scrapie-enriched fraction and in many cases the recovery of activity from the columns was low. Ultrasonicated fractions gave higher recoveries but in all instances it was concluded that no significant purification of the agent had been achieved.

11.5.3 Effect of enzymes on infectivity

Scrapie membrane preparations have been exposed to the action of a

TABLE 11.1

The effect of enzymes on scrapie infectivity.

Enzyme	Loss of infectivity		Reference
	$(-\log_{10}\mathrm{ID}_{50})$	(%)	
Nucleases:			
DNAase	0.0–0.1	0–13	2
RNAase	0.0–0.1	0–13	2
Carbohydrases:			
Glycosidase extract of *Trichomonas foetus*	0.2	16	5
Hyaluronidase	0.3	20	5
Neuraminidase	0.2	16	3
Lipases:			
Pancreatic lipase	0.0	0	3
Phospholipase A + 0.1% Tween 80	+0.2	+16	3
Phospholipase A + 0.1% Tween 80 + 1 M NaCl	0.2	16	3
Phospholipase C	0.0	0	3
Proteases:			
Ficin	0.7	50	1
Papain	1.5	93	3
Pronase	0.8	63	5
Pronase (on acetone-extracted brain)	1.5–4.0	93–>99	5
Trypsin	0.0–1.0	0–90	1, 2, 3
Trypsin (on fluorocarbon-treated brain)	0.8	63	4

(1) Haig and Clarke, 1965; (2) Hunter, unpublished; (3) Hunter and Millson, 1967; (4) Hunter et al., 1969; (5) Millson, unpublished.

number of hydrolytic enzymes; firstly, with the aim of obtaining information on the nature of the agent, and secondly, as a possible means of releasing the agent from membrane fragments. In these studies DNAase, RNAase, phospholipase A and C, neuraminidase, hyaluronidase and glycosidases had little effect on scrapie activity whilst proteolytic enzymes in some instances inactivated the agent by more than 90% (Table 11.1). In one experiment where an acetone powder from scrapie mouse brain was treated with pronase a loss of over 99% of scrapie activity was observed. Whilst there was no significant release of agent from membrane fragments by the use of these enzymes, the effect of the proteolytic enzymes on the agent would suggest that a protein component of infected membranes is necessary for the expression of full scrapie infectivity.

11.5.4 Chemical treatments

Hunter (1965) drew attention to a group of small DNA viruses which have high affinity for tissue components. Procedures used to dissociate these agents were applied to scrapie-infected brain (Hunter and Millson, 1967). Fluorocarbon treatment (arcton-113) at 0–4 °C has been used to remove a large proportion of the lipid and protein from scrapie mouse brain homogenates (Haig and Clarke, 1965; Hunter, 1965). In some instances almost all scrapie activity can be sedimented from the aqueous phase after arcton extraction by centrifuging at 100,000 g for 1 h to give a 10-fold concentration of the agent with respect to protein. Unfortunately, results are variable and in some preparations almost 90% of the infectivity is lost. Treatment of arcton material with 1 M NaCl or 0.05% Triton X-100 or pH 3 buffer failed to release the agent from the readily sedimentable fraction. However a combination of 1 M NaCl + 0.05% Triton X-100 appeared to achieve some dissociation. A similar effect was observed when heated brain (80 °C for 30 min) was extracted with 6 M lithium chloride or 0.5% sodium deoxycholate, but the authors considered that these treatments merely reduced the size of the membrane fragments to which the infective agent was bound.

Several years ago, it seemed to the authors that since it was apparently very difficult to dissociate the infective scrapie agent from its tenacious association with membranes, it might be possible to remove selectively membrane components without affecting the integrity (and hence the activity) of the agent. A number of mild solvents and detergents were used but it was found that they either caused some loss of infectivity or (with the exception of lysolecithin) dissociation of membrane components was limited.

Lysolecithin solubilises a large proportion of membrane proteins and phospholipids from brain tissue (Webster, 1956). Treatment of scrapie mouse brain with this biological surfactant releases approximately 90% of the membrane protein and 80% of the phospholipids into the 100,000 g (1 h) supernatant, whilst the full scrapie activity of the preparation is found in the sediment. It is interesting to note that lysolecithin almost quantitatively precipitates nuclear DNA from whole brain homogenates (Millson and Hunter, unpublished). Removal of the nuclear and myelin fractions from the brain homogenate prior to the lysolecithin treatment renders 98% of the protein and 90% of the phospholipids soluble but leaves full scrapie infectivity still in a sedimentable form. This method can give an almost 100-fold concentration of the agent and preparations containing about 10^{10} infective units per g wet wt of sedimented material. Recently it has been reported that the concentration of agent in the brains of scrapie-affected hamsters in the clinical stage of the disease can be nearly 2 logs higher than that commonly found in clinically affected mice (Marsh and Kimberlin, 1975). Preliminary experiments suggest that lysolecithin treatment of scrapie hamster brain can yield preparations containing 10^{12} infective units per g wet wt.

11.6 Physico-chemical properties

Various physical and chemical methods have been used in studies of the nature of the scrapie agent. As it has not yet been possible to purify the agent these treatments have all been carried out on relatively crude material.

11.6.1 Thermal resistance

The relative resistance of the scrapie agent to heat has been known for many years. Early work in sheep (Stamp, et al., 1959) and goats (Pattison and Millson, 1961) showed that scrapie brain preparations were still infective after heating for 30 min at 100 °C. However, the extent of inactivation was not determined until titration studies became possible in mice.

The heating curve for scrapie agent in mouse brain (Fig. 11.1; Hunter and Millson, 1964b) shows that substantial inactivation is observed near the melting temperature of DNA, a finding confirmed by Mould and Dawson (1970a). Between 3.5 and 4 logs of infectivity are lost by heating at 100 °C for 30 min (Gibbs et al., 1965; Gibbs, 1967; Zlotnik and Rennie, 1967) and trace amounts of activity survive heating at higher temperatures

G. C. Millson et al.

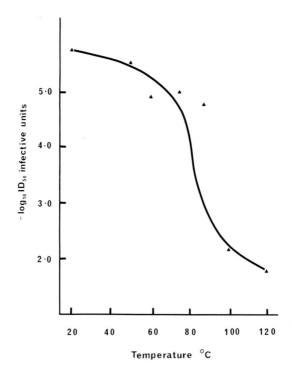

Fig. 11.1 The effect of heat on the scrapie infectivity of affected mouse brain. Samples were exposed to various temperatures for a period of 10 min, and infectivity determined by titration in mice using the intracerebral route of inoculation. For further details see Hunter and Millson (1964b).

(Hunter and Millson, 1964b). Eklund et al. (1963) reported complete in-activation of the agent by heating for 30 min at 100 °C, but in the presence of infusion broth saline some activity survived this temperature. These results indicate that some protection may occur when the agent is heated in crude suspension. Adams (1970) has emphasised that heating of crude brain homogenates at 100 °C results in severe coagulation and the degree of resuspension of heated material is a point to be considered when evalu-ating infective titres.

11.6.2 Effect of hydrogen ion concentration

Both the Chandler and the ME7 strains of scrapie agent are stable over a pH range from 2.1 to 10.5 when exposed for 24 h (Haig and Clarke, 1965;

Mould et al., 1965). However, exposure to 5 N sodium hydroxide results in a dramatic loss of infectivity (Clarke, personal communication).

11.6.3 Effect of alkylating agents

The 'louping-ill vaccine' event (10.4.1) showed that scrapie infectivity was not completely removed by exposure to 0.35% formalin for 3 months (Gordon, 1946). Stamp (1959) later reported that some scrapie infectivity in sheep brain could survive concentrations of up to 3% formalin at 37 °C for 13 days. Studies in goats indicated that treatment of scrapie brain with formalin concentrations of 0.25–20% for 18 h at 37 °C, or 10% formalin or 12% neutral-buffered formalin for periods of 6–28 months produced relatively little loss of infectivity (Pattison, 1965). A similar resistance was observed in mice but no titration studies were made so that it is difficult to determine the degree of inactivation resulting from formalin treatment. However, incubation periods of the disease in mice were much longer when the inoculum consisted of formalin-treated material than when untreated scrapie brain was used: the inactivation of scrapie was probably substantial in some experiments.

Stamp et al. (1959) showed that some scrapie activity was resistant to acetylethyleneimine and concluded that the agent might not contain nucleoprotein. A similar view was expressed by Haig and Clarke (1968) who found that whereas 0.2% β-propiolactone completely inactivated yellow fever virus present in brain tissue, scrapie activity was only reduced by 0.6 $-\log_{10}\mathrm{ID}_{50}$ units. Increasing the concentration to 1% resulted in losses of over 1.0 $-\log_{10}\mathrm{ID}_{50}$ units. Essentially the same degree of inactivation was observed when fluorocarbon-defatted scrapie brain preparations were treated with 1% β-propiolactone.

11.6.4 Effect of organic solvents, salt solutions and detergents

The effects of a wide range of organic solvents, salt solutions and detergents have been observed in studies aimed at dissociating the agent from its membrane location (see Tables 11.2, 3 and 4). Treatments with fluorocarbon, potassium tartrate or sodium cholate had minimal effects on scrapie activity whilst 80% 2-chloroethanol, guanidinium hydrogen bromide and sodium dodecyl sulphate produced a more marked inactivation of the agent. In general, the loss of infectivity is related to the ability of reagents to disrupt membrane structure, with the exception of periodate where the conditions

TABLE 11.2

The effect of organic solvents on scrapie infectivity.

Solvent	Loss of infectivity		Reference
	$(-\log_{10}ID_{50})$	(%)	
Acetone	2.5	>99	9
Acetone and ether	1.8	96	4
Alcoholic iodine	3.0	>99	1
n-Butanol	0.2–1.0	16–91	7, 9
2-Chloroethanol (50%)	1.9	98	7
2-Chloroethanol (80%)	5.0	>99	7
Chloroform	0.7	50	8
Chloroform–butanol	1.7	95	5
Chloroform–methanol (2:1)	1.1–1.6	91–94	7, 9
Chloroform–methanol (2:1; 75°C, 15 min)	2.1	>99	7
Ether	1.0–2.2	90–99	2, 8, 9
Ether (on fluorocarbon-treated brain)	1.6	94	8
Fluorocarbon:			
Arcton 113 1st extraction	0.4	25	5
2nd extraction	0.8	63	5
1st extraction	1.3	92	3
2nd extraction	3.0	>99	3
3rd extraction	3.7	>99	3
on heated brain	2.4	>99	6
Freon 113	0.6	40	8
Methoxyethanol (80%)	1.4	93	7
n-Pentanol	0.7	50	7
Phenol (90%)	4.7	>99	6

(1) Bell et al., 1972; (2) Eklund et al., 1963; (3) Haig and Clarke, 1965; (4) Hunter and Millson, 1964a; (5) Hunter and Millson, 1967; (6) Hunter et al., 1969; (7) Hunter et al., 1971; (8) Lavelle, 1972; (9) Mould et al., 1965.

used were relatively specific for polysaccharides and would not be expected to affect the basic structural integrity of the membrane (Hunter et al., 1969). Whilst Adams et al. (1972) found that scrapie activity was sensitive to a number of oxidising agents, they found that the effect of periodate was less than that observed by Hunter et al. (1969).

TABLE 11.3

The effect of salt solutions on scrapie infectivity.

Reagent	Molar concentration	Loss of infectivity ($-\log_{10}ID_{50}$)	(%)	Reference
Caesium chloride	2.5	1.1–1.8	91–96	2, 8
Guanidinium hydrogen bromide	6.0	5.0	>99	7
Lithium chloride	6.0	0.5–1.5	32–93	4
Lithium chloride	12.0	4.0	>99	3
Lithium diodosalisylate	0.3	1.0	90	7
Potassium permanganate	0.002	2.4	>99	1
Potassium tartrate	1.1–1.4	0.8	63	6
Sodium chloride	1.0	0.9	79	4
Sodium iodate, pH 3.5	0.01	1.1–1.3	91–92	5
Sodium periodate, pH 3.5	0.01	1.5	93	1
Sodium periodate, pH 3.5	0.01	3.6	>99	5
Urea	6.0	2.0	99	3

(1) Adams et al., 1972; (2) Gibbs, 1967; (3) Hunter, unpublished; (4) Hunter and Millson, 1967; (5) Hunter et al., 1969; (6) Kimberlin et al., 1971; (7) Millson, unpublished; (8) Mould et al., 1965.

TABLE 11.4

The effect of detergents on scrapie infectivity.

Detergent	Concentration (%)	Loss of infectivity ($-\log_{10}ID_{50}$)	(%)	Reference
Anionic:				
Lysolecithin, 1st extraction	0.8	0.0–0.1	0–13	4
Lysolecithin, 2nd extraction	0.8	0.1–0.3	13–20	4
Sodium cholate	0.5	0.1	13	1
Sodium deoxycholate	0.5	0.4	25	3
SDS, 1st extraction	0.5	1.2	92	4
SDS, 2nd extraction		2.0	99	4
SDS	5.0	3.0	>99	6
Sodium lauryl sarcosinate	0.5	1.5	93	5
Cationic:				
Cetyl-trimethylammonium bromide	0.5	0.4	25	6
Non-ionic:				
Triton X-100	0.05	1.5	93	3
Tween 80	0.01	0.0	0	2

(1) Collis, unpublished; (2) Hunter and Millson, 1964b; (3) Hunter and Millson, 1967; (4) Hunter et al., 1971; (5) Kimberlin, unpublished; (6) Millson, unpublished.

11.6.5 Determination of operational size

Wilson et al. (1950) were able to show that infectivity from scrapie-affected sheep brain passed a 47 nm but not a 27 nm filter; an observation which was confirmed by Eklund et al. (1963) and Gibbs (1967) using scrapie mouse brain preparations. However, these studies only give an upper limit to the size of the infective agent. An attempt to determine the minimum size (operational size) of the agent was made by Kimberlin et al. (1971) who treated washed microsomal membranes from scrapie-affected mouse brain with ultrasonic vibrations to produce very small membrane fragments. An average of 20% of the infectivity in the sonicates passed through a 50 nm filter and 95% of the activity in the filtrates was still associated with membrane material after banding on tartrate gradients. No detectable activity was recovered from the 20–35 nm filtrates, although a significant amount of lipoprotein material was present. Electron microscopic studies showed the presence of particulate material in the 50 nm filtrate but the material in the 20–35 nm filtrate was amorphous. These results suggest that the limiting size of particles with scrapie activity is approximately 30 nm, corresponding to the thickness of a typical mammalian membrane. Exclusion chromatography of ultrasonicated and filtered membranes on Bio-Gel A50 indicated that the operational size of the agent is at least 5×10^7 daltons.

Pattison and Sansom (1964) and Pattison and Jones (1967) reported that very small amounts of infectivity from goat and mouse brain sources could pass through Visking tubing of 2–4 nm APD under certain conditions. These claims should be set against the problem of establishing absolute integrity of the Visking tubing and also the possibility of contamination when such minute amounts of agent are being examined (see 10.4.2). It should be emphasised that rigorous attempts to verify these claims using both goat and mouse materials failed to detect any dialysable scrapie infectivity (Mould and Dawson, 1968, 1970b).

11.6.6 Irradiation studies

The dose of ionising radiation required for inactivation is far greater for scrapie than for conventional viruses (Alper et al., 1966; Alper and Haig, 1968; Field et al., 1969) and the target size of the agent has been calculated to be approximately 1.5×10^5 daltons. The agent also requires exceptionally high doses of UV-irradiation for inactivation at 254 and 265 nm (Alper et al., 1966, 1967; Haig et al., 1969; Gadjusek and Gibbs, 1968); the D_o

value at 254 nm is 2.4 × 10^5 ergs/mm^2. Moreover, the agent displays an inactivation spectrum which is uncharacteristic of viruses when exposed to monochromatic light at 237, 250 and 280 nm (Latarjet et al., 1970).

The results of the irradiation studies and the relatively high resistance of the scrapie agent to many physical and chemical treatments, particularly to treatments with acetylethyleneimine and β-propiolactone which rapidly inactivate most viruses, has led many workers to consider that the scrapie agent may not contain nucleic acid. This suggestion has stimulated a number of alternative proposals which attempt to explain the coding and replication of scrapie-specific information in terms of polysaccharides (Field, 1966), proteins (Griffith, 1967), basic proteins (Pattison and Jones, 1967) and membrane glycoproteins (Gibbons and Hunter, 1967). Other proposals include the related membrane and the linkage-substance hypotheses (Gibbons and Hunter, 1967; Adams and Field, 1968).

However, since scrapie agents have some virus-like properties in that biologically different strains exist which are capable of replication, it seems more reasonable to seek explanations for the unusual properties of scrapie agents in terms of nucleic acid before considering completely novel alternatives. The most important evidence against the existence of scrapie nucleic acids comes from the UV-irradiation studies which have raised two issues: the high dose required to inactivate scrapie at 254 nm and the unusual inactivation spectrum which shows a greater sensitivity of scrapie to irradiation at 230 nm than at 254 or 280 nm.

The calculated ionising radiation target of scrapie agent is 150,000 daltons, i.e., not much larger than the purified infectious RNA of potato spindle tuber viroid (PSTV) or of satellite tobacco ringspot virus (SAT) which have molecular weights of approximately 80,000. Both purified PSTV and SAT are more resistant to UV-irradiation than nucleic acids of higher molecular weight, with D_o values at 254 nm of approximately 4 × 10^4 ergs/mm^2 (Diener et al., 1974). This figure is within a factor of 10 of the D_o value for scrapie irradiated in crude suspension at the same wavelength. However, Diener (1973) has reported that PSTV irradiated in crude tissue extracts is at least as resistant to UV light as scrapie.

Siegel and Wildman (1954) have shown that the U1 and U2 strains of tobacco mosaic virus (TMV) differ in their degree of resistance to UV-irradiation at 254 nm while the RNA core showed no difference (Siegel et al., 1956). The U2 strain and the purified RNA from both U1 and U2 strains exhibited the expected inactivation spectrum with greater sensitivity to UV at 260 nm than at 230 or 280 nm. However, the intact virion of the U1 strain was found to more readily inactivated at 230 nm than at 254 and 280 nm (Kleczkowski and McLaren, 1967). Some differences in sensitivity have thus been shown to reside in the protein coat of these strains (Streeter and Gordon, 1968). However the atypical response of U1 strain to UV-irradiation probably depends on a particular structural relationship between nucleic acid and protein, which enables energy absorbed by the nucleic acid to be transferred to the protein (Kleczkowski and McLaren, 1967).

Although scrapie is inactivated far less readily by UV irradiation than the U1 strain of TMV, the intimate association of infectivity with the structure of some cell membranes may provide an analogous situation. The exceptional resistance of the agent to UV-irradiation and the unusual inactivation spectrum may be explicable in terms of an agent of small target size closely associated with a membrane protein.

11.7 Possible nature of the scrapie agent

There is a discrepancy in the estimated size of the scrapie agent using different methods. On the one hand the radiation target size is approximately 1.5×10^5 daltons; on the other, the operational size in excess of 5×10^7. A reasonable interpretation of this discrepancy is that scrapie activity is mediated by at least a two-component system (Hunter, 1972); one component being relatively small (the ionising radiation target) and possessing scrapie-specific information, and another, much larger component which may be some part of the host cell membrane and which has an unknown function in expressing the biological infectivity of the scrapie-specific component. If the small component is considered to be a nucleic acid – similar in size to viroid RNA – which is intimately associated with the host cell membrane, many of the unusual features of the agent can be understood more easily, including the high degree of stability of scrapie activity to many physico-chemical treatments (11.6).

It is clear that the possibilities for the purification of such an agent are strictly limited (as indeed seems to be the case), and that unlike orthodox viruses the scrapie agent may not exist as a discrete infective particle with a characteristic morphological structure. This may explain why many electron microscopical studies have failed to identify scrapie virus particles, even when partially purified preparations from scrapie-infected hamster brain, known to contain approximately 10^{12} infective units per gram, (11.5.4) have been examined (Millson et al., unpublished). The apparent lack of a specific immunological response in susceptible animals after experimental infection with the agent may reflect the inability of the host to recognise a scrapie-specific molecule amidst a large complex of normal donor membrane components. Evidence that the tissue source and the strain of mouse from which agent is obtained can considerably affect pathogenesis is understandable in terms of the different membranes with which the agent is associated (Dickinson and Outram, 1973). Furthermore, the controlled nature of agent multiplication which in the cell line appears to be linked to cell division (Clarke and Haig, 1970), and in the mouse may be limited by the number of available replication sites (Dickinson et al., 1972; see 14.5) can also be understood in terms of the two-component system.

11.8 Search for infectious scrapie nucleic acid

Original attempts to isolate infectious nucleic acids from scrapie-affected goat brain were unsuccessful (Hunter, unpublished). Later extractions of affected mouse brain showed that scrapie activity was inactivated by the phenol used in the procedures for isolating both DNA and RNA (Hunter et al., 1969). In response to Diener's suggestion that scrapie may be an animal viroid (Diener, 1972), two groups of workers independently re-examined the possibility of isolating an infectious nucleic acid from scrapie mouse brain. Marsh et al. (1974) employed procedures which have been successfully used for the extraction and concentration of low molecular weight pathogenic RNA from citrus exocortis viroid. They monitored the ability of their methods to extract biologically active nucleic acid by incorporating purified encephalomyocarditis (EMC) virus into the original scrapie homogenate. Under these conditions the recovery of infectious EMC RNA was 7% of that obtained when purified EMC virus was extracted alone. However, no scrapie activity was found in the nucleic acid extract. A similar study by Ward et al. (1974) used mengovirus as a control for the isolation of infectious nucleic acid also confirmed these observations. The lack of success in isolating a pathogenic nucleic acid from scrapie brain preparations led both groups of authors to conclude that scrapie is unlikely to be a viroid. At the same time these findings suggest that a putative scrapie nucleic acid may have little or no infectivity when tested in mice.

11.9 Problems in identifying a scrapie-specific nucleic acid

Radioactively labelled nucleic acid precursors have been used in in vivo studies designed to identify a scrapie-specific nucleic acid. This technique is limited, particularly in respect of the rapid dilution of the injected precursors, the small amount of agent and its apparent slow replication rate (14.3). Nevertheless, evidence of a small labelled DNA fraction in scrapie brain has been reported (Adams, 1972; Hunter et al., 1973). However, these findings must be seen against the background of an increased turnover of nuclear DNA in scrapie mouse brain (Kimberlin et al., 1974) and other biochemical abnormalities which are found in affected brain at the clinical and preclinical stages of the disease (see 13.2 and 3). It is therefore necessary to demonstrate firstly that a putative scrapie nucleic acid is present in tissues such as spleen which, in the preclinical stage of infection, contain high titres of agent but have few biochemical abnormalities (13.3); and secondly, that it is essential for the biological activity of scrapie agent. It has been mentioned earlier (11.8) that a scrapie-specific nucleic acid is unlikely to be biologically active when assayed in the mouse, and therefore a system must be developed whereby it can be reconstituted with membrane components to form the fully biologically active agent.

A number of preliminary, unpublished studies have been carried out to discover ways of disaggregating membranes with scrapie activity and producing a loss of infectivity, and then recombining membrane components to regain biologically active agent. Figs 11.2–5 give some of the results obtained with urea, sodium dodecyl sulphate (SDS), cetyl-trimethylammonium bromide (CTAB) and lithium chloride.

Firstly an increase in titre was observed when urea-treated scrapie membranes were dialysed against either saline or 6 M urea. Membranes which were held for the same period of time in 6 M urea did not show any increase in titre (Fig. 11.2). This finding suggests the removal of 'inhibitory' substances during dialysis. Secondly, when SDS was employed to dissociate scrapie membranes a loss of titre was observed, but recombination of the solubilised membrane components by dialysis against 10 mM Mg^{2+} resulted in a further loss of titre under conditions where approximately 80% of the solubilised membrane protein recombined into ultrastructurally recognisable membranes (Fig. 11.3). Although there are several explanations for this finding, one may be the requirement of the scrapie-specific molecule for binding with some specific membrane components which is blocked by non-specific binding during dialysis against Mg^{2+}. Dialysis against buffer without Mg^{2+} gave very little recombination of membrane components but scrapie activity was consistently higher under these conditions when compared with undialysed samples held in SDS for the same period of

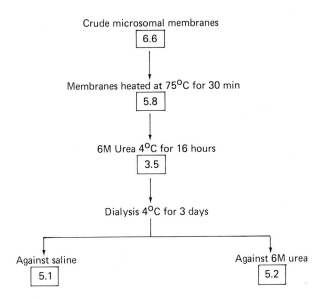

Titres expressed as $-\log_{10} ID_{50}$ units per 0.03 ml inoculated intracerebrally into Compton white mice.

Fig. 11.2. Effect of urea and of subsequent dialysis on the infectivity of membranes isolated from scrapie mouse brain.

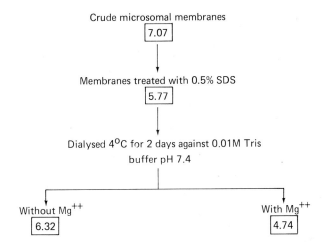

Titres expressed as $-\log_{10} ID_{50}$ units per 0.03 ml inoculated intracerebrally into Compton white mice

Fig. 11.3. Effect of sodium dodecyl sulphate (SDS) and of subsequent dialysis on the infectivity of membranes isolated from scrapie mouse brain.

G. C. Millson et al.

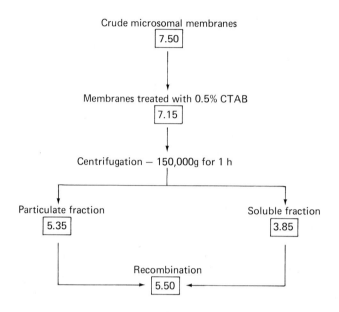

Crude microsomal membranes

| 7.50 |

Membranes treated with 0.5% CTAB

| 7.15 |

Centrifugation — 150,000g for 1 h

Particulate fraction

| 5.35 |

Soluble fraction

| 3.85 |

Recombination

| 5.50 |

Titres expressed as − $\log_{10}ID_{50}$ units per 0.03 ml inoculated
intracerebrally into Compton white mice.

Fig. 11.4. The effect of cetyl-trimethylammonium bromide (CTAB) on the infectivity of membranes isolated from scrapie mouse brain.

time. These results may mean that SDS causes some dissociation of components of infective agent which can be partially reversed by removing SDS.

Thirdly, dissociation of scrapie membranes by the use of CTAB resulted in very little loss of scrapie infectivity (Fig. 11.4). An overall loss of infectivity was observed when the particulate residue and soluble fractions were separated by high-speed centrifugation. Thus, it would appear that components essential for infectivity are present in both fractions. Attempts to recover full scrapie activity by simply recombining the particulate and soluble fractions and by the use of various dialysis procedures have so far been unsuccessful. Finally, treatment of scrapie membranes with 6 M lithium chloride results in a loss of 90% or more of the infectivity. Subsequent dialysis against either 6 M LiCl or saline always produces a slight recovery of titre. High-speed centrifugation of the treated but undialysed samples showed that most infectivity was sedimentable (150,000 *g* for 1 h). However, in contrast to the CTAB experiments, when the soluble fraction is added back to the sediment there is some recovery of titre, particularly

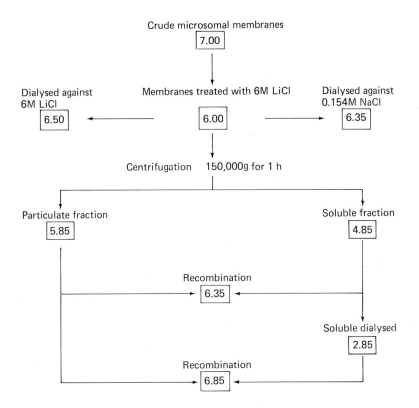

Titres expressed as − $\log_{10} ID_{50}$ units per 0.03 ml inoculated intracerebrally into Compton white mice.

Fig. 11.5. Effect of 6 M lithium chloride on scrapie activity of infected membranes isolated from scrapie mouse brain.

if the soluble fraction is dialysed prior to recombination with the particulate fraction (Fig. 11.5).

However, all of these experiments are subject to the problem of interpreting infectivity titres described earlier (11.3). In the present context, it is not known whether the loss and regain of infectivity is due to a disaggregation and recombination of components essential for infectivity or, alternatively, to a release and/or chemical modification of membrane constituents which alter the pathogenesis of disease in the assay animals and hence the operational titres of infectivity. To distinguish between these two possibilities it is necessary to isolate chemically at least one component and show that the regain of infectivity is critically dependent upon its incorporation into the biologically active complex.

An alternative approach is based on the assumption that scrapie activity is largely dependent on the combination of a scrapie-specific nucleic acid with some normal membrane constituents; i.e., that scrapie activity can be obtained by combining components which in isolation are inactive. Such a demonstration would meet the criteria necessary to assay the putative scrapie nucleic acid.

A number of attempts have been made to combine nucleic acids extracted from scrapie-affected mouse brain with normal membranes, based on the results shown in Figs 11.2–5. So far these attempts have been unsuccessful. There may be many reasons for this failure. One basic reason is perhaps the difficulty of extracting the scrapie-specific nucleic acid; another could be the requirement for a fairly specific association between the putative scrapie nucleic acid and certain membrane constituents, an association which may be blocked by the presence of many competing molecules. Current work is particularly concerned with simplifying the system by testing the ability of lipid and individual protein fractions from normal membranes to combine with certain nucleic acid fractions from scrapie-infected tissues. At the present time, it seems likely that the resolution of the scrapie problem may depend upon rigorous and patient application of the techniques of membrane biochemistry.

References

ADAMS, D. H. (1970) Pathol.–Biol., 18, 559.

ADAMS, D. H. (1972) J. Neurochem., 19, 1869.

ADAMS, D. H., CASPARY, E. A. and FIELD, E. J. (1969) J. Gen. Virol., 4, 89.

ADAMS, D. H. and FIELD, E. J. (1968) Lancet, 714.

ADAMS, D. H., FIELD, E. J. and JOYCE, G. (1972) Res. Vet. Sci., 13, 195.

ALPER, T., CRAMP, W. A., HAIG, D. A. and CLARKE, M. C. (1967) Nature, 214, 764.

ALPER, T. and HAIG, D. A. (1968) J. Gen. Virol., 3, 157.

ALPER, T., HAIG, D. A. and CLARKE, M. C. (1966) Biochem. Biophys. Res. Commun., 22, 278.

BELL, T. M., FIELD, E. J. and JOYCE, G. (1972) Res. Vet. Sci., 13, 198.

CHANDLER, R. L. (1961) Lancet, i, 1378.

CHANDLER, R. L. (1963) Res. Vet. Sci., 4, 276.

CLARKE, M. C. and HAIG, D. A. (1970) Res. Vet. Sci., 11, 500.

DICKINSON, A. G. and FRASER, H. (1969) Nature, 222, 892.

DICKINSON, A. G., FRASER, H., MEIKLE, V. M. H. and OUTRAM, G. W. (1972) Nature, New Biol., 237, 244.

DICKINSON, A. G. and OUTRAM, G. W. (1973) J. Comp. Pathol., 83, 13.

DIENER, T. O. (1972) Nature New Biol., 235, 218.

DIENER, T. O. (1973) Ann. Clin. Res., 5, 268.

DIENER, T. O., SCHNEIDER, I. R. and SMITH, D. R. (1974) Virology, 57, 577.

EKLUND, C. M., HADLOW, W. J. and KENNEDY, R. C. (1963) Proc. Soc. Exp. Biol., 112, 974.

EKLUND, C. M., KENNEDY, R. C. and HADLOW, W. J. (1967) J. Infect. Dis., 117, 15.

FIELD, E. J. (1966) Br. Med. J., 2, 564.

FIELD, E. J., FARMER, F., CASPARY, E. A. and JOYCE, G. (1969) Nature, 222, 90.

GAJDUSEK, D. C. and GIBBS, J. C. (1968) Res. Publ. Assoc. Res. Nerv. Ment. Dis., 44, 254.

GIBBONS, R. A. and HUNTER, G. D. (1967) Nature, 215, 1041.

GIBBS, J. C. (1967) Curr. Topics Microbiol. Immunol., 40, 44.

GIBBS, J. C., GAJDUSEK, D. C. and MORRIS, J. A. (1965) In: (D. C. Gajdusek, C. J. Gibbs, Jr, and M. Alpers, Eds) Natl. Inst. Neurol. Dis. Blindness, Monogr. 2. Slow, Latent and Temperate Virus Infections. (U.S. Government Printing Office, Washington, D.C.) p.195.

GORDON, W. S. (1946) Vet. Rec., 58, 516.

GRIFFITH, J. S. (1967) Nature, 215, 1043.

HAIG, D. A. and CLARKE, M. C. (1965) In: (D. C. Gajdusek, C. J. Gibbs, Jr, and M. Alpers, Eds) Natl. Inst. Neurol. Dis. Blindness, Monogr. 2. Slow, Latent and Temperate Virus Infections. (U.S. Government Printing Office, Washington, D.C.) p. 215.

HAIG, D. A. and CLARKE, M. C. (1968) J. Gen. Virol., 3, 281.

HAIG, D. A., CLARKE, M. C., BLUM, E. and ALPER, T. (1969) J. Gen. Virol., 5, 455.

HUNTER, G. D. (1965) In: (D. C. Gajdusek, C. J. Gibbs, Jr, and M. Alpers, Eds) Natl. Inst. Neurol. Dis. Blindness, Monogr. 2, Slow, Latent and Temperate Virus Infections. (U.S. Government Printing Office, Washington, D.C.) p. 259.

HUNTER, G. D. (1972) J. Infect. Dis., 125, 427.

HUNTER, G. D., KIMBERLIN, R. H., COLLIS, S. C. and MILLSON, G. C. (1973) Ann. Clin. Res., 5, 262.

HUNTER, G. D., GIBBONS, R. A., KIMBERLIN, R. H. and MILLSON, G. C. (1969) J. Comp. Pathol., 79, 101.

HUNTER, G. D., KIMBERLIN, R. H., MILLSON, G. C. and GIBBONS, R. A. (1971) J. Comp. Pathol., 81, 23.

HUNTER, G. D. and MILLSON, G. C. (1964a) Res. Vet. Sci., 5, 149.

HUNTER, G. D. and MILLSON, G. C. (1964b) J. Gen. Microbiol., 37, 251.

HUNTER, G. D. and MILLSON, G. C. (1967) J. Comp. Pathol., 77, 301.

HUNTER, G. D., MILLSON, G. C. and MEEK, G. (1964) J. Gen. Microbiol., 34, 319.

KIMBERLIN, R. H., MILLSON, G. C. and HUNTER, G. D. (1971) J. Comp. Pathol., 81, 383.

KIMBERLIN, R. H., SHIRT, D. B. and COLLIS, S. C. (1974) J. Neurochem., 23, 241.

LATARJET, R., MUEL, B., HAIG, D. A., CLARKE, M. C. and ALPER, T. (1970) Nature, 227, 1341.

LAVELLE, G. C. (1972) Proc. Soc. Exp. Biol. Med., 141, 460.

KLECZKOWSKI, A. and MCLAREN, A. D. (1967) J. Gen. Virol., 1, 441.

MARSH, R. F. and KIMBERLIN, R. H. (1975) J. Infect. Dis., 131, 104.

MARSH, R. F., SEMANCIK, J. S., MEDAPPA, K. C., HANSON, R. P. and RUECKERT, R. R. (1974) J. Virol., 13, 993.

MILLSON, G. C., HUNTER, G. D. and KIMBERLIN, R. H. (1971) J. Comp. Pathol., 81, 255.

MORRIS, J. A., GAJDUSEK, D. C. and GIBBS, C. J. (1965) Proc. Soc. Exp. Biol. Med., 120, 108.

MOULD, D. L., DAWSON, A. MCL. and SMITH, W. (1964) J. Gen. Microbiol., 35, 491.

MOULD, D. L., SMITH, W. and DAWSON, A. MCL. (1965) J. Gen. Microbiol., 40, 71.

MOULD, D. L. and DAWSON, A. MCL. (1968) J. Comp. Pathol., 78, 115.

MOULD, D. L. and DAWSON, A. MCL. (1970a) J. Comp. Pathol., 80, 595.

MOULD, D. L. and DAWSON, A. MCL. (1970b) Res. Vet. Sci., 11, 304.

OUTRAM, G. W., DICKINSON, A. G. and FRASER, H. (1974) Nature, 249, 855.

OUTRAM, G. W., DICKINSON, A. G. and FRASER, H. (1975) Lancet, i, 198.

PATTISON, I. H. (1964) Vet. Rec., 76, 333.

PATTISON, I. H. (1965) J. Comp. Pathol., 75, 159.

PATTISON, I. H. and JONES, K. M. (1967) Vet. Rec., 80, 2.

PATTISON, I. H., JONES, K. M. and KIMBERLIN, R. H. (1969) Res. Vet. Sci., 10, 214.

PATTISON, I. H. and MILLSON, G. C. (1961) J. Comp. Pathol., 71, 350.

PATTISON, I. H. and SANSOM, B. F. (1964) Res. Vet. Sci., 5, 340.

SIEGEL, A. and WILDMAN, S. G. (1954) Phytopathology, 44, 277.

SIEGEL, A., WILDMAN, S. G. and GRINOZA, W. (1956) Nature, 178, 1117.

STAMP, J. T., BROTHERSTON, J. G., ZLOTNIK, I., MACKAY, J. M. K. and SMITH, W. (1959) J. Comp. Pathol., 69, 268.

STREETER, D. G. and GORDON, M. P. (1968) Photochem. Photobiol., 8, 81.

WARD, R. L., PORTER, D. D. and STEVENS, J. G. (1974) J. Virol., 14, 1099.

WEBSTER, G. R. (1957) Nature, 180, 660.

WILSON, D. R., ANDERSON, R. D. and SMITH, W. (1950) J. Comp. Pathol., 60, 267.

ZLOTNIK, I. (1968) J. Comp. Pathol., 78, 19.

ZLOTNIK, I. and RENNIE, J. C. (1967) Br. J. Exp. Pathol., 48, 171.

CHAPTER 12

The pathology of natural and experimental scrapie

H. FRASER

12.1 General introduction

There is now general agreement that the type of disease and clinical symptoms
of scrapie, described in Ch. 10, result from disturbances in the normal
function of the CNS. The only histological lesions which have been found
are confined to the CNS, and they must stem from an underlying biochemical lesion as yet unrecognised. Although the relationship between the
assumed primary biochemical lesion and the histological ones will be shown
to be very precise, it is unknown whether the relationship is direct or whether
there are many intervening steps – the detailed symptoms of the disease
cannot be correlated with the detailed histological lesions.

The microscopic changes are almost entirely degenerative, and are unaccompanied by inflammation. In some circumstances cases of advanced
clinical scrapie occur in which definitive histological lesions in the central
nervous system are minimal or insufficient for diagnostic purposes. With this
exception, the most typical and prominent changes in clinically advanced
cases is the occurrence of vacuoles in the neuroparenchyma which have
long been regarded as the characteristic lesion (Holman and Pattison, 1943;
Zlotnik, 1958a). In natural scrapie large vacuoles in the cytoplasm of
neurons, particularly in the medulla oblongata, pons, mid-brain, and spinal
cord, are the most striking and diagnostically useful lesion. Under some
circumstances, particularly in certain examples of experimental scrapie, but
also to a more restricted extent in the natural disease, the vacuolation may
occur as a spongy change in which vacuoles, relatively smaller in size than

Slow virus diseases of animals and man, edited by R. H. Kimberlin
© *North-Holland Publishing Company 1976*

those in neurons, occur in the neuropil. The only reactive accompaniment of these degenerative lesions is an hypertrophy and possible proliferation of astrocytes (Hadlow, 1961; Pattison and Jones, 1967).

It has become almost axiomatic in descriptions of the pathology of scrapie that there is strict bilateral symmetry of the vacuolar and astrocytic changes, which has prompted certain hypotheses for its pathogenesis: in particular that it more closely resembles toxic than infective processes (Mackenzie and Wilson, 1966) or that the primary lesion is in astrocytes which are intimately and functionally related to the symmetrically distributed vascular system (Pattison and Jones, 1967). In view of this, it is emphasised that the lesions produced by certain scrapie agents can be asymmetrical.

It is clear from the present knowledge of scrapie in mice that there is great variety in the details of the histological lesions, but that, when several biological variables in the host and the agent are accurately controlled, there is great precision and invariability in the lesions. Table 14.2 shows some of the important variables affecting the agent and host upon which this depends. In natural scrapie, as well as in the rest of this group of degenerative encephalopathies (Ch. 15), several of these important variables cannot be controlled, and therefore the reasons why the lesions vary from case to case are relatively uninterpretable, making attempts at classification of the diseases on the basis of their lesions, arbitrary and of little value. To be a valid basis for conclusions about aetiology and pathogenesis, classification of lesions must be based on independent (i.e., non-histological) criteria. It cannot be stressed too strongly that the primary cellular and biomolecular disorders of scrapie are completely unknown. There have been several attempts in this group of diseases to resolve the pathological variation into arbitrary groupings thus implying basically different categories for causation or pathogenesis – contrasting for instance, 'natural' and 'experimental' scrapie, Cheviot and Suffolk scrapie or four arbitrary 'types' of Creutzfeldt–Jakob disease (Zlotnik, 1962a; Parry, 1969; Daniel, 1972). The effects of important factors are so confounded in this sort of arbitrary classification that any real basis of biologically meaningful differences remain unknown. To overcome these problems, inherent in much qualitative and descriptive pathology of naturally occurring disease, it becomes necessary to develop strictly independent and objective quantitative systems for handling the data, to be able to identify the causes of differences in pathology.

12.2 Quantitative pathology in scrapie

12.2.1 Factors affecting the pathology of scrapie

Our knowledge of the pathology is based historically on the naturally occurring disease and various forms of the experimental disease in sheep and goats but their contribution is small compared with that from mice, particularly inbred mice. For example, in our laboratory, studies of the factors responsible for pathological variation have involved 40,000 inbred mice of over twenty strains and crosses, providing over 500,000 measurements of the severity and distribution of different types of lesions.

The effects have been estimated of factors such as host genotype, age and sex, route of inoculation, and others relating to the agent strain, such as passage status, dose, donor genotype and the organ used for preparing inocula. Equivalent studies are not possible in other species or with the natural disease: hence no systematic pathological studies in different sheep breeds with different agents have yet been made.

In inbred mice, if all the relevant variables are rigorously controlled, the pathology of scrapie is extraordinarily precise in comparison with 'conventional' infections both in the development of the lesions and in their details at the terminal stage of the disease. This will be described on a comparative basis in sheep, goats and mice with regard to the various forms of lesion, i.e., vacuolar degeneration (a) in grey matter, (b) in white matter and (c) occurring asymmetrically; cerebral amyloidosis, and glial changes in scrapie, including astrocytic alterations.

12.2.2 Vacuolar degeneration; the 'lesion profile' system

Vacuolation is the most characteristic lesion seen in scrapie with standard staining methods of paraffin sections, and it can be measured using relatively objective standards. This involves scoring the intensity of degeneration in nine areas of grey matter (Fig. 12.1) using a 6-grade (0–5) score. Normal unvacuolated brain is scored '0', a score of '1' is applied to isolated vacuoles, too few to be recognised as pathological. Each score from 2–5 represents roughly the logarithm of the number of vacuoles in a microscope field covering an area of approximately 0.02 μm^2 (Fraser and Dickinson, 1968). The nine defined areas are: (1) dorsal medulla nuclei; (2) cerebellar cortex adjacent to the fourth ventricle; (3) superior colliculus; (4) hypothalamus; (5) central thalamus; (6) hippocampus; (7) septal area; (8) and (9) two areas

270 *H. Fraser*

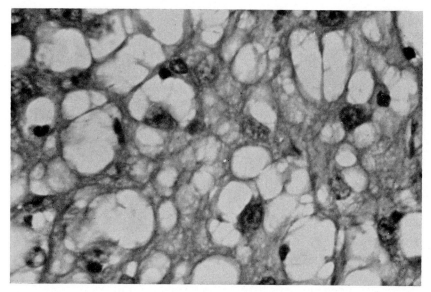

Fig. 12.1 Severe vacuolation (status spongiosus) in cerebral cortex of VM mouse terminally affected with scrapie produced by the ME7 agent. Haematoxylin–Eosin. × 400.

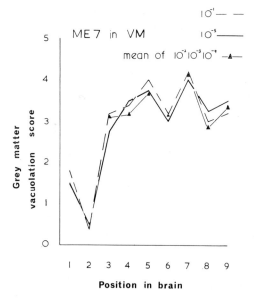

Fig. 12.2 Lesion profiles of ME7 scrapie agent in VM mice injected intracerebrally with
serial 10-fold dilutions of infected brain source. The three intermediate dilutions are
pooled onto a single plot for clarity of presentation.

of cerebral cortex. These were chosen in the light of experience by Dr. I. Zlotnik (Mould et al., 1967) and are still found to be a generally representative array. However, this has subsequently been extended for scoring white matter vacuolation, over a 4-grade scale (12.5).

The precision of the lesion profile system can be seen in Fig. 12.2 which shows that on the basis of 5–8 mice per group there is no change with 100,000-fold differences in the number of infectious units of ME7 agent injected intracerebrally. This accuracy depends on other variables being held constant, such as sex and the route of inoculation but is unaffected by the region of the brain into which the injection is given. (Fraser and Dickinson, 1967; Fraser, 1971; Outram et al., 1973).

12.2.3 *Agent discrimination and host-strain effects*

Several scrapie agents have now been investigated in great detail and some of these have been biologically 'purified' by cloning techniques. This is important because it has become apparent that many scrapie isolates from natural cases contain more than one strain of agent (10.8). The characteristics of a number of more recent isolates are less well known than the cloned agents, and while they may appear to be different, they could still be unresolved mixtures: therefore it is better at this stage to refer to them with the interim designation of 'sources' rather than 'strains or 'agents'.

Many strains of scrapie can be identified and discriminated from each other solely on the basis of the grey matter lesion profile. (Other details for different strains of scrapie are dealt with in Ch. 14.) With some agents (e.g., ME7) the shape of the lesion profile is broadly similar in a variety of mouse strains, although the overall damage in the scored areas is greater in some than others. With other agents there is little resemblance between their profiles in different mouse strains (e.g., between 22A in C57BL and VM mice). It has become operationally convenient, for reasons unconnected with the histopathology, to study the lesion profiles of different agents routinely in two inbred mouse strains, VM ($sinc^{p7}$) and C57BL ($sinc^{s7}$) (Fraser and Dickinson, 1973) and to a lesser extent in their F_1 cross (see Ch. 14), although additional strains have also been used extensively, but in most cases the choice of strain has been unconnected with the histology. As a result, our histological data are widely based in relation to host genotype. Fig. 12.3 shows the radically different lesion profiles for the three agent strains, ME7, 79A and 87A. These agents are representative of ones which differ profoundly on the basis of two further lesions – white matter vacuola-

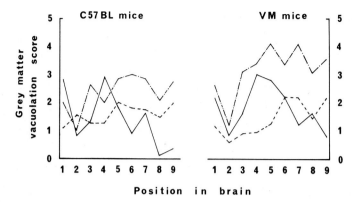

Fig. 12.3 Lesion profiles of ME7, 79A and 87A scrapie agents in C57BL and VM mice
(–·–·–·–, ME7; -------, 79A; ————, 87A).

tion and cerebral amyloidosis (see 12.4 and 5). A further important difference is that the vacuolar degeneration of 87A frequently shows an asymmetrical pattern (12.3). With 87A the amyloid, usually in the form of plaques (12.4) most severely affects those parts of the brain which this agent leaves relatively unvacuolated. In contrast, the far less frequent plaques found in only occasional mice with ME7 agent are present close to, or within, vacuolated areas. Fig. 12.4 shows that the lesion profiles of ME7 and 22C agents only have minor differences from one another though they are easy to distinguish using certain incubation period parameters (Ch. 14). Cerebral amyloidosis has not been observed with either 79A or 22C in the six or seven strains of

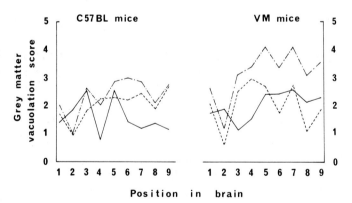

Fig. 12.4. Lesion profiles of ME7, 22A and 22C scrapie agents in C57BL and VM mice
(–·–·–·–, ME7; -------, 22C; ————, 22A).

mice we have used for lesion profile studies. The agent 22A is included in
Fig. 12.4 because it has profound incubation period differences from most
other agents so far isolated (10.6; 14.4.4d), as well as differences in terms of
other pathological parameters. Variation in the age at which clinical signs
develop is completely without effect on the lesion profile with all agents and
host strains in which this has been studied.

12.2.4 The effect of route of infection and other sources of variation on lesion profile

Other factors which influence the lesion profile, sometimes as profoundly
as do host and agent differences, are route of inoculation and the genotype
of the donor used as the source of an inoculum. This emphasises the extreme
caution needed in interpreting lesion profiles and other differences, if such
variables are not controlled (see Table 14.2). Perhaps the most puzzling
finding is the way in which the major variables do not act independently
of one another – they interact (Outram et al., 1973). This means, as an
extreme example, that the occurrence of white matter vacuolation depends
on the route of injection as well as on the agent strain, in some combinations.

The change of lesion profile with different inoculation routes (Fig. 12.5)
provokes speculations concerning pathogenesis and the spread of agent
to and within the CNS, particularly since lesion distribution may reflect
the distribution of agent within the brain (10.8), which in turn might be
influenced by the point of entry to the CNS. Following inoculation into a

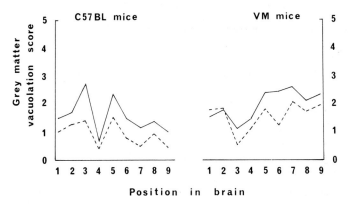

Fig. 12.5. Lesion profiles of 22A scrapie agent in C57BL and VM recipients, following
intracerebral (————) and intraperitoneal (------) inoculation.

site outside the CNS, the early replication is in lympho-reticular or allied cell systems (14.3), but it is not known how agent then reaches the vulnerable nervous tissues; this is still a matter for speculation, even with some acute viral encephalitides (Albrecht, 1968; Blinzinger and Anzil, 1974). The logical choice may appear to be between spread via the vascular or the peripheral nervous systems, but there are as yet no laboratory data to indicate which is more likely.

A much wider question concerns the basis of the effects caused by host genotype and agent strain. How can such radical differences in profile shape and other pathological parameters be explained? An understanding of this may clearly be an integral step in solving the basic molecular problems of this type of disease, although it should not be assumed that the mechanisms associated with different major variables will have a common basis. Some may reflect accessibility to the CNS (Dickinson and Outram, 1973), perhaps with particular agents or sources having selective preference for localising in some regions in certain host genotypes. It must not be forgotten that scrapie does not always produce vacuolar damage accompanying advanced clinical signs (12.7) or that it is often trivial in extent. Similarly in transmissible mink encephalopathy, a marked reduction in vacuolar degeneration has been found in aged mink of the Chediak–Higashi genotype (15.4). This suggests that vacuolation may be a side effect, not in the main stream of the lethal events of pathogenesis.

12.3 Asymmetrical vacuolation

We now have a large number of cases of scrapie in mice in which grey matter vacuolation occurs asymmetrically. This refutes the widely held view that the lesions are always strictly bilaterally symmetrical, which has been stated for both the natural and experimental disease (Pattison et al., 1959; Pattison, 1965a, b; Mackenzie and Wilson, 1966; Pattison and Jones, 1967; Field et al., 1967a). The asymmetry occurs at high frequency with particular agents (e.g., 87A), as well as with some sources where agent typing is at an early stage. Asymmetry of diffuse spongy degeneration is seen sometimes, or the asymmetrical vacuolation can be intensely focal, with intervening areas of otherwise normal neuroparenchyma, in which case it is easier to recognise (Fig. 12.6). The areas in the brain where asymmetrical lesions occur most frequently are the mesencephalic tectum, the geniculate bodies, pretectal area, septal nuclei, cerebral cortex and, most often, in a

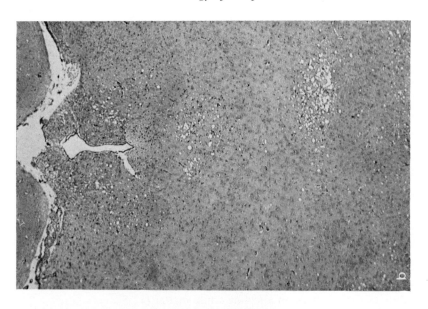

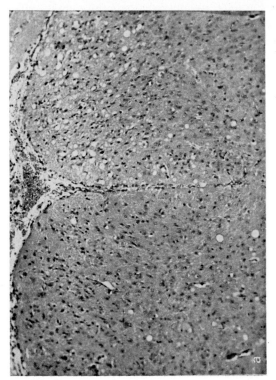

Fig. 12.6. Asymmetrical grey matter vacuolation in C57BL mice infected with 87A scrapie agent: (a) confined to one side of mesencephalic tectum; (b) showing a focal distribution in thalamus. Haematoxylin–Eosin. × 190 and × 80.

TABLE 12.1

Correlation between plaques and asymmetrical vacuolation in VM mice affected with
different scrapie agents. —, absent; +, occasional; ++, infrequent; +++, frequent.

Agent	Plaques	Asymmetry
ME7	+	—
22C	++	+
22C	—	—
87A	+++	+++
87V	+++	++
79A	—	—

variety of sites in the thalamus. Asymmetry has not been identified as a
feature of the vacuolation affecting white matter (12.5).

With new scrapie isolates in mice, usually only the intracerebral route of
inoculation is used, so when asymmetrical lesions were first recognised
we assumed that this was in some way related to the trauma of asymmetrical
injection. It was a surprise, therefore, to find that asymmetrical vacuolation
can also occur after intraperitoneal injection (Bruce and Fraser, unpublished).
Asymmetrical lesions tend to occur with those agents producing the most
severe cerebral amyloidosis in the form of plaques (Table 12.1). An understanding of this association will probably be of great importance for the
wider aspects of pathogenesis. Asymmetrical lesions have been reported in
a case of Creutzfeldt–Jakob disease (Goldhammer et al., 1972) in which
amyloid plaques, similar to those seen in scrapie (Bruce and Fraser, 1975)
are known to occur.

12.4 Cerebral amyloidosis and amyloid plaques

We have recently begun to recognise some of the factors which control
the occurrence of cerebral amyloidosis. Some agents (e.g., 87A, 87V) induce
extensive cerebral amyloidosis in all mouse strains studied, others (e.g.,
ME7, 22A) do so less extensively and only in certain mouse strains, while
some (e.g., 79A, 22C) fail to induce recognisable amyloid in any of a wide
range of mouse strains.

The amyloid occurs predominantly as discrete plaques, but may also

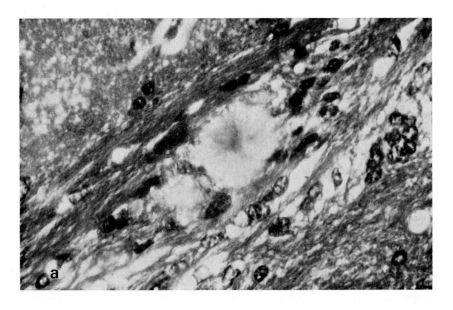

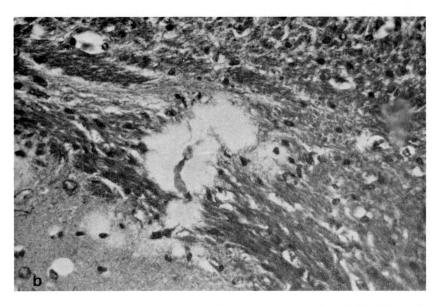

Fig. 12.7. Amyloid plaques in the corpus callosum of VM mice infected with 87A scrapie agent: (a) stellate plaque; Masson's trichrome. × 800; (b) plaque showing association with capillary, Heidenhains. × 800.

adopt a diffuse appearance. Deposits occur in numerous sites in the brain: cerebrum, diencephalon, mid-brain, medulla and cerebellum. Only one group have previously mentioned the existence of plaques in scrapie, both in naturally affected sheep and in mice, but no systematic study was attempted (Beck et al., 1964; Beck and Daniel, 1965b). They mentioned periodic acid–Schiff (PAS) positive, weakly argentophilic plaques in the cerebellum, similar or identical to 'kuru plaques'. The only other identification of amyloid plaques in sheep scrapie (Bruce, personal communication) has been in the fore-brains of Cheviot sheep injected intracerebrally with scrapie from a Blackface sheep, which was infected by close contact with natural cases (Dickinson et al., 1974) (source 104, Table 10.1).

The amyloid in scrapie-infected mice has a wide range of forms; these vary from discrete 'stellate', 'amorphous' and 'giant' plaques, to faint 'shadowy' plaques, or it can appear as 'diffuse' or 'perivascular' deposits, all six categories apparently comprising a continuous range (Fig. 12.7). They are argyrophilic, PAS-positive, congophilic, birefringent with toluidine blue and stain with the fluorescent dye, thioflavine S. Microglia are frequently found surrounding the plaques. The ultrastructural appearance is one of amyloid associated with and surrounded by distended and altered neurites and glial cells (Bruce and Fraser, 1975). These features are also found with classical 'senile plaques' which occur in the brains of aged humans, dogs and rhesus monkeys, as well as in presenile and senile dementia (Wiśniewski and Terry, 1973).

Details of the time sequence in which scrapie-infected mice develop plaques are unknown although they can appear relatively early and long before vacuolation occurs, at least with 87A agent (Bruce and Fraser, unpublished). The age at which scrapie-affected mice are seen to have plaques depends on the incubation period for the particular combination of agent and host strain used (14.4.4) but they are frequently seen in mice well under a year of age. They have never been seen in a large series of extremely old mice, several over 1000 days of age, when these are uninjected with scrapie (Fraser and Bruce, 1973) and senile plaques have not been observed in aged mice elsewhere (Dayan, 1971).

12.5 White matter changes

Scrapie is in no sense a primary demyelinating disease, and in routine histological preparations of brains of affected sheep the white matter is virtually undamaged histologically. However, as neuronal degeneration and loss occur in the grey matter, it can be anticipated that some secondary nerve fibre degeneration will also occur. There are marginal changes in white matter in scrapie such as neutral fat and Marchi-positive lipid in the brainstem and cerebellar white matter (Zlotnik, 1958a; Beck and Daniel, 1965a) and secondary degeneration in long fibre tracts in naturally affected sheep (Wight, 1960; Palmer, 1968) and in mice infected with ME7 agent (Fraser, 1969). However, a quite different type of degeneration, an idiopathic vacuolar degeneration unrelated to secondary degeneration, occurs in scrapie in mice with certain agents. Agent strain and host genotype are the two major factors that control the occurrence, intensity and distribution of this lesion. The lesion appears to be a primary vacuolar degeneration of white matter (Fig. 12.8) which is found particularly with the types of scrapie used in the majority of laboratories working with scrapie. At least two such agents came from the 'drowsy' goat source: 139A (synonym 'Chandler') and 79A (10.6) and they produce severe damage in many regions of white matter in all the twelve mouse strains studied, although with differences between different strains. It is also known that the 'drowsy' source contains another strain of agent, 79V, which produces little or no white matter vacuolation (Fraser, unpublished). Serial transmission through rats from third mouse-passage 'Chandler' agent (Chandler and Fisher, 1963), yielded an agent which does not produce the severe white matter lesion in mice (Pattison and Jones, 1968). This suggests that in the early mouse passages from the 'drowsy' source a mixture of agent strains was present, some producing white matter vacuolation and others, such as 79V, producing none and that the latter replicate preferentially in rats. Surprisingly, significant accumulations of neutral fat have not been found in the white matter of mice terminally affected with 'Chandler' agent, although neutral fat is present in some grey matter areas (Mackenzie and Wilson, 1966).

Many of the factors which can influence the grey matter lesion profile can also influence the profile of white matter vacuolation, although the two parameters seem to be largely independent. With the ME7 agent, white matter vacuolation is usually mild or non-existent but occasionally it is severe, depending on the genotypes of the tissue donor and the recipient, and on the sex of the recipient (Fraser, 1970). When ME7 is passaged in BRVR or in BALB/c mice there is a donor-genotype effect on the white matter lesion, as well as effects due to route of injection and sex of recipient, but the effect of these factors on white matter vacuolation is quite different depending on whether the injection route is intracerebral or intraperitoneal (Tables 12.2 and 3). It should be mentioned that, as these studies used un-

H. Fraser

TABLE 12.2

Factors influencing vacuolation in the white matter in scrapie in BRVR and BALB/c
mice with ME7 agent passaged from BRVR or BALB/c donors. The factors recipient
genotype and sex, donor genotype, inoculation route and organ source of agent are
treated as independent variables in an analysis of least squares and the significance of the
effects between subgroups and of interactions are based on an analysis of variance. The
design is according to the protocol of Dickinson and Outram (1973).

	Cerebellum	Tegmentum
Recipient	× × ×	—
Sex	× × ×	× × ×
Donor	× × ×	× × ×
Route	—	—
Organ	—	—
Recipient × route	×	—
Sex × route	×	× ×
Donor × route	× × ×	× ×
Sex × donor	—	×

−, $P > 0.05$; ×, $P < 0.05$; × ×, $P < 0.01$; × × ×, $P < 0.001$.

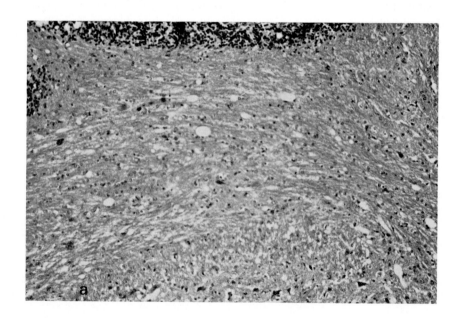

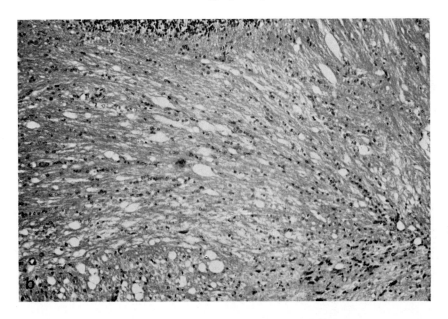

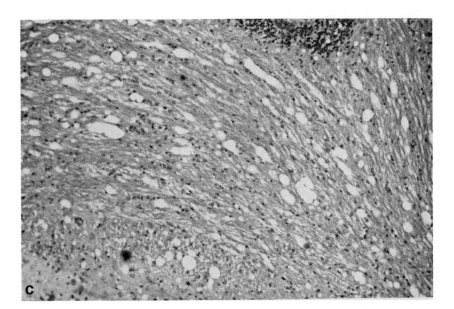

Fig. 12.8. White matter vacuolation in cerebellum: (a) score 1; (b) score 2; (c) score 3. Haematoxylin–Eosin. × 190.

TABLE 12.3

Factors influencing vacuolation in the white matter of ME7 scrapie-affected mice in Table 12.2, treated independently: (a) route difference, and, within the intraperitoneally injected groups, (b) donor difference and (c) sex difference. i.p., intraperitoneal; i.c., intracerebral; cwm, cerebellar white matter; twm, tegmental white matter.

	(a) Routes separately				(b) Donors separately*				(c) Sexes separately*			
	i.c.		i.p.		BRVR		BALB/c		Male		Female	
	cwm	twm	cwm	twm	cwm	twm	cwm	twm	cwm	twm	cwm	twm
Recipient	×××	××	—	—	×	—	—	—	—	—	—	—
Sex	—	—	×××	×××	×××	×××	—	—	—	—	—	—
Donor	—	××	×××	×××	—	—	—	—	××	×	×××	×××
Organ	—	—	×××	×××	—	—	—	—	—	—	×××	—
Donor × organ	××	—	—	××	—	—	—	—	—	—	—	××
Donor × recipient	—	—	—	—	—	—	—	—	—	—	—	×
Donor × sex	—	—	—	×	—	—	—	—	—	—	—	—

* Within intraperitoneal group only.

—, $P > 0.05$; ×, $P < 0.05$; ××, $P < 0.01$; ×××, $P < 0.001$.

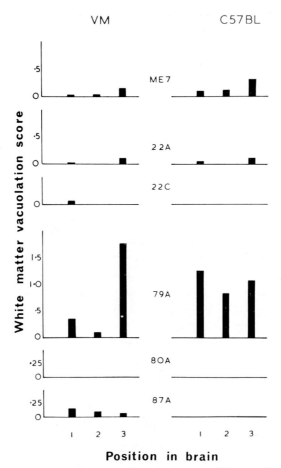

Fig. 12.9. White matter vacuolation in three brain areas (1) cerebellar white matter; (2) white matter in the mesencephalic tegmentum; (3) white matter of the pyramidal tract (at the level of the thalamus) in C57BL and VM mice with different scrapie agents. The intensity of vacuolation is graded in a scale 0–3; the maximum lesion is scored as a 3.

cloned ME7, it is possible that there are two strains present which are very similar in most of their properties but which replicate at different rates in these two genotypes of mice.

Differences in the extent of white matter damage for a limited range of agents is shown in Fig. 12.9. This method is applied routinely to all mouse brains on which grey matter scoring is carried out (12.2.2), and involves scoring the intensity of vacuolation in the cerebellum, tegmentum and pyramidal tract, each scored on a standardised scale of increasing intensity from 0 to 3 (Fig. 12.8). 79A agent always produces severe degeneration but the degree of vacuolation is modulated significantly by the mouse genotype (e.g., VM or C57BL). Similar effects are found with 139A agent. At the other end of the scale, 22C does not produce any demonstrable white matter changes, having now been tested in 12 inbred strains. Just as some agents tend to produce more white matter vacuolation than others, mouse strains tend to vary consistently, so that some (e.g., BALB/c and A2G) are more liable to incur severe white matter damage (Table 12.2; Fraser et al., 1974).

12.6 *Natural scrapie*

The experimental variety and degree of control with inbred mice obviously has no possible equivalent in the natural disease. It is therefore quite inappropriate to attempt to categorise meaningfully differences in the natural disease in sheep when the many sources of variation are uncontrolled. For this reason, many of the published accounts, based on small samples from intrinsically heterogenous or undefined populations, are now uninterpretable in important respects. It may appear that basic differences exist between the relatively wide range of natural cases and the very narrow experimental spectrum of agents which have been used in sheep (10.6), or even between

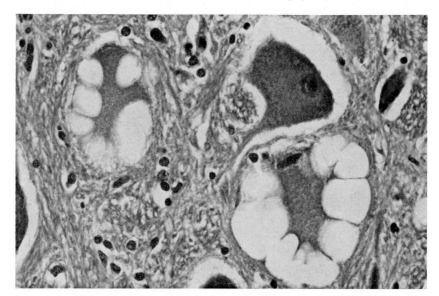

Fig. 12.10. Vacuolation of neurons in mid-brain of Cheviot sheep affected with scrapie. Haematoxylin–Eosin. × 190.

different breeds of sheep, but it would be premature to interpret the bio-logical basis of such differences, particularly when they are founded on relatively small samples. It is a formidable task to assess which strains of agent are involved in the natural disease (10.8).

Although naturally occurring scrapie has been recognised in sheep for several centuries, it has only recently become recognised in goats (10.4.2). As no essential pathological differences between the two species have been found, description of the natural disease will be confined to sheep (Holman and Pattison, 1943; Zlotnik, 1958a). The most prominent lesion is vacuolation of nerve cells in the medulla, pons and mesencephalon which consists of single or multiple vacuoles causing ballooning of nerve cells. The surrounding cytoplasm usually shows various stages of degeneration (Fig. 12.10). Occasionally this vacuolation extends to emergent nerve fibres. Vacuolated neurons and other forms of neuronal degeneration occur throughout the brainstem, most frequently in the facial, cuneate, gracile, ambiguus, dorsal motor vagus, superior olivary, abducens and red nuclei, the reticular forma-tion, the nucleus of the spinal tract of the fifth nerve, and sometimes in the Purkinje cells of the cerebellar cortex. Intravacuolar, eosinophilic, protein-containing spheroids, 2–12 μm in diameter as well as other granules and bodies of various sizes, are often present (Holman and Pattison, 1943; Palmer, 1957a, b; Zlotnik, 1957a). In addition to overt neuronal vacuolation, there is often interstitial spongy degeneration in the same areas, which is sometimes particularly prominent in mid-line regions where it may extend rostrally to include diencephalic and telencephalic non-cortical regions. In addition to vacuolation, other deteriorative changes may occur, including overt necrosis.

The existence of vacuolated neurons in apparently healthy sheep has been emphasised by many workers, but whereas in natural scrapie the vacuolation is widely recognised as severe, in apparently 'healthy' sheep the vacuolation is trivial (Zlotnik, 1958a; Zlotnik and Rennie, 1957, 1958). Surveys of the number of vacuoles in serial sections of the medulla oblongata established that scrapie-affected sheep, mainly Suffolks, had a significantly greater number than non-scrapie animals of the Blackface, Welsh Mountain, Rambouillet, Hampshire, Columbia breeds and crossbreds of Southdown, Hampshire, Shropshire and Merino (Zlotnik and Rennie, 1958; Zlotnik, 1957b, 1958a). It must be appreciated also that in histologically severe scrapie other regressive changes accompany the overt vacuolar degeneration whereas the vacuoles in 'healthy' animals are found in otherwise normal neurons and neuroparenchyma (Zlotnik and Rennie, 1957). In the absence

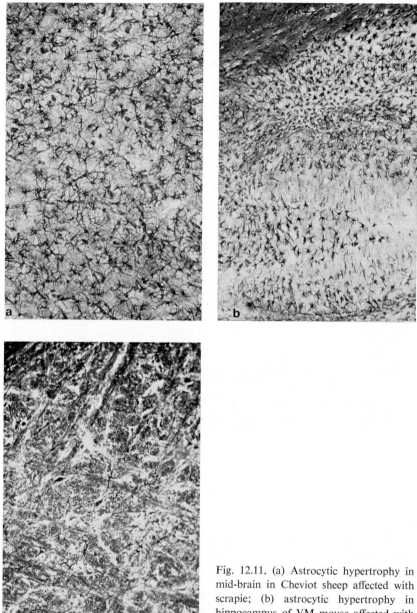

Fig. 12.11. (a) Astrocytic hypertrophy in mid-brain in Cheviot sheep affected with scrapie; (b) astrocytic hypertrophy in hippocampus of VM mouse affected with scrapie; (c) eutrophic astrocytes in mid-brain from healthy, young adult Cheviot sheep. Cajal. × 190.

of other independent tests, severe vacuolation has to be used as the final diagnostic criterion of natural scrapie, but 'scrapie' diagnosed clinically, even by persons with wide experience of the disease, is not always accompanied by diagnostic lesions, which is obviously an unsatisfactory situation. This dilemma is borne out by the results of some experiments where one has the independent information that the sheep have been injected with scrapie and have developed symptoms after a predictable interval (10.7.2 and 12.7).

The regressive neuroparenchymal changes are accompanied and perhaps preceded by an enlargement of astrocytes, as demonstrated by impregnation with metallic gold (Fig. 12.11). This lesion has very limited diagnostic value, firstly because of the non-specific nature of astrocytic response and hypertrophy in general neuropathology, secondly because of technical problems and allied interpretative difficulties, and thirdly because of the difficulty in distinguishing marginal hypertrophy. In contrast the neuronal vacuolation and associated neuroparenchymal degeneration are diagnostically specific, technically easy to demonstrate, and usually severe, at least in the natural disease within the limits of current experience.

It is appropriate at this point to emphasise two important aspects of the pathology in mice which are not typical of the natural disease: (1) murine cerebral amyloidosis which is probably of fundamental significance (Fraser and Bruce, 1973; Bruce and Fraser, 1975). In view of its frequent occurrence in mice injected with isolates from several field cases of scrapie in sheep (10.8), it is surprising that it has only been reported from the natural disease by one group of workers (Beck et al., 1964). (2) In the natural disease (and in experimental forms in sheep as far as is known at present) vacuolation is almost exclusively confined to the grey matter, any white matter changes being trivial. However, the three scrapie agents (79A, 139A, ME7) which together have been studied experimentally more than any others, produce severe vacuolation in white matter in mice, though with ME7 only in some mouse strains (12.5), but very little experimental work has been done with these agents in sheep. Reasons are given elsewhere (10.6) for considering that the first two may not be of sheep origin, whereas ME7 appears to be the commonest agent in sheep (10.8).

12.7 Experimental infection of sheep, goats and other laboratory animals

The principle lesions produced experimentally are broadly similar to those in the natural disease, consisting of vacuolar degeneration, astrocytic hypertrophy and in some cases cerebral amyloidosis, and again without any manifestations of immunological or inflammatory reaction. Some of the implied differences between natural scrapie and that produced by experimental passage into other hosts can simply arise from uncontrolled strains of agent and natural mixtures of strains so that the frequent assertion that 'the agent' becomes 'modified' by passage into different host species, becomes unsupportable upon closer examination (10.11).

The experimental disease has been studied extensively in Cheviot sheep in an experiment designed to select breeding lines of different susceptibility following subcutaneous injections of SSBP/1 scrapie (Dickinson et al., 1968; 10.7.2). It is now established that this scrapie source contains different strains of agent (10.6). The most complete pathological study of SSBP/1 in Cheviot sheep was by Zlotnik (1958b), who confirmed an earlier observation that the vacuolation of neurons in the medulla is usually less severe than generally recognised in the natural disease.

The frequency of vacuolation and degeneration of neurons in the pons is similar to or may exceed that in the medulla, while the mesencephalon is less severely affected. Astrocyte proliferation and myelin destruction do not occur. The mild nature of the lesions in SSBP/1-infected Cheviots is borne out by a recent study (Dickinson and Rennie, personal communication) of four Cheviot sheep severely affected with scrapie. Counts of the number of vacuoles in serial sections of the medulla and pons, modified from the method of Zlotnik (Zlotnik, 1957b, 1958a; Zlotnik and Rennie, 1957, 1958) showed the number to be very low and sometimes within the range previously reported for healthy animals (Table 12.4).

In contrast, experimental infections in the goat, using various agent sources, have usually resulted in a more severe vacuolation (Zlotnik, 1961, 1962b; Pattison, 1965a, b) although, in this species too, some cases have very few vacuoles (Pattison et al., 1959). Hadlow (1961) conducted the most extensive study, using numerous goats of mixed breeding. The neuroanatomical distribution of the lesions was remarkably constant, despite the use of various injection routes and inocula prepared from two different sources. The thalamic nuclei were the most severely affected, although regressive changes were widespread elsewhere, including the septum, corpus striatum, cerebellar cortex and brainstem, but not the cerebral cortex. The most prominent changes were vacuolar and other forms of degeneration and loss of neurons, including status spongiosus. White matter changes were not seen. A prominent astrocytic response, described as hypertrophy with

TABLE 12.4

Neuronal vacuolation in serial sections in healthy and scrapie-affected sheep

	Breed (number of sheep examined)	Age	Status	Range of numbers of vacuoles (number of serial sections)	
				Medulla (54)	Pons (18)
A	Cheviot (4)*	1–4 year	severe clinical scrapie following SSBP/1 injection (s.c.)	5– 173	24– 88
B	Cheviot (15)	4½ months–2½ years	clinical scrapie following SSBP/1 injection (i.c.; s.c.)	150– 748	18–509
C	Suffolk (7); Cheviot (1); Swaledale (2)	not stated	clinical scrapie occurring naturally	1185–4266	—
D	Cheviot (15)	½–1 year	healthy	0– 28	0
E	Border Leicester cross (2); Cheviot (8)	matched with C	healthy	0– 57	—
F	Blackface (21)	1 year	healthy	3– 76	—
G	Welsh Mountain (10)	not stated	healthy	0– 39	—
H	Cross-bred hoggs (10)	1 year	healthy	0– 27	—
I	Columbia (10)	not stated	healthy	0– 21	—
J	Rambouillet (8)	not stated	healthy	0– 18	—
K	Hampshire (15)	not stated	healthy	6	—

* Selected as examples of cases of severe scrapie in which the pathological confirmation could not be made.
References: Zlotnik, 1957b, 1958a (C and E); Zlotnik, 1958b (B and D); Zlotnik and Rennie, 1958 (F–K); Dickinson and Rennie, personal communication (A).

slight proliferation, accompanied these changes in some regions, but in others such as the thalamus, mesencephalon and cerebellar cortex, this was disproportionately more severe than the concomitant neuronal damage. In a study by Zlotnik (1961) of goats infected with goat- or sheep-passaged SSBP/1 scrapie or natural goat scrapie, the most severely affected area was the tegmentum, and he emphasised the regularity of the pattern of damage. On the other hand Pattison examined four goats infected with a single primary scrapie source from a naturally infected Welsh Mountain sheep and reported differences between them in the distribution of lesions (Pattison, 1965b).

Scrapie has also been transmitted to a variety of laboratory animals besides mice, these include rats and golden hamsters (Zlotnik, 1963; Chandler and Fisher, 1963; Zlotnik and Rennie, 1965) and Chinese hamsters, voles and gerbils (Chandler and Turfrey, 1972). The pathology in these species – vacuolation accompanied by astrocytic hypertrophy – is generally similar to that in other species and in the degenerative encephalopathies as a whole. The usefulness of these species is completely outweighed for most purposes by the elegant degree of control permitted by the use of genetically defined inbred mice. However, the possibility must be borne in mind that some strains of scrapie may produce disease more readily in species other than mice, which may not even be 'susceptible' to all strains present in sheep (10.4 and 8). Also titres of agent in hamster brain can reach 100-fold higher levels than occur in mice (Kimberlin and Marsh, 1975; Marsh and Kimberlin, 1975).

12.8 Glial changes

There is now general acceptance amongst experimental pathologists working with scrapie that astrocytic hypertrophy is a normal part of the encephalopathic process (Fig. 12.11). The question remains open, however, whether this hypertrophy is accompanied by proliferation, although many workers have interpreted the histology in this way (Hadlow, 1961; Zlotnik, 1962a; Eklund et al., 1963, 1967; Lampert et al., 1971; Joubert et al., 1974), while others explicitly exclude proliferation (Pattison and Jones, 1967; Field, 1969). There are considerable interpretative problems in attempting to identify changes in numbers of any neuro-ectodermal cells using routine methods, as the identification of particular cell-types is notoriously difficult. There are elegant selective staining procedures for astrocytes, using heavy metals, but these are capricious and unreliable, especially for large-scale

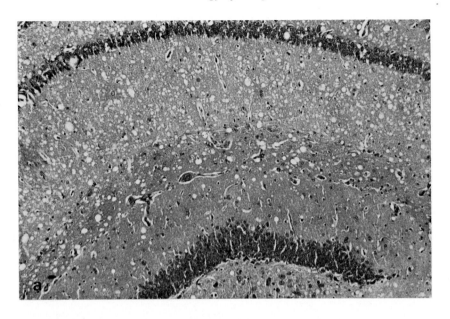

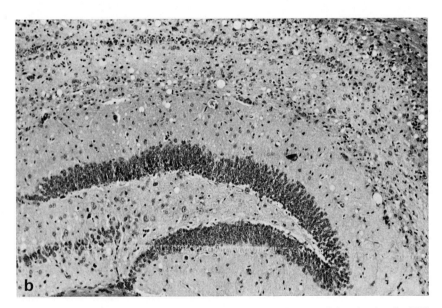

Fig. 12.12. (a) Vacuolation in hippocampus unaccompanied by glial activation in C3H mouse infected with ME7 agent. (b) Glial activation in hippocampus in C57BL mouse infected with ME7 agent. Haematoxylin–Eosin. × 190.

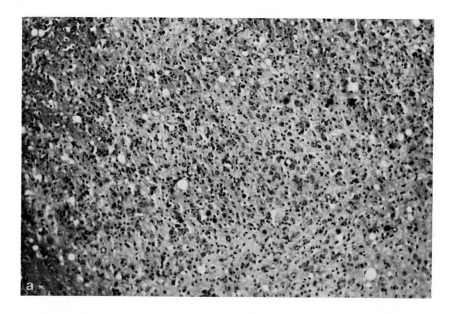

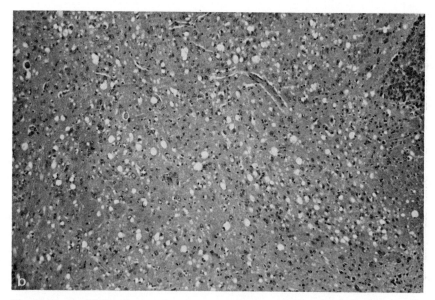

Fig. 12.13. (a) Glial activation in thalamus in C3H mouse infected with 22A scrapie agent; (b) vacuolation in thalamus unaccompanied by glial activation in BALB/c mouse infected with ME7 agent. Haematoxylin–Eosin. × 190.

routine quantitative studies. Astrocytes in the normal mouse brain fail to be impregnated with gold in the most common (Cajal) method for astrocytes and therefore such a technique is quite inappropriate to identify astrocytic proliferation in what would, on other grounds, be the species of choice. One report mentioned a 3-fold increase in the number of astrocytes staining with gold in mouse scrapie caused by 'Chandler' agent, but it was concluded that this resulted from astrocytic hypertrophy rather than from an increase in number (Giorgi et al., 1972). In view of the inability to stain eutrophic astrocytes, to derive a baseline, it is difficult to see how such data can be interpreted as hypertrophy occurring either with or without cell proliferation.

The astrocytic response is restricted to those brain regions affected by the degenerative, vacuolating process: this has been revealed using an agent (125A) which damages certain regions of the brain severely, leaving other regions unaffected (Fraser and Bruce, unpublished). The astrocytic response appears to precede the vacuolation by about 2 weeks (Eklund et al., 1963; Pattison and Jones, 1967), though this has been studied with few agent sources and not in inbred mice. In mice, the encephalopathic process is sometimes, but by no means always, accompanied by a severe gliosis, particularly in the thalamus and hippocampus (Figs 12.12 and 13). In our

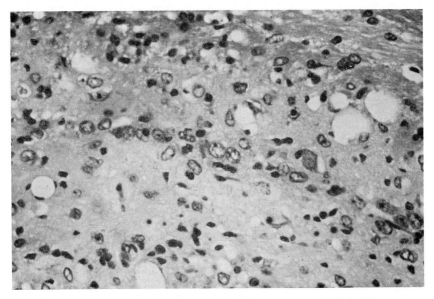

Fig. 12.14. Swollen astrocytic nuclei resembling Alzheimer type II glia in hippocampus of VM mouse infected with 22A agent. Haematoxylin–Eosin. × 800.

studies this has been recognised in about 0.5% of cases but detailed analysis has not yet been started. The cause of this gliosis and any variation due to host or agent strains is not yet understood but its occurrence is difficult to explain and demonstrates the importance of withholding conclusions until they can be based on widely representative samples, with various combinations of route, agent and host. The type of cells in this gliosis varies; microglia and astrocytes can both apparently increase in number, and swollen astrocytic forms including Alzheimer type II cells are also present (Fig. 12.14). Unfortunately microglial staining is rarely successful in mice, so the problem of analysing this gliosis in quantitative terms is still impracticable. Studies of DNA synthesis in scrapie mouse brain do not support the suggestion that cell proliferation is occurring but gliosis may have been absent in these studies (13.3 and 4).

Tissue culture studies of scrapie have yielded a line of cells in which 139A agent replicates: these cells have been described alternatively as astrocytic (Haig, 1970) or as mesodermal (Haig and Clark, 1971; Haig, 1972). As criteria appropriate for their identification were not given, any conclusion concerning their type or origin, based on gross morphology alone, must be recognised as entirely speculative.

Eventual identification of cells involved in these gliotic processes in scrapie, and clear-cut evidence as to whether the astrocytes proliferate, could be very important for basic understanding of this disease. A possible way ahead may be the use of specific neuro-ectodermal cell markers, by cell-specific labelled antibody, such as the astrocyte-specific protein (Bignami et al., 1972).

12.9 Lesions similar to scrapie

The emphasis in this chapter has been on the number of variables which can have precise effects on the pathology of scrapie, and it must be acknowledged that, in an operational sense, we do not know the limits of this variability. Examples have been cited in which the widely accepted degenerative lesions of scrapie are trivial, and it has been pointed out that the underlying lethal biomolecular dysfunction is unknown.

A number of disease processes bear a close resemblance to certain aspects of scrapie, either in broad biological terms or in the narrower context of histopathology. TME, kuru and Creutzfeldt–Jakob disease are generally accepted as homologues of scrapie, but the extent of interrelationships between their causal agents is unknown (Ch. 15). A likelihood that other neurological diseases of unknown aetiology, in particular Alzheimer's disease, might also fall into this category is suggested by similarities in pathology, notably the occurrence of cerebral amyloidosis, although not all scrapie agents produce it, and it has not been reported in TME.

The idea that neural ageing might be caused by scrapie-like processes has also been suggested (Field, 1967; Gajdusek, 1972) on the basis of similarities in the progressive degenerative changes, astrocytic enlargement, vacuolar changes, and cerebral amyloidosis reputed to accompany senescence in certain species. This hypothesis can become a self-fulfilling concept because the limits of the so-called 'senile process' are undefined, and there are only a limited number of ways that nervous tissue can degenerate.

The compound cuprizone is a chelating agent and amine oxidase inhibitor, which produces extensive status spongiosus in the white matter and brain stem of mice and a profound gliosis with Alzheimer type II metaplasia (Carlton, 1969), demyelination followed by remyelination (Blakemore, 1972) and changes in hepatocyte mitochondria and smooth endoplasmic reticulum (Suzuki and Kikkawa, 1969). It has been suggested that this range of brain lesions closely resembles those produced by 'Chandler' agent in mice (Pattison and Jebbett, 1971). A detailed study of brains of RIII mice

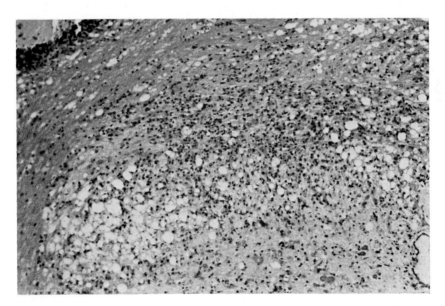

Fig. 12.15. Vacuolation and malacia and severe gliosis in white matter of cerebellum in RIII mouse, produced by cuprizone. Haematoxylin–Eosin. × 190.

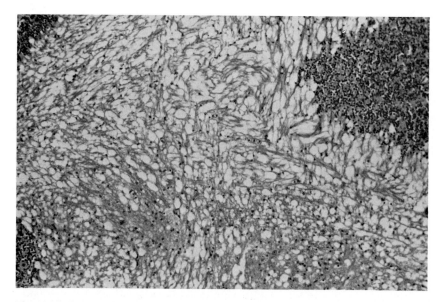

Fig. 12.16. Severe white matter vacuolation in cerebellum of VL mouse, produced by hexachlorophene. Haematoxylin–Eosin. × 190.

killed at intervals following cuprizone at various dose-levels failed to justify the conclusion that the lesions closely resemble any of the range of changes seen in mouse scrapie with a wide variety of agents. At least in these RIII mice, in addition to vacuolation, the cuprizone-induced lesion is an intensely cellular one (Fig. 12.15), quite unlike the lesions associated with scrapie (Fraser, unpublished). Prolonged hexachlorophene intoxication also produces profound spongy degeneration in white matter, ultrastructurally an intra-myelinic oedema (Lampert et al., 1973), and at the light microscopic level this hexachlorophene lesion in VL mice (Fig. 12.16) much more closely resembles the lesion induced in white matter by some scrapie agents (e.g., 139A, 79A) than do the cuprizone lesions.

12.10 Ultrastructural studies

The majority of the dozen or so electron microscopic studies have used mice and only a few have been on the natural disease (Bignami and Parry, 1971, 1972a, b; Narang, 1974a; Carrier et al., 1973). No common unifying ob-servation or hypothesis has emerged either to answer the questions posed by the qualitative optical studies or to provide a clue to the basic pathogenic process. The narrowly based sampling techniques, inherent in electron microscopic methodology, coupled with the immense range of morphological findings – degeneration products, particles, vesicles, etc. – have not been conducive to penetrating interpretations. Few of the studies have been primarily concerned with pathology – the search for an elusive 'scrapie virus' particle having been a major objective.

Studies in mice terminally affected either with 'Chandler' agent and another source isolated from sheep (Lampert et al., 1971) have found focal neuronal swellings with curled membrane fragments and vacuoles formed in the neuritic processes, containing granulo-filamentous material. An interesting finding, also reported in the other degenerative encephalopathies (Lampert et al., 1972) is the coalescence of vacuolar distentions in adjacent neurons and neurites, and the presence in these distentions and in cleared parts in neighbouring cells, of vesicular and curled membrane fragments. Adjacent vacuoles sometimes fuse by an apparent dissolution of the intervening membranes. This process has been reported to occur between neuronal processes (Lampert et al., 1971) and between astrocyte processes (Field and Raine, 1964). Vacuolation also occurs within both pre- and post-synaptic portions of neural processes in which membrane fragments also occur. The extracellular space is not the site of the vacuoles which would appear to be essentially intracellular in origin.

Hypertrophy and alteration in astrocyte cytoplasm can be seen ultrastructurally, but there is no consistency on the question of proliferation (Field and Raine, 1964; Field et al., 1967b; Chandler, 1967, 1968; Lampert et al., 1971). No specific change is associated with oligodendrocytes, although there is a report of 60-nm diameter branched convoluted tubules with dense bodies in them (Field and Narang, 1972). It has been implied that rats and mice may differ with regard to microglial involvement, with only rats having microglial activation, some cells containing pleomorphic, plate-like inclusions (Field et al., 1967b), although this conclusion may well be confounded by agent-strain differences.

Many of the reports refer to various particles, the commonest being rod-shaped particles 35–50 nm in diameter, found both in murine scrapie (David-Ferreira et al., 1968) and the natural disease (Bignami and Parry, 1971, 1972a). The undoubtedly similar 35 nm tubules described by Lampert et al. (1972) are suggestive of virus particles and are frequent in post-synaptic terminals. The 15–26 × 75 nm rod-like particles, also in post-synaptic terminals in natural scrapie (Narang, 1973, 1974b, c) may be identical to these. A variety of other particles, also of unknown significance, have been

H. Fraser

Fig. 12.17. Electron micrograph of stellate amyloid plaque in SM mouse produced by 87A scrapie agent. × 7500.

described in several publications (Field and Raine, 1964; Chandler, 1967; Field and Narang, 1972; Narang, 1973).

There is one ultrastructural study of the 'stellate' type of amyloid plaque which occurs with certain scrapie agents (12.4; Bruce and Fraser, 1975), showing that they possess the characteristics (Fig. 12.17) typical of senile (or neuritic) plaques (Wiśniewski et al., 1973). On the basis of silver studies it has subsequently been suggested that the scrapie plaques are identical with senile plaques (Wiśniewski et al., 1975). This conclusion needs confirmation with independent methods, particularly in view of its important implications for an infectious aetiology of the human dementias in which senile plaques are a prominent and characteristic lesion.

In conclusion it must be said that these studies have not amounted to a more precise definition of the disease than the earlier optical microscopy had done, both failing to distinguish primary from secondary changes. The relative importance of glial and neuronal changes, the questions of cellular proliferation, migration, mitosis and cell death, and the significance of the membranous and vesicular changes remain unknown. The putative viral nature of the variety of particles found can only be regarded as speculative.

12.11 Concluding comments

Infections usually activate host defences which reverse or slow the progress of infection, or with some diseases, contribute to the manifestations of disease by an immunological over-reaction (see Ch. 9). However, there are examples of infectious processes in which reactivity on the part of the host is absent, but these can usually be explained in terms of immunological tolerance or immunosuppression. There remains this group of four diseases in which there are neither homeostatic reactions on the part of the host, nor at present an explanation for this absence (scrapie, Creutzfeldt–Jakob disease, kuru and transmissible mink encephalopathy). In them, although transmission of 'foreign' informational molecules occurs, the host does not mobilize homeostatic mechanisms either against the molecular structure of the agent, its replication products, or the consequent progressive and lethal disorders of cellular physiology.

The only specific morphological alterations in scrapie occur in association with the CNS tissues, and are essentially degenerative and slowly progressive. Although the clinical disease to some extent reflects this neuroectodermal degeneration, it has never been possible to show a precise relationship between the neurological signs and the type, distribution or intensity of lesions. In view of this it is cautionary to introduce two areas where lack of knowledge exists. Firstly, although a great many pathological studies have concentrated on the CNS, very few have been made on the motor, sensory or autonomic peripheral nervous systems. Secondly, a primary biochemical, molecular or cellular lesion for scrapie has not been identified (see Ch. 13), and in view of the widespread replication and accumulation of the agent in non-neural tissues such as in the lympho-reticular organs, there is a probability that important biochemical lesions occur throughout many cells and tissues in the body. It is against this background of ignorance of the primary processes of scrapie that the relative importance of any histo-pathological changes must be viewed.

Although the visible lesions probably reflect only remotely the sequence of events leading to them, their study has been important in furthering our understanding of this group of diseases even though a definition of the disease based on the histological lesions may be too narrow. The important contribution comes from the realisation that the variation in pathology can precisely reflect many of the biological factors which interact to control pathogenesis, and that these can be resolved with carefully applied quantitative methods.

Kuru, Creutzfeldt–Jakob disease, TME and scrapie share many features in their clinical, pathological, biological and transmission properties (see Ch. 15). However, in natural scrapie, the vacuolation is virtually confined to brainstem neurons, whereas in some of its experimental forms and in the other three natural encephalopathies, a spongy vacuolation, or status spongiosus, predominates particularly in the fore-brain. In kuru, TME and Creutzfeldt–Jakob disease vacuolation of nerve cell perikarya is not an outstanding feature of the pathology. It is not necessary to regard the two forms of vacuolar degeneration in grey matter as discrete or unrelated phenomena. They almost certainly reflect different manifestations of the same basic lesion. Ultrastructural studies have failed to arrive at a single view regarding the exact localisation of vacuolation in the neuropil, but the weight of evidence suggests that the lesion is a neuronal and intraneuritic one (Lampert et al., 1971). To what extent vacuolation in the white matter is formed by the same basic defect is open to speculation, although this must appear likely.

Astrocytic hypertrophy is a common reactive response to any alterations in the homeostatic balance in the neuroparenchyma, and in scrapie, predictably, there is a profound enlargement of the astrocytic soma. The limiting technical problems of astrocytic staining have already been discussed (12.8).

The effects and interactions of the major biological variables on the glial reactions have not been studied adequately. It was proposed that scrapie is a 'primary astrocytic disease' (Pattison, 1965a) on the basis of two types of observation: (a) that organ-cultured nervous tissue from scrapie brains possesses an enhanced growth ability analogous to the transformation of cells by oncogenic viruses (Field and Windsor, 1965; Gustafson and Kanitz, 1965; Haig and Pattison, 1967; Webb, 1967) and (b) that the earliest histological lesion in scrapie is an hypertrophy of astrocytes, as revealed by gold chloride (Eklund et al., 1967) and suggests a benign neoplastic change (Field, 1967). Any support for this heterodox view must centre upon a demonstration that the cells growing in culture are astrocytes and that these altered properties of astrocytes are not similarly present in other known situations of astrocytic reactivity, but such a demonstration has not been made. Astrocytes respond rapidly to brain injury, and it is therefore predictable that an alteration in them occurs at an early stage of the disease. Astrocytic alterations in scrapie are almost certainly reactive in nature.

An alternative hypothesis has been proposed, that microglial cells, or brain macrophages, occupy the central position in pathogenesis, and that mesodermal cells (including reticulo-endothelial and endothelial cells) are responsible for agent replication in vitro, and also in the spleen and lympho-reticular tissues in vivo. This hypothesis is based on rather unsubstantial evidence that cultures of 'scrapie mouse brain' cells give rise to foci of mesodermal cells (Haig and Clark, 1971; Haig, 1972); that the scrapie lesion is accompanied by an increase in the microglial population, for which there is only contradictory evidence available (Field, 1967), and that capillary endothelial cells are prominent in explant cultures of scrapie brain (Haig and Pattison, 1967). At present such observations must be treated with the greatest caution, and any interpretation based on them recognised as speculative.

Another school of thought exists which maintains that scrapie is a 'system degeneration' (Beck et al., 1964). Although there are differences in the clinical presentation of scrapie, it has not been found that the histopathological features can be associated with any particular neurological abnormalities although this may reflect the lack of appropriate clinical tests in laboratory animals and livestock. The 'system degeneration' view of scrapie, with implications regarding aetiology, is that the pathology is consistent with a 'spontaneous' degeneration in two major anatomico-physiological systems in the CNS: the cerebellar and hypothalamo-neuro-hypophyseal systems. Selected clinical occurrences have been invoked to support this, but the wide variability in the distribution, intensity, symmetry and type of lesions associated with different scrapie agents and genotypes of host, as well as with other biological variables, conflict with the restricted view of scrapie as basically due to system degeneration. The phenomenon of system degeneration suggests an analogy with some putative genetic diseases rather than with transmissible diseases. There is one report of a deliberate study which fails to confirm a system degeneration in experimental scrapie in sheep (Field, 1969). The term 'Wallerian degeneration' has been used to describe fibre break-down in the spinal cord, cerebellar peduncles and optic tract in scrapie (Palmer, 1968). This was interpreted as a 'dying-back' process, which is inconsistent with secondary or 'Wallerian degeneration'. A low-grade secondary degeneration is a predictable outcome of the neuronal destruction of scrapie. In the mouse there is quite severe and wide-spread degeneration of nerve fibres in the CNS with at least two, biologically quite different, agents commonly in experimental use. With ME7 agent the distribution of fibre degeneration is haphazard and does not involve any particular neuro-anatomical systems and is widespread throughout the grey matter (Fraser, 1969). In the white matter as a whole, large diameter fibres display evidence of damage, but this may reflect the particular agent (ME7) and mouse strains used in these studies. With 'Chandler' agent, in non-inbred mice, lipid breakdown products occur in various sites, but surprisingly, not in the white matter areas where vacuolation is severe (Mackenzie and Wilson, 1966).

The inbred mouse work with scrapie has shown that vacuolation can occur asymmetrically, even following intraperitoneal injection, and agent strain is the most important variable determining this. Of particular importance is the finding that those agents associated with asymmetry are also those liable to induce a high frequency of cerebral amyloid plaques. Much more work is needed with these plaque-producing agents because, although the amyloid deposits take a variety of forms, some bear a close resemblance to human senile plaques even though no comparable structures have been identified in extremely old mice uninfected with scrapie (12.4).

The pathology of scrapie is predominantly degenerative, and no homeostatic host response can resist the destructive progress of the disease. However, synthesis of a new protein, amyloid, which is characterised as a complex of immunoglobulin light chains (Glenner et al., 1971), must be recognised as a departure from the implicit generalisation that the disease is a totally degenerative process. Systemic amyloidosis is recognised as a terminal reaction of an exhausted immunoresponsive system, or the result of a disorganised immunoglobulin synthesis of certain effector cell malignancies. That such a protein occurs in some forms of the degenerative encephalopathies offers important possibilities towards understanding their pathogenesis. A re-examination of immune responses in scrapie now seems worthwhile using pathological criteria to choose the most likely combinations of agent and host for immunological study.

References

ALBRECHT, P. (1968) Curr. Topics Microbiol. Immunol., 43, 43.

BECK, E. and DANIEL, P. M. (1965a) In: (D. C. Gajdusek, C. J. Gibbs, Jr, and M. Alpers, Eds) Natl. Inst. Neurol. Dis. Blindness, Monogr. 2, Slow, Latent and Temperate Virus Infections. (U.S. Government Printing Office, Washington, D.C.) p. 85.

BECK, E. and DANIEL, P. M. (1965b) In: (D. C. Gajdusek, C. J. Gibbs, Jr, and M. Alpers, Eds) Natl. Inst. Neurol. Dis. Blindness, Monogr. 2, Slow, Latent and Temperate Virus Infections. (U.S. Government Printing Office, Washington, D.C.) p. 203

BECK, E., DANIEL, P. M. and PARRY, H. B. (1964) Brain, 87, 153.

BIGNAMI, A., ENG, L. F., DAHL, D. and UYEDA, C. T. (1972) Brain Res., 43, 429.

BIGNAMI, A. and PARRY, H. B. (1971) Science, 171, 389.

BIGNAMI, A. and PARRY, H. B. (1972a) Brain, 95, 319.

BIGNAMI, A. and PARRY, H. B. (1972b) Brain, 95, 487.

BLAKEMORE, W. F. (1972) J. Neurocytol., 1, 413.

BLINZINGER, K. and ANZIL, A. P. (1974) Lancet, ii, 1374.

BRUCE, M. E. and FRASER, H. (1975) Neurobiol. Appl. Neuropathol., 1, 189.

CARLTON, W. W. (1969) Exp. Mol. Pathol., 10, 274.

CARRIER, H., JOUBERT, L., LAPRAS, M. and GASTELLU, J. (1973) Bull. Soc. Sci. Vet. Med. Comp., 75, 101.

CHANDLER, R. L. (1967) Res. Vet. Sci., 8, 98.

CHANDLER, R. L. (1968) Br. J. Exp. Pathol., 69, 52.

CHANDLER, R. L. and FISHER, J. (1963) Lancet, 2, 1165.

CHANDLER, R. L. and TURFREY, B. A. (1972) Res. Vet. Sci., 13, 219.

DANIEL, P. M. (1972) J. Clin. Pathol., 25, suppl. (R. Coll. Pathol.) 6, 97.

DAVID-FERREIRA, J. F., DAVID-FERREIRA, K. L., GIBBS, C. J. and MORRIS, J. A. (1968) Proc. Soc. Exp. Biol. Med., 127, 313.

DAYAN, A. D. (1971) Brain, 94, 31.

DICKINSON, A. G. and OUTRAM, G. W. (1973) J. Comp. Pathol., 83, 13.

DICKINSON, A. G., STAMP, J. T. and RENWICK, C. C. (1974) J. Comp. Pathol., 84, 19.

DICKINSON, A. G., STAMP, J. T., RENWICK, C. C. and RENNIE, J. C. (1968) J. Comp. Pathol., 78, 313.

EKLUND, C. M., HADLOW, W. J. and KENNEDY, R. C. (1963) Proc. Soc. Exp. Biol. Med., 112, 974.

EKLUND, C. M., KENNEDY, R. C. and HADLOW, W. J. (1967) J. Infect. Dis., 117, 15.

FIELD, E. J. (1967) Dtsch. Z. Nervenheilk., 192, 265.

FIELD, E. J. (1969) Int. Rev. Exp. Pathol., 8, 129.

FIELD, E. J. and NARANG, H. K. (1972) J. Neurol. Sci., 17, 347.

FIELD, E. J. and RAINE, C. S. (1964) Acta Neuropathol., 4, 200.

FIELD, E. J., RAINE, C. S. and JOYCE, G. (1967a) Acta Neuropathol., 8, 47.

FIELD, E. J., RAINE, C. S. and JOYCE, G. (1967b) Acta Neuropathol., 9, 305.

FIELD, E. J. and WINDSOR, G. D. (1965) Res. Vet. Sci., 3, 130.

FRASER, H. (1969) Res. Vet. Sci., 10, 338.

FRASER, H. (1970) Proc. VIth Int. Congr. Neuropathol., (Masson et Cie, Paris) p. 897.

FRASER, H. (1971) Ph.D. Thesis, University of Edinburgh.

FRASER, H. and BRUCE, M. E. (1973) Lancet, i, 617.

FRASER, H., BRUCE, M. E. and DICKINSON, A. G. (1974) Proc. VIIth Int. Congr. Neuropathol., Budapest, 1-7th Sept. 1974, in press.

FRASER, H. and DICKINSON, A. G. (1967) Nature, 216, 1310.

FRASER, H. and DICKINSON, A. G. (1968) J. Comp. Pathol., 78, 301.

FRASER, H. and DICKINSON, A. G. (1973) J. Comp. Pathol., 83, 29.

GAJDUSEK, D. C. (1972) Adv. Gerontol. Res., 4, 201.

GIORGI, P. P., FIELD, E. J. and JOYCE, G. (1972) J. Neurochem., 19, 225.

GLENNER, G. G., TERRY, W., HARADA, M., ISERSKY, C. and PAGE, D. (1971) Science, 172, 1150.

GOLDHAMMER, Y., BUBIS, J. J., SAROVIA-PINHAS, I. and BRAHAM, J. (1972) J. Neurol. Neurosurg. Psychiat., 35, 1.

GUSTAFSON, D. P. and KANITZ, C. L. (1965) In: (D. C. Gajdusek, C. J. Gibbs, Jr, and M. Alpers, Eds) Natl. Inst. Neurol. Dis. Blindness, Monogr. 2, Slow, Latent and Temperate Virus Infections. (U.S. Government Printing Office, Washington, D.C.) p. 221.

HADLOW, W. J. (1961) Res. Vet. Sci., 2, 289.

HAIG, D. A. (1970) Proc. VIth Int. Congr. Neuropathol. (Masson et Cie, Paris) 856.

HAIG, D. A. (1972) Report IInd Int. Congr. Virol., Budapest (Karger, Basel) p. 208.

HAIG, D. A. and CLARKE, M. C. (1971) Nature, 234, 106.

HAIG, D. A. and PATTISON, I. H. (1967) J. Pathol., 93, 724.

HOLMAN, H. and PATTISON, I. H. (1943) J. Comp. Pathol., 53, 231.

JOUBERT, L., BONNEAU, M. and CARRIER, H. (1974) Rev. Med. Vet., 125, 647.

KIMBERLIN, R. H. and MARSH, R. F. (1975) J. Infect. Dis., 131, 97.

LAMPERT, P., GAJDUSEK, D. C. and GIBBS, C. J. (1972) Am. J. Pathol., 68, 626.

LAMPERT, P., HOOKS, J., GIBBS, C. J. and GAJDUSEK, D. C. (1971) Acta Neuropathol., 19, 81.

LAMPERT, P., O'BRIEN, J. and GARRETT, R. (1973) Acta Neuropathol., 23, 326.

MACKENZIE, A. and WILSON, A. M. (1966) Res. Vet. Sci., 7, 45.

MARSH, R. F. and KIMBERLIN, R. H. (1975) J. Infect. Dis., 131, 104.

MOULD, D. L., DAWSON, A. MCL., SLATER, J. S. and ZLOTNIK, I. (1967) J. Comp. Pathol., 77, 393.

NARANG, H. K. (1973) Res. Vet. Sci., 14, 108.

NARANG, H. K. (1974a) Acta Neuropathol., 28, 317.

NARANG, H, K, (1974b) Acta Neuropathol., 29, 37.

NARANG, H. K. (1974c) Neurobiology, 4, 349.

OUTRAM, G. W., FRASER, H. and WILSON, D. T. (1973) J. Comp. Pathol., 83, 19.

PALMER, A. C. (1957a) Vet. Rec., 69, 1318.

PALMER, A. C. (1957b) Nature, 179, 480.

PALMER, A. C. (1968) Vet. Rec., 82, 729.

PARRY, H. B. (1969) In: (C. W. M. Whitty, J. T. Hughes and F. O. MacCallum, Eds)

Virus Diseases and the Nervous System. (Blackwell Scientific, Oxford and Edinburgh) p. 99.

PATTISON, I. H. (1965a) In: (D. C. Gajdusek, C. J. Gibbs, Jr, and M. Alpers, Eds) Natl. Inst. Neurol. Dis. Blindness, Monogr. 2, Slow, Latent and Temperate Virus Infections. (U.S. Government Printing Office, Washington, D.C.) p. 249.

PATTISON, I. H. (1965b) Vet. Rec., 77, 1388.

PATTISON, I. H., GORDON, W. S. and MILLSON, G. C. (1959) J. Comp. Pathol., 69, 300.

PATTISON, I. H. and JEBBETT, J. N. (1971) Res. Vet. Sci., 12, 378.

PATTISON, I. H. and JONES, K. M. (1967) Res. Vet. Sci., 8, 160.

PATTISON, I. H. and JONES, K. M. (1968) Res. Vet. Sci., 9, 408.

SUZUKI, K. and KIKKAWA, Y. (1969) Am. J. Pathol., 54, 307.

WEBB, H. E. (1967) Proc. R. Soc. Med., 60, 698.

WIGHT, P. A. L. (1960) J. Comp. Pathol., 70, 70.

WISNIEWSKI, H. M., BRUCE, M. E. and FRASER, H. (1975) Science, in press.

WISNIEWSKI, H. M., GHETTI, B. and TERRY, R. D. (1973) J. Neuropathol., 32, 566.

WISNIEWSKI, H. and TERRY, R. D. (1973) Progr. Neuropathol., 2, 1.

ZLOTNIK, I. (1957a) Nature, 179, 737.

ZLOTNIK, I. (1957b) Nature, 180, 393.

ZLOTNIK, I. (1958a) J. Comp. Pathol., 68, 148.

ZLOTNIK, I. (1958b) J. Comp. Pathol., 68, 428.

ZLOTNIK, I. (1961) J. Comp. Pathol., 71, 440.

ZLOTNIK, I. (1962a) Acta Neuropathol., suppl., 1, 61.

ZLOTNIK, I. (1962b) J. Comp. Pathol., 72, 366.

ZLOTNIK, I. (1963) Lancet, ii, 1072.

ZLOTNIK, I. and RENNIE, J. C. (1957) J. Comp. Pathol., 67, 30.

ZLOTNIK, I. and RENNIE, J. C. (1958) J. Comp. Pathol., 68, 411.

ZLOTNIK, I. and RENNIE, J. C. (1965) J. Comp. Pathol., 75, 147.

CHAPTER 13

Biochemical and behavioural changes in scrapie

Richard H. KIMBERLIN

13.1 Introduction

Early studies of the biochemical abnormalities in scrapie-affected animals were carried out using sheep and goats, before the experimental form of the disease in mice became available in the early 1960s. Some of these early studies involved estimating various brain constituents in an attempt to find biochemical correlates of the known histopathological lesions of scrapie which occur entirely within the CNS (12.1). Many measurements of CSF, blood and urine constituents were also carried out (see Kimberlin, 1969a, for references) with the particular aim of finding a simple diagnostic test for scrapie. Most of this work gave negative results and little has been done since on the biochemistry of the disease in sheep and goats (see Parry and Livett, 1973). The main exceptions are some histochemical studies of glycosidase activity in scrapie-affected brains. Of particular interest is the presence of intraneuronal 'droplets' of β-glucuronidase activity in naturally and experimentally infected sheep brain. These droplets have a pronounced brick-red colour when suitably stained, and despite the uncertainty of the role of increased β-glucuronidase activity in pathogenesis (see later), their presence may be a useful diagnostic test of sheep scrapie in addition to the usual histopathological lesions (Mackenzie et al., 1968a; Mackenzie, 1971).

A great deal of work has been carried out on the biochemical changes in scrapie-affected mice and to a lesser extent in other rodents. The main reason for this is the unsuitability of many conventional techniques for studying the nature and pathogenesis of the disease. For example, the present

Slow virus diseases of animals and man, edited by R. H. Kimberlin
ⓒ *North-Holland Publishing Company 1976*

inability to identify the agent in cells and tissues by electron microscopic or immunological methods has left only two main approaches to the study of pathogenesis. One is based on measurements of infectivity (14.3 and 4) and the other on studies of the biochemical and physiological events in disease development. The latter approach has attracted many workers simply because it is not dependent on extremely time-consuming biological assays of infectivity. It is, however, fraught with many difficulties as the following discussion will show, and despite the large amount of work carried out, very little is known of the pathogenesis of scrapie in molecular terms.

Much of the subject matter of this chapter has been reviewed elsewhere (see Kimberlin, 1969a, 1976a, b). It is therefore unnecessary to describe all the studies that have been carried out. Instead, the discussion will be limited to a critical account of the problems involved in this kind of study and to a review of selected aspects of the field where extensive studies have been made.

13.2 *Biochemical changes in clinically affected mice*

Since scrapie is a disease of the CNS and all the known histopathological abnormalities occur in brain, the great majority of biochemical determinations have been made on this tissue. Furthermore, most studies have employed a strain of scrapie transmitted to mice by Chandler (the 'Chandler' strain; see 10.6) which gives a similar incubation period in most of the commonly used mouse strains.

In the clinical stage of scrapie produced by the Chandler strain there is a progressive decrease in body weight which is accompanied by a general reduction in the size of many organs, e.g., liver, spleen and submaxillary salivary gland (Kimberlin, 1969b). There is also a decrease in brain weight which is accompanied by alterations in the content of water, protein, lipid, carbohydrate, nucleic acid, Na^+, K^+ and inorganic phosphate (see Kimberlin, 1969a, for references). However, even in the terminal stages of disease most of these changes are relatively slight ($\pm 10\%$) and it is interesting that many biochemical functions are scarcely altered at all; for example the respiratory activity of brain homogenates, protein and RNA metabolism (see Kimberlin, 1969a).

Some of the changes that occur can be related to some extent to the histopathological abnormalities present. It is reasonable to suggest that the altered Na^+ and K^+ concentrations are due to an altered function of astrocytes, and the changes in major brain constituents probably reflect, in part, the vacuolation (Pattison and Smith, 1963) and nerve fibre degeneration seen histologically (Fraser, 1969). Furthermore, these degenerative changes are possibly brought about by the increased activity of several lysosomal hydrolases which include acid proteinase, RNAase, DNAase and a number of glycosidases (13.3).

Although these abnormalities give some indication of the type of disease process in scrapie, it will be appreciated that the onset of clinical disease represents the end of the line in so far as pathogenesis is concerned. Therefore it is almost impossible to evaluate the significance of changes at this late stage when many secondary events complicate the picture.

With one exception, which will be discussed later, all the biochemical abnormalities found outside the CNS have been studied only in clinical scrapie. Hence the significance of an increased β-glucuronidase activity in spleen (Little and Adams, 1971), and of an increased incorporation of [^{3}H]thymidine and [^{14}C]glucosamine into a DNA–polysaccharide complex in the same tissue (Adams et al., 1969) is also difficult to assess.

It is therefore imperative to study biochemical changes in scrapie at much earlier times if their significance in disease pathogenesis is to be revealed. The following is a brief synopsis of the most detailed investigations which meet this requirement.

13.3 Biochemical changes in scrapie-infected mice before the onset of disease

A decreased activity of an enzyme which hydrolyses ATP directly to AMP occurs about half-way through the incubation period of Chandler scrapie in affected brain. This enzyme is present in the mitochondrial, microsomal and nuclear fractions of brain but the decrease in activity in scrapie mice occurs mainly in nuclei. The supernatant fraction contains little apparent enzyme activity. However, the supernatant from scrapie brain does contain an inhibitor which in 'mix-match' experiments will reduce the enzyme activity of normal nuclei to the same level that is found in scrapie nuclei. Preliminary characterisation of this inhibitor suggests it is a protein (Hunter et al., 1972b).

Some interesting changes in polyamine metabolism have been described in scrapie brain. During the first half of the incubation period (after intra-cerebral infection) the concentrations of spermine and spermidine decrease and, then in the latter half, they progressively increase. The latter increase correlates with the progressive development of astrocyte hypertrophy but the significance of the earlier change is not known. These changes in polyamine concentration are accompanied by increased rates of $[^{14}C]$putrescine incorporation into both spermine and spermidine (Giorgi, et al., 1972). In contrast to the situation in brain, the concentration of spermine and spermidine in spleen is higher during the first half of the incubation period (Giorgi et al., 1972). This is the only early change reported in spleen. In view of the rapid rate of agent replication in this tissue, even after intracerebral inoculation of Chandler agent (Hunter et al., 1972a) it is tempting to suggest a possible association between increased polyamine concentration and agent replication in spleen. However, more extensive studies must be carried out to establish this point.

Several studies have been made of glycoprotein metabolism in scrapie brain. Briefly, a substantial increase in the rate of incorporation of radioactively labelled L-fucose and D-N-acetyl mannosamine into brain glycoproteins has been described. These increases are progressive, starting about half-way through the incubation period and reaching twice the control levels during the clinical stage of the disease (Hunter and Millson 1973; Suckling and Hunter, 1974a). It would appear that this is a general increase in precursor incorporation involving the majority of proteins found in all the subcellular fractions examined; the significant exception is the proportionately higher labelling of low molecular weight glycoproteins in the soluble fraction of scrapie brain (Hunter et al., 1972b). Some related studies have shown that an increased activity of galactosyl- and N-acetyl glucosaminyl-transferase is present in scrapie brain when measured in vitro using endogenous acceptors. However, these changes did not occur until the early clinical signs of scrapie developed. Much earlier increases in the activity of galactosyl- and fucosyl-transferases can be detected using exogenous acceptors, and in the clinical stage of the disease these activities reach twice the normal values (Suckling and Hunter, 1974b).

It is interesting to note that a number of hydrolytic enzymes have an increased activity in scrapie brain, including several glycosidases. A progressive increase in enzyme activity, starting about half-way through the incubation period, has been reported for β-glucuronidase, N-acetyl-β-D-glucosaminidase, acid DNAase and acid proteinase (Millson, 1965; Mould et al., 1967; Mackenzie et al., 1968a; Kimberlin et al., 1971). However, in the clinical stage of the disease a much broader group of 11 glycosidases has been studied, some of which also show increased activity in the second half of the incubation period (Millson and Bountiff, 1973). The characteristic profile in mice showing clinical signs consists of 4 enzymes with a marked increase in activity, namely, β-glucuronidase (GUR), N-acetyl-β-D-glucosaminidase (GAM), *N*-acetyl-β-D-galactosaminidase (GALAM) and α-mannosidase at pH 4.1 (MAN 4.1); two others, α-fucosidase and β-xylosidase are moderately increased in activity and five remain unaltered, that is α- and β-glucosidases, α- and β-galactosidases and α-mannosidase at pH 6.0. Although subsequent studies have indicated that these enzymes are not primarily involved in scrapie pathogenesis, the profile of enzyme changes has been a useful parameter for comparing scrapie and TME in hamsters (Kimberlin, 1973; Kimberlin and Marsh, 1975).

Finally, there is a substantial increase (2–3-fold in clinically affected mice) in the rate of nuclear DNA synthesis in scrapie brain, which is first detectable at about half-way through the incubation period (Kimberlin and Hunter, 1967; Adams, 1972; Kimberlin, 1972). The rate of DNA synthesis in mitochondria or in other extranuclear sites is not generally increased (Kimberlin, 1972; Adams, 1972) but there are reports of an increased rate of incorporation of [^{3}H]thymidine and of [^{14}C]glucosamine into a DNA–polysaccharide complex in mouse brain (Adams et al., 1969, 1970). More recently, the presence of a small single-stranded DNA in scrapie brain has been claimed (Adams, 1972). However, none of these extranuclear abnormalities have been studied earlier than the clinical stage of the disease. The increased rate of nuclear DNA synthesis is probably related to the increased rates of histone synthesis and histone acetylation which also occur before the onset of clinical signs (Caspary and Sewell, 1968a, b). However, it is interesting that the elevated rate of DNA synthesis in scrapie occurs mainly in a metabolically labile fraction of nuclear DNA which is also found in normal brain (both young and adult); in other words, the increase is not specifically associated with the turnover of cells (Kimberlin et al., 1974b).

Most of the abnormalities described in this section are progressive, and first occur at the beginning of the second half of the incubation period. This appears to be highly significant because the onset and progression of histopathological lesions follows a similar time course (Table 13.1). Indeed, various authors have established a close correlation between the development of either astrocyte hypertrophy or vacuolation (or both) and changes in the following:- acid proteinase, GAM, spermine and spermidine (Mould et al., 1967; Mackenzie et al., 1968a; Giorgi et al., 1972). In another study, a close correlation was found in the time of onset and rate of development of vacuolation, astrocyte hypertrophy, the increase of activities of GAM, GUR and acid DNAase and the increased rate of DNA synthesis. This study was carried out on mice infected intraperitoneally in which the rate of increase of infective agent was also determined in brain. So close was the time of

TABLE 13.1

Temporal relationships of the known events taking place in mice after intracerebral inoculation of high doses of the Chandler strain of scrapie. This table only gives a very
approximate guide to the time when various changes are taking place. The incubation
period for Chandler scrapie after intracerebral inoculation in most mouse strains is
16–20 weeks. For other details see text.

Event	Stages in the incubation period when each event occurs			
	1st quarter	2nd quarter	3rd quarter	4th quarter
Increase in infectivity:				
Spleen	*****			
Brain	*****	*****	*****	*****
Behavioural changes		*****	*****	*****
Histological changes:				
Vacuolation			*****	*****
Astrocyte hypertrophy			*****	*****
Biochemical changes:				
Polyamine metabolism				
Spleen	*****	*****	*****	*****
Brain	*****	*****	*****	*****
ATP → AMP inhibitor			*****	*****
Glycoprotein synthesis			*****	*****
Glycosidases			*****	*****
Nuclear DNA synthesis			*****	*****
Histone metabolism			*****	*****

onset of all these changes that they appeared to form part of a single disease
process which was initiated when the concentration of agent in brain
exceeded $6.0 \log_{10}$ units of infectivity/20 mg of brain (Kimberlin et al., 1971).

13.4 Biochemical changes in different cell populations in scrapie brain

A general problem in evaluating any biochemical change in diseased brain
is that of associating the change observed with individual populations of
cells in this very complex tissue. This problem is particularly great in
diseases like scrapie because they involve both 'proliferative' changes
associated with astrocyte hypertrophy and 'degenerative' changes such as

vacuolation and nerve fibre degeneration (Ch. 12). Histochemical techniques have been used to correlate enzyme changes with histological lesions. For example, hypertrophied astrocytes can be detected as easily by their increased content of oxidative enzymes (glutamate dehydrogenase, lactate dehydrogenase, NADH diaphorase and NADPH diaphorase) as they can by Cajal staining (Buening and Gustafson, 1971b). The distribution of increased GAM activity also correlates quite well with astrocyte hypertrophy (Mackenzie et al., 1968a; Buening and Gustafson, 1971b). Both groups of authors have observed a mainly paravascular association of GUR activity in mouse scrapie and Mackenzie et al. (1968a) emphasised that this distribution was rather different from that observed for GAM activity. In fact the increase in GUR tends to be associated more with areas of vacuolation than with astrocyte hypertrophy. This association is also apparent in affected sheep brain since vacuolation and increased GUR activity are commonly found within neurons (Mackenzie et al., 1968a). However, the anatomical complexity of brain, coupled with the diffuse localisation of many histochemical reactions, makes a precise association of enzymes with cells difficult; the intraneuronal droplets of GUR seen in sheep scrapie is a striking exception in the present context.

Some attempt to overcome this problem have been made in two further studies. Chandler and Smith (1968) applied a histochemical test for GAM at the ultrastructural level and found that some GAM activity was associated with medium-sized neurons in mouse scrapie. Buening and Gustafson (1971a) studied the histochemical localisation of GAM and GUR in primary cell cultures of normal and scrapie mouse brain on glass coverslips. Primary cells can be attached to glass much more easily from scrapie-affected mouse brain than normal adult brain (Field and Windsor, 1965; Gustafson and Kanitz, 1965; Haig and Pattison, 1967; Pattison et al., 1971), which is why the scrapie cultures had a higher number of cells with GAM and GUR activity. However, the main point is that enzyme activity was associated with fibroblastic-type cells, probably blood capillary pericytes, microglia or oligodendroglia. No activity was detected in astrocytes and it was considered doubtful if neurons would survive in such an in vitro system (Buening and Gustafson, 1971a).

Two other techniques are currently available to examine the occurrence of biochemical abnormalities in specific brain cells, namely, the bulk separation of different cell types using density gradient centrifugation, and autoradiography. Cell separation has not so far been applied to study the biochemical changes in scrapie brain and autoradiography has only been used in one case; to study the increased rate of nuclear DNA synthesis in affected mouse brain (Kimberlin and Hunter, 1967). It was anticipated that this abnormality might be associated with the astrocyte response in scrapie, particularly as some authors have emphasised the occurrence of an astro-cytosis (i.e., an increase in the number of these cells), in addition to astrocyte hypertrophy (12.8). However, this prediction was not fulfilled. An increased number of labelled glial nuclei was detected autoradiographically throughout the brains of mice infected by an intracerebral route. Moreover, the increase occurred very early in the incubation period of scrapie; but it did not correlate with the astrocyte response, which came several weeks later (Kimberlin and Anger, 1969). More importantly, it did not occur in mice infected by an intraperitoneal route of inoculation. In this case the number of labelled glial nuclei remained the same in both normal and scrapie animals at all stages in the incubation period even though the increased rate of [^{3}H]thymidine incorporation, measured biochemically, occurred after intra-peritoneal infection as it did using an intracerebral route (Kimberlin et al., 1971). It was therefore concluded that the increased number of labelled cells seen autoradiographically soon after intracerebral infection was some kind of local response caused by the injection of biochemically altered brain tissue in the inoculum.

13.5 *The generality of biochemical abnormalities in scrapie*

These findings raise the general point that the inoculation of a suspension of normal brain is not necessarily an adequate control for infection with a scrapie brain homogenate. This point is relevant to the great majority of studies on scrapie since the intracerebral inoculation of clinically affected brain is widely used to achieve a short incubation period. It is, therefore, important when searching for early events in pathogenesis to confirm any abnormalities using a peripheral route of inoculation when non-specific effects due to the inoculation of biochemically altered brain are less likely to occur. An alternative tissue source of agent might be considered to minimise the problem: particularly spleen which has a relatively high con-centration of agent long before any of the degenerative changes of scrapie are apparent (Hunter et al., 1972a).

It is well known that many different strains of scrapie agent exist (10.6). These can be distinguished using biological criteria and also by histo-pathological lesion profiles (see 12.2.3 and 14.4.4d). The latter method is particularly relevant to the present discussion since different agent strain–

mouse strain combinations produce a very wide spectrum of damage in scrapie, in terms of distribution and severity of vacuolation. It was stated earlier in this Chapter that most studies of the biochemistry of scrapie-infected brain have been carried out with the Chandler strain of agent. This strain is unusual in that it commonly produces a great deal of white matter status spongiosus, whereas an agent such as ME7 usually gives minimal white matter damage, depending on the mouse strains used and 22C has not been found to produce any white matter damage in a wide variety of mouse strains (12.5). It is therefore important to consider the generality of a bio-chemical change in terms of different agent strain–mouse strain combinations.

There are no published reports where this point has been investigated experimentally. However, some preliminary studies have shown that 3 other agent strains give a broadly similar glycosidase profile as the Chandler agent. Of particular interest is the finding that strain 79A in BALB/c mice, which produces extensive white matter damage, gives a glycosidase profile identical to that found with 22C in BALB/c mice which produces virtually no white matter vacuolation at all (Kimberlin and Dickinson, unpublished).

The point about the generality of a particular biochemical change applies equally well to other variables in scrapie. The route of inoculation has al-ready been mentioned; other variables include the strain of mouse with respect to the *sinc* gene (which controls the length of incubation period of mouse scrapie; 14.5) and the species of host.

In one histochemical study, an increased activity of GAM was confirmed in mice, and detected in rats but was absent from goats and sheep. Similar studies of GUR showed that the increased activity in scrapie mice was paravascular whereas in affected sheep and goats it was mainly within neurons (Mackenzie et al., 1968a). Biochemical measurements of glycosidase activity in scrapie-affected rats have revealed the same large increase in the activity of GAM, GALAM and GUR as found in mice but MAN 4.1, which is also much increased in mice, remains unaltered in rats (Kimberlin et al., 1976). The profile in scrapie-affected hamsters differs markedly from rats and mice in that only GUR and β-xylosidase show a substantial increase in activity. However, the increased rate of incorporation of DNA and glycoprotein precursors is similar in both mice and golden hamsters (Kimberlin and Marsh, 1975).

There is, however, a major difference between scrapie-affected mice and hamsters in the time of occurrence of the various abnormalities. In mice, the enzyme changes and the increased incorporation of glycoprotein and DNA precursors first occur at the same time as the histopathological lesions:

long before the clinical disease develops. In hamsters, the histological lesions first occur preclinically but the biochemical changes do not arise until late in the clinical disease in spite of the very high concentrations of agent present in brain at earlier times. This comparison between species is, there-fore, very revealing and suggests that many of the changes found in mouse scrapie are not fundamental to scrapie pathogenesis (Kimberlin and Marsh, 1975; Marsh and Kimberlin, 1975).

13.6 Biochemical changes in brain in other pathological states

It has already been stressed that mice inoculated with normal tissue are not necessarily adequate controls for scrapie-infected mice. Consequently, it is important to investigate other conditions which produce brain damage in order to identify those abnormalities which are specific for scrapie and those which are produced by a variety of unrelated causes.

Two attempts to investigate this problem have been made. In the first, a histochemical study of glycosidases in infected mouse brain was made in which scrapie was compared with louping ill, yellow fever and infection with *Herpes simplex:* conditions with histopathological lesions quite different from scrapie (Mackenzie et al., 1968b). The viral encephalitides all resembled scrapie in producing an increased GUR activity, although the distribution was not the same in each infection. However, an increased activity of GAM and GALAM was found only in scrapie suggesting some degree of specificity with these two glycosidases.

The second approach was based on the observation that mice fed on a diet containing cuprizone (biscyclohexanoneoxaldihydrazone) developed histo-pathological lesions resembling those induced by the Chandler strain of scrapie, particularly white matter vacuolation, gliosis and astrocyte hyper-trophy (Pattison and Jebbett, 1971). It should be appreciated that white matter vacuolation is not produced by all scrapie agents (12.5) and, that ultrastructurally, cuprizone-induced vacuolation is different from that found in scrapie and the related subacute spongiform encephalopathies (Gonatas, 1970; Adornato and Lampert, 1971). However, cuprizone administration to mice produces a series of biochemical changes in brain which are remarkably similar to those produced by Chandler strain of scrapie (Kimberlin, et al., 1974a). The profile of 11 glycosidases was almost identical in the two con-ditions, both qualitatively and quantitatively. In addition, cuprizone induced increases in the activity of acid DNAase and in the rate of incorporation of [^{3}H]thymidine into DNA which were very similar to those found in scrapie.

The only major differences between cuprizone toxicity and scrapie infection (in terms of brain biochemistry) was the lack of any increased incorporation of glycoprotein precursors in cuprizone-fed mice. These results clearly indicate that many of the biochemical abnormalities of scrapie are not specific for this disease and hence are unlikely to be of primary significance in scrapie pathogenesis; at the same time, they suggest that cuprizone toxicity could be a useful control for some of the non-specific effects of Chandler scrapie in future experiments.

Some recent studies have shown that a severe white matter vacuolation and a generalised astrocyte hypertrophy can also be induced in golden hamster brain by feeding cuprizone. In view of the very different glycosidase profiles in hamster and mouse scrapie, it was of interest to investigate the glycosidase profile of cuprizone toxicity in hamster brain. The results showed, on the one hand, a marked disimilarity to both scrapie and cuprizone in mice, and on the other, a general similarity to scrapie in the hamster. It would appear therefore, that cuprizone could be used as a control for some of the non-specific changes produced by scrapie in hamster brain (Kimberlin et al., 1976).

13.7 Behavioural changes in scrapie

A number of behavioural tests have been applied to scrapie-affected mice in the hope of detecting early brain damage which could provide clues to the pathogenesis of disease. Savage and Field (1965) found that the emotional responsiveness of scrapie mice, as measured by the open-field test, decreased progressively during the second half of the incubation period and in the clinical stage of the disease. Heitzman and Corp (1968), using an emergence test, found abnormal behaviour even earlier in scrapie-affected mice. In fact, significant differences were found after about one-third of the incubation period had elapsed following intracerebral infection. More recently, a detailed study has been made of the drinking and feeding habits of mice infected with scrapie (Outram, 1972). In particular, it was found that a progressive decline in polydipsia induced by giving mice sucrose or saline solutions takes place early in the incubation period of scrapie. This phenomenon has been observed with three different agent strains using two routes of inoculation and several strains of mice some of which differed at the *sinc* locus.

Not only do these results establish the generality of the drinking response in scrapie but there was a good correlation between the time of onset of the response and the length of incubation under a very wide variety of conditions. This correlation and the early occurrence of the drinking response in mice suggests that it is caused by some primary interaction between the agent and cells in the CNS. It is worth emphasising that this behavioural change first occurs at a time when the concentration of agent in brain is relatively low and the histopathological lesions of scrapie have scarcely begun to develop.

13.8 Concluding remarks

Table 13.2 summarises, very briefly, some of the more important abnormalities found in scrapie brain so far; and the likelihood of their involvement in scrapie pathogenesis is indicated by the number of 'yes's' given. It seems highly likely that most of these abnormalities are secondary effects, although at least two findings, those concerning polyamine metabolism and the inhibitor of ATP conversion to AMP, should be investigated further. The abnormality in polyamine metabolism found in spleen is uniquely interesting as the only early change reported in this tissue. If this finding is confirmed it may well be that an altered polyamine metabolism reflects an important interaction between agent and cells outside the CNS which is, perhaps, concerned with agent replication. However, much more work needs to be done before one can usefully speculate on the significance of this abnormality.

Even though we still have no certain knowledge of the primary bio-chemical events in the pathogenesis of scrapie, some important general points about the problem have been established by the studies carried out so far.

First, the necessity for establishing the generality of a biochemical response to scrapie infection under a wide variety of conditions has been stressed. This is particularly true in relation to the *sinc* gene which controls the incubation period of scrapie in mice. Secondly, the question of suitable controls has been raised and the value of examining other CNS conditions has been discussed as an important way of testing the specificity of a particular biochemical change in scrapie. The close similarity of some aspects of brain damage caused by scrapie infection and cuprizone toxicity is intriguing; it suggests that there may only be a limited number of ways the CNS can respond to traumatic agents and it raises the question of what is the common denominator which underlies a similar response induced by such apparently diverse factors as scrapie and cuprizone.

The third point is that it is important to be able to associate biochemical

TABLE 13.2

Brief summary of the most studied biochemical changes in scrapie brain and their possible significance in scrapie pathogenesis. The spaces in the table signify not known. A biochemical change which is likely to be of fundamental importance in scrapie pathogenesis should have a row of 6 YES's opposite it. For other details see text.

Biochemical change	Mice				Hamsters	
	Preclinical onset	Progressive	Absent from other CNS infections	Absent from cuprizone toxicity	Preclinical onset	Progressive
Polyamine metabolism	YES	YES				
ATP → AMP inhibitor	YES	YES			NO	YES
Glycoprotein synthesis	YES	YES		YES		
Histone synthesis	YES	YES				
Nuclear DNA synthesis	YES	YES		NO	NO	YES
DNAase	YES	YES		NO		
N-acetyl-β-D-glucosaminidase	YES	YES	YES	NO	NO	NO
N-acetyl-β-D-galactosaminidase	YES	YES	YES	NO	NO	NO
β-Glucuronidase	YES	YES	NO	NO	NO	YES
Mannosidase, pH 4.1	YES	YES		NO	NO	NO
α-Fucosidase	YES	YES		NO	NO	NO
β-Xylosidase	YES	NO		NO	NO	NO

changes in scrapie brain with specific cell types. The brain is such a complex
organ that the interpretation of biochemical abnormalities is very difficult
without this information. However, it should be emphasised that there is
no knowledge of which of the cell populations in brain contain the agent;
this is another question which needs to be investigated.

Finally, it is apparent that even in the early clinical stage of the disease
few of the biochemical abnormalities are particularly striking and many of
the metabolic functions of brain proceed more or less unimpaired (13.2).
This is an important point because it suggests that the primary biochemical
events in scrapie are relatively subtle and hence will be difficult to detect.

A number of general features in the development of the disease support
this view. For instance, the pathogenesis of scrapie is an extremely well
regulated process which is why the length of incubation period can be
measured with great precision despite the long periods of time involved
(14.4.2a). Several pieces of evidence suggest that the long incubation period
of scrapie is due to the initial delay before agent multiplication begins in the
CNS (which varies with the conditions of infection) and then to the slow
rate of multiplication within this tissue. However, it is significant that despite
the prolonged replication process, the onset of clinical disease is charac-

DIAGRAM 13.1.

Possible causal relationships in the pathogenesis of clinical scrapie.

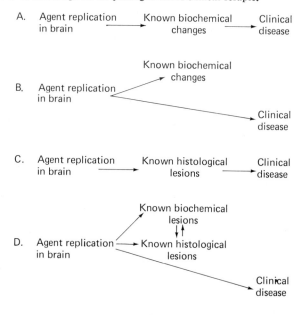

teristically very sharp, at least in rodents. This situation suggests that onset of disease is triggered by some threshold event, probably directly related to the attainment of a certain critical concentration of agent in brain. Before this critical concentration is reached it would appear that an equilibrium situation exists in which damage caused by the agent is corrected by some 'repair' or 'compensatory mechanism'; in other words, the clinical disease develops when the compensatory mechanism can no longer cope with the ever increasing amount of agent-induced damage. If this assessment is correct, then the task of identifying preclinically the subtle biochemical process involved in pathogenesis will clearly be difficult.

It must not be forgotten, of course, that biochemical and histopathological changes occur in mouse scrapie before the onset of clinical disease. Therefore with these abnormalities the compensatory process could be said to be failing. However, it has been shown earlier in this chapter that many of the known biochemical lesions are not likely to be of fundamental importance to the pathogenesis of disease, despite the fact that several develop progressively as the concentration of agent increases; this is represented in Diagram 13.1 as situation B rather than A.

The significance of the histopathological lesions has been discussed in detail elsewhere (see Ch. 12). In the normal course of events one would reasonably suppose that histological lesions lead directly to the development of clinical disease, particularly if immunological and ultrastructural techniques showed the presence of agent in damaged cells (see Diagram 13.1C). However, such demonstrations have not been made in scrapie, and because of the unusual nature of the disease it may be unwise to make any assumptions of this kind. Indeed there is good reason to question a simple causal relationship between the known histopathological lesions of scrapie and the clinical disease (see Chs 12 and 15; Kimberlin, 1976a, b). Hence, it may be better to summarise the importance of the known histopathological lesions of scrapie by means of Diagram 13.1D.

The uncertainty about the significance of histological lesions makes any studies of biochemical abnormalities difficult to interpret, unless they can also be detected before any abnormal histology is apparent. Hence the need to find biochemical changes at even earlier times in the incubation period. The behavioural studies of scrapie-infected mice indicate that biochemical changes are there to be found very early in the incubation period. The altered drinking response in scrapie mice is particularly interesting because it not only satisfies the basic requirements for a primary change but it is also an aspect of animal behaviour that is relatively well understood in basic neurophysiology. It is early events of this kind which seem to be the promising ones to explore in order to understand the pathogenesis of clinical scrapie.

References

ADAMS, D. H. (1972) J. Neurochem., 19, 1869.

ADAMS, D. H., CASPARY, E. A. and FIELD, E. J. (1969) J. Gen. Virol., 4, 89.

ADAMS, D. H., CASPARY, E. A. and FIELD, E. J. (1970) Arch. Ges. Virusforsch., 30, 224.

ADORNATO, B. and LAMPERT, P. (1971) Acta Neuropathol., 19, 271.

BUENING, G. M. and GUSTAFSON, D. P. (1971a) Am. J. Vet. Res., 32, 953.

BUENING, G. M. and GUSTAFSON, D. P. (1971b) Am. J. Vet. Res., 32, 959.

CASPARY, E. A. and SEWELL, F. M. (1968a) Biochem. J., 108, 37P.

CASPARY, E. A. and SEWELL, F. M. (1968b) Experientia, 24, 793.

CHANDLER, R. L. and SMITH, K. (1968) Res. Vet. Sci., 9, 228.

FIELD, E. J. and WINDSOR, G. D. (1965) Res. Vet. Sci., 6, 130.

FRASER, H. (1969) Res. Vet. Sci., 10, 338.

GIORGI, P. P., FIELD, E. J. and JOYCE, G. (1972) J. Neurochem., 19, 255.

GONATAS, N. K. (1970) Proc. 6th Int. Congr. Neuropathol. (Masson et Cie, Paris) p. 49.

GUSTAFSON, D. P. and KANITZ, C. L. (1965) In: (D. C. Gajdusek, C. J. Gibbs, Jr, and M. Alpers, Eds) Natl. Inst. Neurol. Dis. Blindness, Monogr. 2, Slow, Latent and Temperate Virus Infections. (U.S. Government Printing Office, Washington, D.C.) p. 221.

HAIG, D. A. and PATTISON, I. H. (1967) J. Pathol. Bacteriol., 93, 724.

HEITZMAN, R. J. and CORP, C. R. (1968) Res. Vet. Sci., 9, 600.

HUNTER, G. D., KIMBERLIN, R. H. and MILLSON, G. C. (1972a) Nature, New Biol., 235, 31.

HUNTER, G. D. and MILLSON, G. C. (1973) J. Comp. Pathol., 83, 217.

HUNTER, G. D., MILLSON, G. C. and HEITZMAN, R. J. (1972b) Ann. Inst. Pasteur, 123, 571.

KIMBERLIN, R. H. (1969a) Biochem. J., 114, 20P.

KIMBERLIN, R. H. (1969b) Res. Vet. Sci., 10, 392.

KIMBERLIN, R. H. (1972) J. Neurochem., 19, 2767.

KIMBERLIN, R. H. (1973) Biochem. Soc. Trans., 1, 1058.

KIMBERLIN, R. H. (1976a) Merrow Monogr., Microbiol. Ser., (Newcastle-upon-Tyne, U.K.) in press.

KIMBERLIN, R. H. (1976b) Sci. Progr., in press.

KIMBERLIN, R. H. and ANGER, H. S. (1969) J. Neurochem., 16, 543.

KIMBERLIN, R. H. and HUNTER, G. D. (1967) J. Gen. Virol., 1, 115.

KIMBERLIN, R. H. and MARSH, R. F. (1975) J. Infect. Dis., 131, 97.

KIMBERLIN, R. H., COLLIS, S. C. and WALKER, C. A. (1976) J. Comp. Pathol., in press.

KIMBERLIN, R. H., MILLSON, G. C., BOUNTIFF, L. and COLLIS, S. C. (1974a) J. Comp. Pathol., 84, 263.

KIMBERLIN, R. H., MILLSON, G. C. and MACKENZIE, A. (1971) J. Comp. Pathol. 81, 469.

KIMBERLIN, R. H., SHIRT, D. B. and COLLIS, S. C. (1974b) J. Neurochem., 23, 241.

LITTLE, K. and ADAMS, D. H. (1971) Res. Vet. Sci., 12, 202.

MACKENZIE, A. (1971) Ph.D. Thesis, University of Reading, U.K.

MACKENZIE, A., MILLSON, G. C. and WILSON, A. M. (1968a) J. Comp. Pathol., 78, 43.

MACKENZIE, A., WILSON, A. M. and DENNIS, P. F. (1968b) J. Comp. Pathol., 78, 489.

MARSH, R. F. and KIMBERLIN, R. H. (1975) J. Infect. Dis., 131, 104.

MILLSON, G. C. (1965) J. Neurochem., 12, 461.

MILLSON, G. C. and BOUNTIFF, L. (1973) J. Neurochem., 20, 541.

MOULD, D. L., DAWSON, A. MCL., SLATER, J. and ZLOTNIK, I. (1967) J. Comp. Pathol., 77, 393.

OUTRAM, G. W. (1972) J. Comp. Pathol., 82, 415.

PARRY, H. B. and LIVETT, B. G. (1973) Nature, 242, 63.

PATTISON, I. H., CLARKE, M. C., HAIG, D. A. and JEBBETT, N. J. (1971) Res. Vet. Sci., 12, 478.

PATTISON, I. H. and JEBBETT, N. J. (1971) Res. Vet. Sci., 12, 378.

PATTISON, I. H. and SMITH, K. (1963) Res. Vet. Sci., 4, 269.

SAVAGE, R. D. and FIELD, E. J. (1965) Anim. Behav., 13, 443.

SUCKLING, A. J. and HUNTER, G. D. (1974a) Biochem. Soc. Trans., 2, 1124.

SUCKLING, A. J. and HUNTER, G. D. (1974b) J. Neurochem., 22, 1005.

CHAPTER 14

The pathogenesis of scrapie in mice

George W. OUTRAM

14.1 Introduction

'I realise now that I shall not live to see the leucocyte problem solved'

Virchow, 1898

The reader who has reached here via the previous chapters will be well prepared to expect problems and paradoxes in the study of the pathogenesis of scrapie in mice. Small mammal models are often very welcome in the study of diseases, and the successful passage of scrapie agents to mice in the early 1960s occasioned high hopes of a swift breakthrough in what had proved to be a somewhat intractable problem in the natural hosts (Ch. 10). In the event, however, and despite a large amount of work, mouse models have served mainly to confirm the unorthodox nature of the disease and have not so far provided a much needed theory capable of encompassing all the known facts and observations. This immediately presents problems to the writer of a section such as this. An adequate theory, once described, provides him with a structure for assembling and briefly discussing a large corpus of experimental observations. In the absence of such a theoretical framework he has the choice of either using the most obvious (but inappropriate) models derived from another field (in this case virology), or of painstakingly setting out the operational details of the experimental procedures and steadfastly refusing to put any fundamental interpretation upon them. The former could be entertaining but would be misleading, while the latter would inevitably be exceedingly dull – however honest. This chapter must, therefore, be something of a compromise. It is a basic premise of what follows that the repeated failures of the virological and what might be called the 'molecular

Slow virus diseases of animals and man, edited by R. H. Kimberlin
© *North-Holland Publishing Company 1976*

biological' approaches to scrapie to present an adequate theory of the disease have a two-fold consequence: first, that much of the basic experimental work must henceforth be much more consciously 'operational' in design and reportage with findings being described in phenomenological terms rather than virological ones, and second, that there is an ongoing need to extend further the study of biological variation in the disease with a view to discerning its nature in terms that are appropriate to itself, rather than in terms imposed by, say, molecular virology. These points will become clearer later.

Mouse scrapie work during the past 14 years falls into several categories which have been employed in various combinations and with different emphases in about half a dozen laboratories.

(1) The study of the neuropathology (Ch. 12).

(2) Attempts to characterise the agent by physico-chemical methods (Ch. 11).

(3) Attempts to analyse biochemical and physiological changes in infected hosts (Ch. 13).

(4) Attempts to detect very early behavioural changes (Ch. 13 and 14.4.2).

(5) Attempts to study the dynamics of pathogenesis by titration methods (14.3).

(6) Attempts to uncover the fundamental nature of the disease by the study of biological variation (14.4).

This chapter will be confined mainly to a consideration of categories 5 and 6. The interested reader should refer to other chapters in this volume and the recent review by Kimberlin (1976) for further details.

One of the unfortunate consequences of the long incubation periods in this and similar diseases, is that it can take several years to discover a methodological error. Two such errors have been common in the past decade. The first has been too much concentration upon a narrow range of models (usually the one which gives the shortest possible incubation period) with the assumption that it can be used to derive valid generalisations about 'scrapie in the mouse'. The second has been an emphasis upon the clinical phase of the disease because it is when the most obvious pathological events are occurring. As will be seen, both these errors expose the investigator to the danger of being deflected from fundamentals by the lure of conspicuous secondary disease phenomena.

14.2 An outline of scrapie pathogenesis in mice

In lieu of an introductory discussion of the theoretical nature of scrapie disease, which it is not yet possible to give, this section will outline a tentative generalisation of the course of scrapie pathogenesis in the mouse at a phenomenological level, beginning with a brief reminder of the chief anomalies of this disease in terms of conventional virology.

Although scrapie (in many of its numerous forms, 14.4.4d) can certainly be passed from sheep to mouse, mouse to mouse, and from mouse to sheep (Ch. 10) by injecting infected tissues, the most strenuous efforts have never revealed an infectious particle (by electron microscopy), although filtration and irradiation–inactivation techniques give size values in the order of 7–40 nm for the infectious unit; no antibody responses (or anomalies) have been detected in infected hosts; the agent does not appear to infect tissue cultures or cause plaques in them; infectivity is resistant to a wide range of 'viricidal' treatments including heat (e.g., 160 °C dry heat for 1 h), UV-irradiation specific for nucleic acids, many enzymes including lipases, proteases and nucleases, and 20% formol; infectivity titres of only about 10^8 infectious units per gram of brain tissue; remarkably regular incubation periods lasting months or years, during the whole or large part of which agent is replicating and passing between various tissues, mostly without causing overt damage; lesions appear to be confined to the CNS where they take the form of a relatively benign-looking, non-inflammatory activation of astrocytes accompanied by various degrees and patterns of distribution of grey- and white-matter vacuolation. The direct causes of either clinical signs or death are unknown.

The object of the study of pathogenesis is to trace all the interactions which occur between pathogen and host with a view to identifying the crucial ones which will permit an understanding and control of the disease. It generally involves questions of entry routes, sequences of infected tissues (cells, organelles), the nature of damage sustained, the nature of host defences, the identification of 'target' tissues whose infection leads to serious damage and perhaps death. It will be seen that very little unequivocal knowledge of this sort has been obtained for scrapie in the mouse. The reasons for this are at least three-fold: the agent leaves no 'trail' as it proceeds on its inexorable course through the host, and the only method of detection and assay depends upon the development of disease in susceptible hosts; secondly, incubation periods are so long that experimental progress is slow; and thirdly, there has been much deflection of research effort into the hope that molecular–virological techniques would provide a short-cut to the answers.

'I gotta use words when I talk to you' (*Sweeney Agonistes*, T. S. Eliot)

– and in highly operational sciences, words can be very misleading. The reader is advised therefore to accept the following generalised statement of scrapie pathogenesis in the mouse with the utmost caution, using it only as a rough guide to the subject and trying not to import too many virological connotations.

Scrapie does not seem to be maternally transmitted in mice (Dickinson, 1967; Field and Joyce, 1970; Clarke and Haig, 1971b) and most instances of lateral contagion are probably traceable to oral ingestion by licking oozed inocula, fighting or cannibalism (Dickinson et al., 1964; Pattison, 1964). Experimental infection is therefore generally by injection (i.c., i.p. or s.c.) although oral dosing and scarification have also been used and can be effective. It would seem that, irrespective of route of infection, the earliest rises in titre occur in organs of the lympho-reticular system such as the spleen and lymph nodes, and can begin at any time between a day or two to many months after injection according to the host–agent strain model that is used (14.4.4). It is not yet possible to distinguish between titre rises due to accumulation and replication. Not until, or only shortly after, the onset of detectable titre in the CNS do the characteristic CNS lesions develop. The mode of transport of agent between tissues is unknown although both haematological and neurological routes have been suspected. There is considerable evidence that after an i.c. injection of the normal standard doses, the mouse is killed by agent which remains and replicates in the brain, with relatively little contribution being made by the considerable amount of agent which is replicating simultaneously in peripheral organs (14.4.4b).

After an apparently irreducible minimum value, incubation periods are inversely proportional to dilution when using relatively untreated inocula. However, the slopes of the titration curves vary considerably in different models and information of this kind may be expected to shed much light on pathogenic processes in the future (14.4.4a). Related to the above, and also suggesting that some sort of site limitation is involved, is the observation that plateau titres seem to occur in several tissues – including possibly the CNS. Whether this represents a static or dynamic equilibrium is an open question. A perspective on these facts is given by the estimated ratios of infectious units to cell number, which is in the order of 1/50 in the spleen and 1/1 in the brain (but see 14.3.3).

The period of clinical signs varies according to the model but generally occupies a mere 5–10% of the incubation period. It often begins with some rather ill-defined behavioural abnormality such as hyperexcitibility, and then passes to a condition of 'definite' disease. The details of this also vary considerably, although they are remarkably consistent within particular model systems (14.4). A rough behavioural generalisation is that models showing predominantly fore-brain lesions tend to be somnolent, while those with hind-brain lesions show locomotor incoordination (Ch. 12). The clinical period has been much studied and is characterised by a plethora of histopathological and biochemical changes, some of which are very important for agent-strain identification (Chs 12 and 13, and 14.4.2), and an 'explosive' phase of agent replication in the CNS which may finally flatten into a brief plateau (14.3). It seems very likely, however, that many of these changes are secondary effects consequent upon a loss of the homeostatic control which the host has managed to sustain in the face of the primary upset of function during earlier phases of the disease, and as such are probably of little use for understanding or control. Recently there has been an increased emphasis upon studies during the pre-clinical phase (Chs 12 and 13). When these findings have been checked for generality in a range of different host–agent models they may provide a means for identifying and investigating the primary interaction between agent and host.

Despite considerable numbers of investigations, no trace of an involvement or abnormality of conventional immune responses (including interferon) has been detected in scrapie-infected mice (Chandler, 1959; Haig and Clarke, 1965; Gibbs et al., 1965; Gardiner, 1966; Clarke and Haig, 1966; Clarke, 1968; Katz and Koprowski, 1968; Gardiner and Marucci, 1969; Worthington and Clarke, 1971; McFarlin et al., 1971; Worthington, 1972; Porter et al., 1973), although there is good evidence that they have some kind of 'inactivation' mechanism (14.6).

The most important discovery regarding the fundamental nature of the interaction between host and pathogen has come from the study of variation, in which it has been shown that despite a very wide variety of disease patterns (in terms of length of incubation period and lesion profile) the replication of all scrapie agents is under the over-riding control of a single gene in the mouse. This gene (called *sinc*, 14.5), or its immediate product, seems to provide the replication site for all known scrapie agents, and something similar to it may exist in sheep (Ch. 10). Its control of replication sets the pace for pathogenesis in both the CNS and in peripheral organs, although in the latter case other genes may also have important contributory effects on pathogenesis (14.6). The normal function of *sinc*, and the tissues and cells in which it is active, are unknown although techniques for converging on this question are being devised (14.5 and 6).

The most obvious, and at first sight, unproblematic method of studying the pathogenesis of scrapie is to try to trace the course of pathogenesis by titrating tissues in infected animals. Accordingly, the rest of this chapter will consist of, first, an account of such titration methods and a critique of their interpretation, second, an outline of the considerable amount of data obtained by more indirect methods, and third, some of the phenomeno-logical observations and speculations derivable from these data.

14.3 Study of the dynamics of pathogenesis by direct titration techniques

Scrapie titrations are a laborious and expensive procedure, requiring many months and sometimes thousands of mice to perform. Only one group of workers have had the single mindedness to attempt a full-scale investigation of pathogenesis by this technique, others confining themselves to more limited questions and often using various titration short-cuts. Furthermore there are limitations to the value of titration techniques while there are no independent methods of detecting and assaying infected tissues: it is no longer possible simply to assume that 'operational titre' and 'functional titre' are the same (see 14.3.4).

14.3.1 A classical titration

In a classical and much-quoted experiment, Eklund, Kennedy and Hadlow undertook in 1962 the only full-scale investigation of the dynamics of scrapie titres in the organs of mice using the then recently isolated 'Chandler' agent (10.6). Interim and final accounts were published (Eklund et al., 1964, 1965, 1967), and the following outline of this work is given to illustrate the types of procedure that are used in scrapie experiments and to emphasise the need to know all operational details before results can be assessed.

Methods   The tissues selected for examination and the intervals at which they were taken are shown in Table 14.1 and Fig. 14.1. Organs and tissues were removed from groups of 3 mice at 13 intervals throughout the incuba-tion period after a s.c. injection in the left thigh. The titre of injected material was independently estimated to be $10^{5.7}$ i.c. LD_{50} units. Inocula for titration were prepared from weighed pools of whole organs, made up to a 10^{-1} suspension in physiological saline containing 10% normal rabbit serum. Titrations were performed by injecting groups of 6 weanling Swiss mice i.c. with 0.03 ml of serial ten-fold dilutions through to 10^{-8}. They were terminated

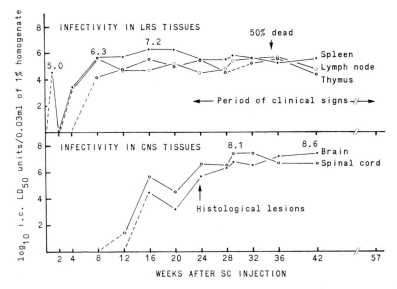

Fig. 14.1. Infectivity titres in various organs at intervals after a s.c. injection of Chandler agent in random-bred mice. Data abstracted from Eklund et al. (1967) to show some of the major features of scrapie pathogenesis in this particular model of the disease. Random-bred mice were injected s.c. in the thigh with $10^{5.7}$ i.c. LD_{50} units/0.05 ml of Chandler agent. The figures on the graphs represent estimated $\log_{10}$ i.c. LD_{50} titres of agent present in the entire organ at a particular point in time.

after one year and Reed–Muench i.c. LD_{50} titres calculated. Estimates of total infectious units in original organs were made. Groups of 3 inoculated and 3 uninoculated control mice of the same age were killed for CNS histology, looking especially for astrocytic activation and cell-vacuolation. A close watch was kept throughout for the onset of clinical signs (details unspecified).

Results After one week agent was found only in the spleen where it reached a titre of 10^5 i.c. infectious units in the whole organ, but at two weeks no infectivity was detected in any of the organs and tissues tested. (No test was performed at three weeks). Four weeks after injection, agent was detected again in spleen (and peripheral lymph nodes totalling about $10^{4.4}$ i.c. LD_{50} units in each organ). By 8 weeks quite high titres were detected also in the thymus and submaxillary glands. It was not until 12 weeks after injection, however, that agent was detected in the CNS (spinal cord). First infectivity in brain was in the 16th week group with a titre of 4.4 i.c. LD_{50} units/0.03 ml.

G. Outram

TABLE 14.1

Approximate titres of agent in various tissues at intervals after subcutaneous injection in the thigh. Simplified data from Eklund et al. (1967) showing approximate i.c. LD_{50} titres of agent in 0.03 ml of 10% homogenate prepared at intervals after a s.c. injection of 0.03 ml of Chandler agent ($10^{5.7}$ i.c. LD_{50} units/0.05 ml) in random-bred mice. Blood clot, serum and testis were also examined but no scrapie activity was ever detected. Uterus, liver and kidney were found to contain low titres in the latter part of the incubation period.

Weeks after injection:	1	2	4	8	12	16	20	24	28	29	32	36	42
Mortality (%):									5	7	25	61	73
Spleen	5	*	4	6	6	6	6	6	6	6	6	5	6
Peripheral lymph nodes			3	6	5	5	5	5	5	5	6	6	5
Thymus				4	5	6	5	5	5	—	5	6	5
Submaxillary salivary gland				6	6	7	5	6	6	6	6	3	3
Lung						4	3	3	3	2	3	2	4
Intestine					2	2	3	5	6	5	5	5	5
Spinal cord					1	6	5	7	7	7	7	7	7
Brain						4	3	6	6	7	7	7	7
Femoral bone marrow								2	3	5	4	5	4

* Spaces indicate no activity detected by assay animals when killed one year after s.c. injection.

Clinical signs began in the 23rd week and were associated with the appearance of the earliest histological changes – astrocytic activation in the brain described as 'hypertrophy and proliferation' although no quantitative data were supplied. Vacuolar and degenerative changes in neurons and ground tissue were never conspicuous and seemed to be later than the astrocytic changes. No inflammation or demyelination were ever observed.

Groups of mice were taken at both the 28th and 29th weeks because of a change in selection policy: up to the 28th week mice were chosen at random from clinically negative individuals and thereafter only from clinically positive ones. No infectivity was detected in blood clot, serum or testis at any time. The salivary gland appeared to be unique in showing a marked decline in titre towards the end of the incubation periods.

Authors' interpretations The authors interpreted their findings in virological terms as follows. After s.c. injection, agent is selectively adsorbed by the spleen during the first week. It then enters an intracellular non-infective phase preparatory to replication of new 'virus'. Thereafter it begins to infect a variety of other organs – perhaps disseminated in blood lymphocytes (at a low level of infectivity). The data for salivary glands suggest that these could be the source of agent for lateral infection (when it occurs) and perhaps for the late appearance of agent in the intestine. When the agent eventually gains access to the CNS (beginning with the spinal cord) it first infects astrocytes causing enlargement and proliferation which leads to a secondary malfunction of neurons as manifested in vacuolation, clinical signs and death. The absence of an immune response is more likely to be due to the slow surreptitious nature of the agent's infection of the host's defence mechanisms than to some supposed physico-chemical property of the agent itself which prevents the recognition of foreignness.

14.3.2 Comments in the light of present knowledge

This series of titrations is probably the most frequently quoted of all mouse scrapie experiments and the broad observational outlines of pathogenesis which it presents for this and similar models is no doubt substantially correct. However, there are two main comments to be made in the light of subsequent findings in mouse scrapie.

First, as to the generality of the findings. It will be shown later (14.4.4) that there are combinations of host- and agent-strain which show very different patterns of pathogenesis from Chandler agent in Swiss mice. These differences certainly relate to absolute timing, and may possibly extend to the order of organ infection. There are a number of more recently discovered, biologically important mouse scrapie models which badly need to be examined by a full-scale titration of this type.

The second comment stems from the authors' assumption that all mice showed essentially the same pattern of agent dynamics (i.e., that the sample of three animals at each stage was large enough to be representative) and that the very wide range in incubation periods (23–57 weeks, with two healthy 'survivors' being discarded at 64 weeks) was to be accounted for in terms of 'variation in individual susceptibility to clinical disease'. Whatever the causes of this variation (and many possible candidates suggest themselves in 14.4) it seems more likely that the wide range in incubation periods is a reflection of a similar range of differences in the dynamics of titre rises. Several of the inflections in the titration curves including: the apparent loss of infectivity during the 2nd week; the dip in brain and cord titres between the 16th and 20th week; the rise in titre coincidental with the change in selection policy between the 28th and 29th week; and perhaps even the apparent decline in salivary gland titres in the longer-living mice – are all likely to be indications of variation in titre dynamics in the small sample taken. It is doubtful therefore if much weight can be put upon the finer details in any attempt to reconstruct the course of pathogenesis.

Another area of doubt is the validity of speculations regarding the relationships between the various histological lesions and the onset of clinical signs. This area is still under investigation (Ch. 12) and the wide variation in clinical syndromes shown by this model (Eklund et al., 1963), make it unlikely that any strict relationship could have been established in this experiment.

14.3.3 Some subsequent, smaller-scale titrations

Several workers have subsequently used sequential titration methods to study the dynamics of agent in brain and spleen, often employing some type of titration short-cut in order to economise on numbers of mice and time. Two types of simplified titration have been used, both with considerable potential for exacerbating even further the interpretive difficulties that are inherent in scrapie titrations. One method seeks to reduce the numbers of mice by estimating titre from the incubation period of a single dilution and reading the value from a previously constructed dose–response curve (valid provided the operational details of the experiments are compatible); the other attempts to reduce the amount of time required by terminating the titration at some point after the bulk of cases have occurred but well before all mice have shown clinical signs and then, depending upon histological criteria, to determine incidence (valid for the quicker models provided that the time of onset of histological lesions always has the same relationship to incubation period). It is unfortunate that, owing to ignorance of all the possible sources of variation, published reports have not always given sufficient operational details for their validity to be assessed. This is an important point since some of the 'contradictions' between the findings of different workers may be a consequence of the tendency still to think in unitary terms of 'scrapie in the mouse' and a failure to appreciate how much variation there can be between different models.

Mouse scrapie models can usefully be divided into two groups: those showing 'operationally short' incubation periods and those showing 'operationally long' ones (14.4.4c and d). This is because the incubation period of scrapie in the mouse is under the over-riding control of the *sinc* gene, the

TABLE 14.2

The effects of strain of agent, *sinc* genotype, route of injection and sex of recipient on incubation period. Representative incubation periods in various scrapie models obtained by pooling results from experimental replicates using weanling mice in groups of at least 6, injected i.c. or i.p. with 0.02 ml of lightly centrifuged supernate (500*g*, 10 min) of 1% homogenates (high-speed teflon-in-glass) of individual terminal mouse brains (C57BL or VM) prepared in sterile physiological saline (no additives). These findings are based upon approximately 4000 mice in about 400 experiments performed during the past 12 years.

Injection route:	i.c.														i.p.							
Agent group:	22A			ME7											22A	ME7						
Agent:	22A	104A	87V	ME7	79A	22C	80A	22L	79V	80V	58A	87A	51C	125A	22A	ME7	79A	22C	79V	58A	87A	51C
VM $sinc^{p7}$																						
♀	195	334*	299	322	276*	439	436	216	251	327	306	586	504*	586*	306	514	422*	540*	370			
♂	206	331*	273	326	309	442	442	247*	290*	318	316	614	543*	519*	319	520	441*	540	435	519		
F1 $sinc^{s7p7}$																						
♀	542>700	566*	244	246	252	252	197	283*	290*	265	226	586*	421*	356*	696*	345	309	327	390	373*	604	
♂	588>700		246	236	263	265	220	290*	265	210	323	429*	651*	345	336	323	390	380			658	
C57BL $sinc^{s7}$																						
♀	453	606*	447*	159	152	177	177	166	228	152	342	326*	333*	601	239	201	234	307*	236*	471	440*	
♂	470	539*	162	157	182	185	172	234	188	167	351	329*	319*	593	238	207	225	340	255*	499	471*	

* S.E. >2½% of the incubation period.

two alleles of which are capable of switching the incubation period of any agent from a 'short' type of incubation period to a 'long' one (see Tables 14.2 and 3), or in other words, any agent can show either an 'operationally short' or an 'operationally long' incubation period according to the *sinc* genotype of the host (Dickinson, 1975; 14.4.4d). In two fully reported experiments, Dickinson and his colleagues investigated the dynamics of the ME7 agent in the brain and spleen of mice homozygous for either the s7 allele (operationally short model) or the p7 allele (operationally long model) (Dickinson et al., 1969; Dickinson and Fraser, 1969a). They found that in the model showing long incubation periods (ME7 agent in VM *sinc*^{p7} mice) there was a considerable delay (weeks) in the onset of detectable titre rises in the brain and spleen when compared with the model showing short incubation periods. Subsequent work with other operationally long models (e.g., 22A agent in C57BL *sinc*^{s7} mice) has shown that the onset of titre rises in the spleen may not occur for many months after injection (Dickinson et al., 1975b). These findings raise the question of where the agent could be during this long period, and whether the spleen always plays the same leading role in pathogenesis.

An ambitious attempt by Mould and his colleagues to trace the details of the dynamics of the ME7 agent in the spleen of C57BL mice after i.c. injection foundered because of the use simultaneously of a stock of assay animals which were known to be segregating for the *sinc* gene and the titration short-cut method of early termination with histological analysis (Mould et al., 1970). The same stock of mice had also been supplied to E. J. Field and his co-workers in Newcastle with the consequence that several of their findings must also be held in doubt – e.g., the report of an 'eclipse phase' (Field et al., 1971) and of modified polyamine metabolism in scrapie mice (Giorgi et al., 1972).

Similar spleen dynamics to those obtained by Eklund et al. (1967) were obtained by Clarke and Haig (1971a) using the Chandler agent i.p. in Compton random-breds, but without any sign of a 'non-infectious' phase. The details of their findings, together with the effects of splenectomy upon incubation period (Fraser and Dickinson, 1970; Clarke and Haig, 1971a; 14.4.41) were used to suggest that titre rises in the spleen were due to replication and not accumulation; that the initial rate of replication was similar to rates of cell division in the spleen; and that there was a direct relationship between spleen size and agent titre.

Hunter and his colleagues have recently reported results of sequential titration experiments using Chandler agent in Compton mice which show no eclipse phase in the spleen (Hunter et al., 1972); that the onset of clinical disease is more or less coincidental with an explosive arithmetical rise in brain titre (Hunter, 1972; Kimberlin, 1975), and that there is a terminal brain plateau phase (Hunter, 1974; Kimberlin, 1975). Some of these findings are based upon unpublished experiments described in passing and therefore few of the operational details are given. It is worth considering the possibility that a terminal plateau could be an artifact of the spread of end-points in a random-bred population.

14.3.4 The relationship between 'operational' and 'functional' titres in scrapie

It is necessary at this point to consider some theoretical problems regarding the interpretation of titration values which have been referred to above. As a quantifying technique, titration in entire hosts gives numerical values which we, at present, have to assume are directly proportional to the number of infectious entities which are functional in the tissue being tested. The term 'functional' is used to include all agent present which is significant for the donor animal, either in terms of its further replication which, as a process, might damage host functions, or as infectivity which will not replicate further in that location but may produce damage by its presence. However, there are some circumstances in which the 'operational' estimate of titre may have little relationship to the 'functional' titre. Some of these circumstances are known to apply to scrapie, others are suspected, while some need to be considered as theoretical possibilities. Obviously this situation is not exclusive to scrapie but applies to any disease for which there is not a variety of independent techniques for conducting assays; and it is also helpful in this respect to have unrelated methods for identifying infected tissues.

Particularly with scrapie, where we have fewer frames of reference than with conventional viruses, the results from a single assay system must be interpreted with caution because of three possibilities: (1) a tissue may contain undetected agent, (2) agent present may be more or less degradable according to the biological, chemical or physical condition of accompanying tissue components, or the presence of biologically active substances used in preparation of the inoculum, and (3) the tissue components present or the preparative technique may cause a re-routing of the course of pathogenesis at a systemic or cellular level.

Some of the agent functionally present in a tissue may be undetected because it is operationally 'labile' using present techniques. This may seem surprising in view of the emphasis on the extreme stability which scrapie agent can display but, especially if this stability comes from some form of structural protection, not all the functional infectivity may be in this stable form, and the proportion of labile agent could vary with the stage of pathogenesis or between different tissues. There is now good evidence that there are processes capable of degrading scrapie activity, at least in the peripheral systems of mice (14.4.4j). This means that experimental treatments of scrapie homogenates, designed to produce direct changes in titre from which to deduce something about the physico-chemical nature of the agent, may in fact cause changes in estimated titre either by changing degradability of agent by the assay animal, or by bringing about a re-routing of the pathogenic process.

What, for instance are we to suppose is the relationship between the 'operational' and 'functional' titre estimates using an i.c. route, when one considers that perhaps 95% (or more) of the inoculum passes immediately into the blood stream (Mims, 1960)? Assuming that this also means that 95% of the 'functional' titre escapes into peripheral systems it could result in at least a 20-fold error in estimates, which may not be important when comparing inocula, but becomes very relevant when trying to relate the number of infectious units to other parameters, i.e., the number of cells in an infected organ.

Two final points: one practical way of guarding against some of these potential sources of error (pending the development of independent ways of detecting and assaying scrapie agents) would be to duplicate titration estimates using both the i.c. and the i.p. routes; when a policy of early termination of titrations is adopted, care needs to be taken not to assume that a termination date based upon the use of untreated material will also be appropriate for treated material, since it is possible to have prolonged incubation periods without an equivalent loss of titre (Dickinson and Taylor, personal communication).

All the titration experiments described in this section, together with many reported in Ch. 11, are subject to interpretative qualification in the light of the above possibilities and need to be scrutinised with them in mind (see 11.3 and 9).

14.4 Indirect methods of studying the dynamics of the pathogenesis of scrapie in mice

In view of the problems outlined above regarding the use of titration techniques for investigating scrapie pathogenesis, it is obviously desirable to have other, perhaps less direct, ways of studying pathogenesis. The tracing of biochemical, histological and physiological changes in infected animals is one such approach, but they are limited by the difficulty of distinguishing primary from secondary effects (see Chs 12 and 13). Recognition of the dangers of generalising from a too restricted sample of the scrapie spectrum points to the possibility of another strategem, namely, the deliberate study of variation itself. This approach is well exemplified by classical genetics and is the logical choice for the most economical mode of investigating any biological problem which is suspected of being well outside the range of currently understood phenomena.

14.4.1 The scrapie 'black-box'

If it is accepted that the large number of negative findings reported for scrapie using conventional techniques is tantamount to an indication that this disease belongs to a new class (and is not simply a slightly aberrant virus that is difficult to investigate), then scrapie becomes a classic example of a 'black-box' phenomenon. This is a picturesque but useful notion for describing any system which displays regular relationships between arrays of INPUTS and OUTPUTS, the details of which can be used to make a phenomeno-logical analysis of the possible CONTENTS of the 'black-box'. The study of variation in mouse scrapie is just such an approach: it attempts to discover all the major variables in the disease in order (a) to be able to design properly controlled experiments, and (b) to try to delineate the main features of the disease in its own terms as a preparation for a more specialised attack upon these main features when they have been identified. A corollary of this is that there can be little value in the type of 'one-off' experiment performed by groups unfamiliar with scrapie procedures, no matter how technically competent they are otherwise.

It will, however, only be possible to give the very broadest outlines of the findings obtained by the investigation of scrapie variation in the mouse, since the body of information is now very large, highly operational, and much of it unpublished. A brief account will first be given of the OUTPUTS from the scrapie 'black-box', then an indication of the current CONTENTS of the box and finally an annotated listing of the major INPUTS which influence or are suspected of influencing the OUTPUTS. These are summarised in Fig. 14.2.

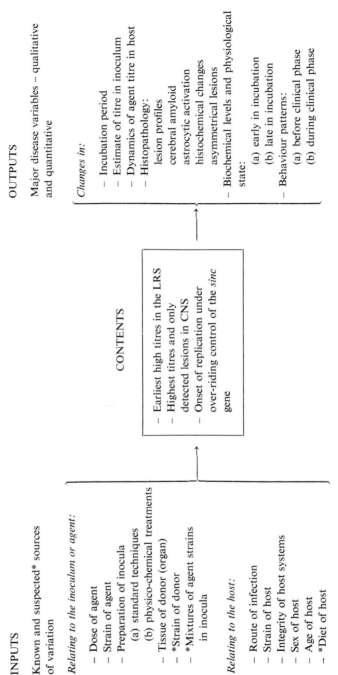

INPUTS

Known and suspected* sources
of variation

Relating to the inoculum or agent:

- Dose of agent
- Strain of agent
- Preparation of inocula
    (a) standard techniques
    (b) physico-chemical treatments
- Tissue of donor (organ)
- *Strain of donor
- *Mixtures of agent strains
  in inocula

Relating to the host:

- Route of infection
- Strain of host
- Integrity of host systems
- Sex of host
- Age of host
- *Diet of host

CONTENTS

- Earliest high titres in the LRS
- Highest titres and only
  detected lesions in CNS
- Onset of replication under
  over-riding control of the *sinc*
  gene

OUTPUTS

Major disease variables – qualitative
and quantitative

Changes in:

- Incubation period
- Estimate of titre in inoculum
- Dynamics of agent titre in host
- Histopathology:
    lesion profiles
    cerebral amyloid
    astrocytic activation
    histochemical changes
    asymmetrical lesions
- Biochemical levels and physiological
  state:
    (a) early in incubation
    (b) late in incubation
- Behaviour patterns:
    (a) before clinical phase
    (b) during clinical phase

Fig. 14.2 The mouse–scrapie 'black-box'

14.4.2 Outputs and measurable parameters

The outputs in the natural disease show a wide range of variation which is very difficult to make sense of; mouse models offer an opportunity for analysis of both qualitative and quantitative outputs.

14.4.2a Changes in incubation period

The incubation period of any disease usually refers to the length of time between infection and the onset of clinical signs. However, 'definite' clinical signs in scrapie are generally preceeded insidiously by rather variable 'preliminary' signs which differ from one model to another, so that an end-point has to be rather carefully defined. A standard method has been in use in these laboratories for many years (Dickinson and Mackay, 1964) and the following is an outline.

The assessment of the clinical status of infected animals is begun several weeks before the expected time of onset of disease by experienced personnel who examine the animals at weekly intervals, in the early afternoon on a day when there has been no disturbance by husbandry routines. The scoring depends on considerable acquaintance with normal behaviour of each mouse strain, some agent–host combinations being more difficult to score than others.

As far as possible, cages are arranged to contain more than one treatment group, and display only coded information in order to reduce scoring bias. Once advanced clinical signs appear, the animals are observed on a daily basis in order to pin-point the end of incubation and obtain fresh material for histology. Animals are killed either two weeks after the onset and per-sistence of certain predefined positive signs or in extremis if this is sooner. Scored in this manner, the incubation periods of all but limiting dilutions show a reasonably normal, non-skewed distribution which permits the application of simple statistical techniques.

Incubation period values obtained by this technique using various com-binations of agent strain, host genotype and route of injection are shown in Tables 14.2 and 3. They show a remarkable consistency within each experi-mental model and this fact is one of the most significant features of the pathogenesis of scrapie. Biologists usually associate such clock-work regularity with genetically controlled developmental processes, and it suggests that the pathogenesis of scrapie may be linked to something of the kind (see 14.7).

14.4.2b Changes in titre estimates and sequences of organ infection
Several INPUTS can influence titre estimates, with or without a consonant change in incubation period. This is a complex and little understood matter as the previous section illustrates and much work remains to be done (14.4.4b, h, j and l).

14.4.2c Histological and biochemical changes
Histological and biochemical changes are described in Chs 12 and 13. Particular attention may be drawn to the following: the great specificity of lesion type and pattern of distribution according to model suggests that some type of 'targeting' is in operation; the absence of any kind of hypersensitivity reaction (inflammation or demyelination) in the CNS, despite many months of infection, suggests some positive 'switching-off' mechanism of these reactions; there is a clear need to concentrate more upon changes during the preclinical phases of the disease when there is less danger of deflection of research effort by secondary effects of disease; and finally, only those biochemical changes whose incidence in different models has been checked are likely to be of much use in the study of pathogenesis (as a first approximation, those of wide generality are most likely to be primary ones).

14.4.2d Behavioural changes
Long before the onset of clinical signs, and as early as the first quarter of the incubation period (13.7), subtle behavioural changes begin to occur in infected mice. These include a decreased responsivity in open-field testing (Savage and Field, 1965), abnormal behaviour in emergence tests (Heitzman and Corp, 1968), changes in food consumption and a progressive decline in polydipsic responsiveness to saline or sucrose solutions (Outram, 1972). Only the last of these has been checked for generality. Such behavioural changes could provide a means of identifying the primary CNS lesion, and the evidence is consistent with an upset of catecholamine-mediated communication (Outram, 1972).

14.4.3 Contents of the scrapie black-box

The nature of the basic interaction between pathogen and host is still unknown. However, a few major features have been established: there is always a high infection of the lympho-reticular system; highest titres and most probably the primary lesions occur in the CNS, and the timing of the onset of replication and hence length of incubation period is under the major control of the *sinc* gene. Numerous other potential CONTENTS of the box have been suggested but their validity or generality is in too much doubt for inclusion at this time.

14.4.4 Inputs; sources of biological variation in patterns of pathogenesis

The following are the major sources of variation in patterns of pathogenesis that have so far been detected in mouse scrapie. Individually and collectively they throw a suggestive light upon the nature of the contents of the scrapie black-box but much more analysis of them remains to be done.

14.4.4a Dose of agent
Progressive dilution of the inoculum causes prolongation of incubation periods in a highly regular manner with an increasing number of 'survivors' at the higher dilutions. Except in respect of timing, dilution does not seem to affect other outputs. However, the slopes of the dose-response curves and the value of the limiting dilution vary quite considerably from one model to another according to other input variables such as agent strain, route of infection, host genotype, strain and organ and donor, and the nature of the preparation of inocula. These data (still sketchy in some important models) suggest that there are some interesting dynamic relationships resulting from interactions between replication rate, site limitation and at least one process of agent 'inactivation' (14.4, 4j).

14.4.4b Route of infection
The route of infection affects the length of incubation period and other time-related outputs (Ch. 13); lesion profiles (Outram et al., 1973b) and estimated titre (Hunter and Millson, 1964; Mould and Dawson, 1968; Dickinson, 1975; Dickinson and Fraser, 1975; Carp, personal communication). The evidence suggests that when relatively untreated inocula are used for an i.c. injection, the 'short' incubation periods compared with those obtained using peripheral routes are due to agent being able to remain and replicate in its 'target' organ without having to undergo pathogenic steps in peripheral organs (including the lympho-reticular system). Conversely, agent injected by any of a variety of peripheral routes is obliged to undertake a number of other pathogenic steps before gaining access to its 'target' organ. However, there is evidence that inocula which have been heat treated contain agent which cannot remain and replicate in the brain even after an i.c. injection but has to undergo extra pathogenic steps in peripheral systems (Dickinson and Fraser, 1969b;

G. Outram

TABLE 14.3

The effects of strain of agent, route of injection and sex of recipient on incubation period in various mouse strains all homozygous for the s7 allele of *sinc*. For experimental detail see legend to Table 14.2.

Injection route:	i.c.						i.p.			
Agent group:	22A	ME7					22A	ME7		
Agent:	22A	ME7	79A	22C	58A	87A	22A	ME7	79A	22C
C57BL ♀	453	159	152	177	152	342	601	239	201	234
♂	470	162	157	182	167	351	593	238	207	225
RIII	429	167	136	166		360		248	189	
	432	164	140	175		404*		247	195	
BALB/c	470	172	161	185				266	227	267
	470	169	166	198				256*	232	272
VL	430	158	151					250	205	
	437	160	155					256	221	
A2G	464	173	160				585	234	197	
	451	173	166					245	203	
C3H	429	169	138	174					195	250
	433	172	141	176				232	202	246
BRVR	467	171	147	191				236		
	493	171	145	179				224		
BSVS	457	172	146		163	363				
	472	161	144*		159					
SM	438	174	148		159	355			204	
	428*	174	150		160	379			207	
DBA/2		157	145					217		
		158*	152					216		
129		157	142					211		
		156	149					214		
NZB		179						244		
		170						253*		

* S.E. $> 2\tfrac{1}{2}\%$ of the incubation period.

Dickinson and Taylor, personal communication). This could apply to other kinds of treatment and lead to spurious estimates of 'titre-loss' (14.3). Interactions between route, age of host, and the effects of drug treatments on incubation period and titre estimates, indicate that pathogenesis in the periphery may involve some mechanism of agent 'inactivation'.

14.4.4c Strain of mouse host

The strain of mouse host affects the length of incubation, the type of lesion and shape of lesion profile (Ch. 12) and some biochemical changes (Ch. 13). Investigations of incubation periods in a wide variety of mouse genotypes (some of which are reported in Tables 14.2 and 3) have found only one gene which has a major effect, the rest being trivial in comparison. This gene called *sinc* is discussed more fully in 14.5. The small range of the effects of other genes can be seen by comparing RIII and BALB/c mice as hosts. Another gene with a relatively small but significant effect is *Dh* which causes spleen-lessness in heterozygotes and prolongs the i.p. incubation period of ME7 from 226 to 282 days but has no effect on i.c. incubation periods (Dickinson and Fraser, 1972). This suggests that there is probably only a short biochemical pathway between *sinc* gene function and agent replication (see 14.5).

14.4.4d Strain of agent

Much work in these laboratories has been devoted to analysing field isolates of natural and experimental scrapie in sheep and goats into their constituent agent strains (see Chs 10 and 12). As many sheep with natural scrapie are infected with more than one strain of agent, the process of isolating separate agents in mice is a long one involving serial passage in particular genotypes of mice, at present principally in either C57BL or VM mice because of their difference in *sinc*, and then serial cloning of the agents by passing only from cases injected with limiting dilutions of infectivity. Scrapie agents cannot be typed by the conventional morphological, serological or plaque-formation techniques of microbiology, one has to use instead the peculiar biological features which have emerged during the study of scrapie variation itself. Some scrapie workers have even been hesitant to accept the existence of agent strains because unfamiliar methods have had to be used for their discrimination. Criteria which have to be used include: the incubation periods of standard supernates after i.c. and i.p. injection in C57BL *sinc*^{s7}, VM *sinc*^{p7} and (C57BL × VM)F$_1$ *sinc*^{s7p7} mice (Dickinson, 1975) and the lesion profiles in these genotypes (Fraser and Dickinson, 1973; Ch. 12). As well as these grey- and white-matter profiles there are now an increasing number of other histological markers for discriminating agent strains, including the occurrence of cerebral amyloid (Fraser and Bruce, 1973; Bruce and Fraser, 1975) and asymmetrical lesions (Fraser et al., 1975; Ch. 12). Ancillary help can often be obtained when necessary, from the details of pre-clinical and clinical behaviour patterns such as over-eating, body-weight changes, hind-limb incoordination, somnolence, etc. (Outram, 1972 and unpublished).

'Purification' of agents by serial passage using animals given only end-point dilutions takes many years and much testing at each step. What is especially remarkable about this is the stability of some agents so obtained. For example the ME7 agent has identical properties whether passaged either 8 times in C57BL *sinc*^{s7} or 6 times in VM *sinc*^{p7} mice. Other agents are entirely stable when passaged in a particular genotype, for example 22A passaged 6 times in VM mice, but when 'back-passaged' in another genotype they are liable to change slowly and progressively, though this change does not necessarily start at the first back-passage and has never been observed to restabilise after only one back-passage – 22A back-passaged in C57BL shows these features. Whether we are dealing in such cases with mutational changes is an open question, as is the possibility of recombination, and marker systems are being developed with the aim of attacking this problem.

Agents are named according to the system advocated for viral nomenclature, using numerals and capital letters, which signify, respectively, separate isolates and different strains present in these isolates. As can be seen from Tables 14.2 and 3, they fall into two major groups: the ME7 and the 22A group according to their relative incubation periods in the *sinc* genotypes.

The molecular basis of these agent-strain differences is still unknown although some type of 'informational molecule' must clearly be involved. Whether or not this is a nucleic acid must still be regarded as an open question.

14.4.4e Mixtures of agents

Natural sources of scrapie and their early passages in mice frequently contain mixtures of agents (Ch. 10). Unrecognised mixtures of agents with very different physico-chemical properties are a potential source of erroneous interpretations of experiments designed to test the 'inactivation' effects of various treatments (Ch. 11). It has been shown that an 'operationally slow' agent can exercise some form of interference with the pathogenesis of a subsequently injected 'operationally quick' agent, causing various degrees of prolongation of incubation periods of the latter even to the extent of complete elimination of its contribution to the death of the host (Dickinson et al., 1972, 1975a; 14.5). This effect of one agent upon the pathogenesis of another is attributable to competition for the *sinc*-controlled replication site, and its degree depends upon numerous other experimental variables such as routes of injection, relative doses and the timing of the injections with respect to each other.

It is not known if such agent competition plays any role in the pathogenesis of 'natural' mixtures, but it could clearly play at least a contributory part in some cases where there are difficulties of inter-specific passage. In the case of artificial, in vitro mixtures, it is suspected that interference phenomena may occur, but only when there are great discrepancies in titre (Dickinson, personal communication).

14.4.4f Organ or infected tissue of donor
In general this does not seem to be a major source of variation except that the operational titre varies according to tissue and there is no way of knowing how this is related to the 'functional' titre (14.3). Only slight differences in lesion profile have been reported when comparing agent from brain and spleen (Fraser and Dickinson, 1967; Outram et al., 1973b) so that if the infective unit of scrapie contains some obligatory host-component one has to assume on present data that this is the same in both tissues.

14.4.4g Strain of donor
Similar remarks can be made about this potential source of scrapie variation. A report that the ME7 agent passaged in BALB/c and BRVR mice undergoes some subtle modification through association with strain-specific tissues (Dickinson and Outram, 1973; Outram et al., 1973b) is being elaborately checked-out at present. If it is established that donor-tissue components can specifically influence incubation period length and lesion profile this would clearly be of great importance in our understanding of the nature of host–pathogen interactions.

14.4.4h Preparation and treatment of inoculum
No systematic investigation of the effects of different preparative techniques (degree of fragmentation and dispersal; use of filtration and centrifugation; addition of substances such as serum, antibiotics, solubilisers and buffers) and their relationship with route of injection has been published. However, numerous incidental observations strongly suggest that such procedures and additives could greatly influence pathogenesis in certain circumstances.

It has already been sufficiently emphasised that attempts to make deductions about the nature of the agent by using physico-chemical treatments of the inoculum as a means for modifying pathogenesis (e.g., changes of incubation period and/or titre estimates) contain many traps for the unwary, since the physical and chemical changes in the inoculum could influence pathogenesis by changing host responses rather than by direct action on the agent. For normal purposes, therefore, it would seem wisest to use minimally manipulated materials, e.g., lightly centrifuged (500*g* for 10 min) supernates prepared in simple physiological saline.

14.4.4i Sex of recipient
It can be seen from Table 14.2 that frequently (but not invariably) male mice have longer incubation periods than females. The fact that the order can be absent or reversed in some models shows that the phenomenon is not due to observer bias. The biological basis of this effect is not known and an attempt to modify and extend it by treating male and female mice with sex hormones during the early stages of incubation period had no significant result (Outram, unpublished). These findings have some analogies with observations of earlier and higher incidence in female mammals of auto-immune phenomena and of diseases known or suspected to have auto-immune overtones and sequelae (e.g., multiple sclerosis).

14.4.4j Age of recipient
The age of the recipient at the time of injection is known to affect incubation periods but not lesion profiles. The effect depends, however, upon route of injection. After i.c. injections there is only a slight progressive shortening of incubation period from birth onwards into adult life which in the case of ME7 in C57BL mice is at the rate of about 1 day in 20. After an i.p. route of injection, however, some quite unexpected things happen during the pre-weaning period. It has been shown in several widely different models that i.p. injections of 10^{-2} inocula which are 100% lethal in young adult mice, kill only about 80% of neonatal mice, which also show greatly prolonged incubation periods (Outram et al., 1973a). Furthermore, an investigation of i.p. titre estimates using animals between birth and weaning for the assay, shows that there is a progressive increase in operational values as the mice get older (Dickinson personal communication). However, some recent unpublished work in neonatal mice shows them to be very susceptible to agent injected by various s.c. routes. It looks therefore as though the peri-toneal cavity of the mouse has a capacity for 'inactivating' scrapie which becomes relatively inefficient as the animal gets older. The remarkable in vitro 'resistance' of scrapie to inactivation makes it a real question to ask what happens to the agent which is injected i.p. into neonates. Preliminary investigations have shown that it is possible to confer normal adult suscep-tibility on neonatal mice by i.p. injections of living spleen cells from syngeneic adults (Outram, unpublished).

14.4.4k Diet of recipient

Prolongation of i.p. incubation periods has been achieved in mice fed for much of their early life on a diet of maize only (Dickinson, personal communication). Such a diet is likely to be deficient in several essential amino acids and vitamins which may individually or in combination lie at the root of the effect. However, the fact that incubation periods are prolonged suggests analogies with increased longevity of rats and mice on such diets (McCay et al., 1935; Jose and Good, 1973) and decreased immune responsiveness (Walford et al., 1973) which, in the case of scrapie, seems to be paradoxically associated with prolonged incubation periods (see below).

14.4.4l Integrity of host systems

A number of treatments which may be said to upset the integrity of the host have been shown to affect pathogenesis as reflected in changed incubation periods. The best known of these is the effect of adult splenectomy (Fraser and Dickinson, 1970; Clarke and Haig, 1971a) which prolongs i.p. but not i.c. incubation periods. Similar in effect is the possession of the $Dh/+$ genotype which results in spleenless mice (Dickinson and Fraser, 1972). On the other hand, adult thymectomy and ancillary treatments which were profound enough to abolish normal skin graft reactions throughout the time of the incubation period had no marked effect on i.p. incubation periods (McFarlin et al., 1971).

Integrity can also be modified by drugs, although there is often a question as to the exact nature of the drug damage. Prolonged incubation periods together with the occurrence of survivors from what would normally be 100% lethal doses have been obtained by treating recipients with large doses of anti-inflammatory steroids around the time of i.p. injection of agent. The effect has been found with prednisone acetate (Outram et al., 1974) and dexamethazone (Outram, unpublished). The doses required for this steroid effect are rapidly reduced with age, so that something very similar to the neonatal age-effect can be produced in week-old mice by only a single injection of prednisone acetate (50 mg/kg) or dexamethazone (10 mg/kg) a day or two before. Preliminary experiments have shown that a single injection of prednisone acetate followed four days later by a single injection of cyclophosphamide in adults also prolongs the incubation period of i.p. injected scrapie given the day after the cyclophosphamide, although neither of these drug injections is effective alone (Outram, unpublished). A highly immunosuppressive course of cyclophosphamide alone has been shown not to influence i.p. incubation periods (Worthington and Clarke, 1971; Outram, unpublished).

Preliminary findings with two drug treatments which seem to reduce incubation periods may be mentioned. Kimberlin (personal communication) has shown that slightly reduced i.p. incubation periods can be produced using immunostimulatory substances such as a methanol extraction residue from B.C.G., and Fraser (personal communication) obtained shorter i.c. incubation periods of the ME7 agent in BALB/c mice by the addition of thyroxine to the drinking water.

Clearly, most of these treatments have multiple effects upon the host, but it cannot escape notice that they all have in common some modification of the lympho-reticular system. In the paradoxical fashion one has come to associate with scrapie, treatments which are generally 'immunosuppressive' tend to prolong incubation periods, while those which enhance immune responsiveness tend to shorten them. Many of the treatments are also known to have effects upon inflammatory responses, which relates more obviously to the recent report of prolonging effects of injections of Arachis oil (Outram et al., 1975) which is thought to contain an extremely potent anti-inflammatory component (Long and Martin, 1956).

14.5 The significance of the sinc gene for scrapie pathogenesis

The name 'sinc' is an acronym for 'scrapie incubation' and the designations of the two known alleles are 's7' (producing short incubation periods with the ME7 group agents) and 'p7' (producing prolonged incubation periods with the ME7 group agents). The normal function and linkage group of sinc are unknown. It is possible that a similar gene has been found in sheep (Ch. 10).

The range of properties of the alleles of sinc in controlling scrapie incubation periods in various models can best be seen by consulting Table 14.2. Two characteristics are outstanding: first, the phenomenon of reversal in which ME7- and 22A-group incubation periods change places in terms of being operationally 'quick' or 'slow' according to the sinc genotype of the host; and second, the wide range of differences in degree of genetic dominance according to agent – extending to a condition of overdominance in the case of the s7 allele with 22A and 104A agents. Single-gene overdominance has not been otherwise recognised in mammals and should not be confused with 'hybrid vigour' (or heterosis) which occurs in hybrids inheriting a more effective combination of many genes than that possessed by either of the parents. Single gene overdominance has, however, been reported in some micro-organisms where it has been shown to be due to inter-allelic complementation – a condition in which both alleles at a locus contribute monomeres to a multimeric functional product, which in the heterozygote gives rise to a heteromere with a functional efficiency outside the range of either parental homomere (Zimmerman and Gundelach, 1969). Mechanisms for assembling multimeric structures for monomeres are not understood, so that the possibility of free-association models in which hybrids produce both types of homomere as well as the heteromere must be considered.

Since the *sinc* gene has such an over-riding control over the length of scrapie incubation – to the virtual exclusion of other genes – it is likely that there are only a few biological steps separating gene action from agent replication, depending almost exclusively upon some steps between gene duplication and the action of its most immediate products. Clearly it would be useful to know the normal function of this gene.

An attempt has been made to construct a model of gene–agent interaction based on inter-allelic complementation (Dickinson and Meikle, 1971; Dickinson, 1975). Assuming a free-association model for heteromere assembly, one of the necessary conditions is that there must be a severe limit on the number of sites of this *sinc* action – otherwise there would be no overdominance because the heterozygote would contain sufficient numbers of the homomeres to satisfy any strain of agent. With s7 overdominance with 22A, for instance, the heteromere derived from s7p7 would be less efficient than the s7 homomere in supporting agent replication with the consequent very prolonged incubation periods.

Several lines of evidence are compatible with the site limitation which the free-association model requires: the long incubation periods despite the long periods of agent replication; lower limits to the length of incubation period of several months; plateau titres in several organs; the marked effects of surgical and genetic splenectomy on incubation periods. Severe limitation of the numbers of replication sites would lead to two predictable results: agents should compete for such sites, and the dose–response curves in normal as compared with splenectomised mice injected by the i.p. route should differ so that they would cross at the level of dilution upon which the absence of the spleen would have no effect. Evidence in support of both of these is beginning to emerge, although much remains to be done, especially on the dynamics of agent replication in splenectomised animals (Dickinson, unpublished).

Three experiments demonstrating the ability of agents to compete for replication sites have been published, using the i.c. and i.p. routes and demonstrating an apparent loss in titre due to such competition. The general design of these competition experiments was to precede the injection of an operationally quick agent with an injection of an operationally slow one using intervals which would allow time for the quick one to kill the host unless there had been some form of competition for the means of replication. Identification of the agent which actually killed the mouse was to be done by examining lesion profiles (Ch. 12).

Dickinson et al. (1972) showed that when an i.c. injection of 22A in VM *sinc*^{p7} mice was preceded 5 or 9 weeks (but not one week) earlier by an i.c. injection of 22C agent, the incubation period of the 22A was prolonged by about 30 days ($P = 0.001$), which is equivalent to about a five-fold loss of dose. In another experiment using i.p. routes of injection, 22C in RIII *sinc*^{s7} mice was preceded by 22A given 100, 200 or 300 days before. The expected incubation periods of the two agents in this strain of mouse were in the region of 230 days for 22C and 550 for 22A. All the control mice which received preliminary injections with normal VM brain instead of 22A showed incubation periods of about 230 days and 22C-type lesion profiles. All the mice which had received both injections, however, died about 550 days after the first (22A) injection with 22A-type lesion profiles, suggesting that the pathogenic progress of the 22C had been at least severely retarded, and probably even prevented entirely by the prior injection of 22A (Dickinson et al., 1975). Titration (i.c.) of an operationally quick agent in mice already infected i.c. with an operationally slow agent shows prolonged incubation periods at all doses and an apparent loss of about 90% of the titre in the experimental conditions which obtained (Dickinson, 1975).

These findings are evidence for the existence of a remarkably small number of *sinc*-controlled replication sites which have a slow turnover rate and are either replaced only infrequently or the occupying agent has priority of access to them as they are formed. It is not possible to say at present if these relatively rare sites are the sum total of active *sinc* sites or merely a surplus of such sites not already blocked by some natural host component of which the scrapie agent is an analogue.

The lack of a fundamental understanding of the relationship between gene and agent does not preclude the experimental usefulness of *sinc* geno-types. Apart from agent identification and purification already described, the properties of *sinc* can be used in cellular engineering procedures to detect the type of tissue and cell in which the gene is active. Testing genetic chimeras with a variety of agents should help pin-point the cell-type involved. The cultivation of these cells in vitro could open up an entirely new chapter in scrapie research.

14.6 The possible role of the lympho-reticular system

An examination of the sources of variation in mouse scrapie (INPUTS) shows that many have in common a possible involvement of the lympho-reticular system (LRS). This rather imprecise histological expression is meant to refer to all tissues which are common to extraneural organs containing high operational titres of scrapie, and therefore includes elements as diverse as all types of leucocytes, phagocytic cells, stem cells and nervous elements. Attempts to restrict interest to macrophages, mesothelial cells, sympathetic nerves, lymphocytes and lymphoblasts, although useable as 'working hypotheses', are premature.

The main reasons for regarding the LRS as having a primary role in peripheral pathogenesis may be summarised as follows: it is the site of the earliest detectable rises in titre; increased i.p. incubation periods are observed in splenectomised and genetically spleenless mice; there is decreased susceptibility to i.p. injected agent in neonatal mice with evidence for some 'inactivation' mechanism; reduced susceptibility to i.p. injected scrapie is found in mice injected with high doses of glucocorticosteroids and prolonged incubation periods are noted after s.c. injections of Arachis oil. Other observations which may also be construed in this way are: the combination of high agent titres in the LRS with an absence of humoural immune responsiveness; the absence of inflammatory responses in the CNS after many months of infection; the occurrence of cerebral amyloid in some scrapie models; the demonstration that lymphocytes can become sensitised to scrapie-modified tissues as indicated by changed electrophoretic mobility in a cytopherometer (Field and Shenton, 1972); decreased levels of circulating polymorphonuclear neutrophils after scrapie injection (Licursi et al., 1972; Carp et al., 1973; Dickinson et al., 1974); the sex-related differences in incubation period; increased susceptibility to scrapie of animals after immune stimulation or thyroxin; the prolonging effect of a deficient diet upon incubation period; and the apparent 'targeting' of scrapie agents or their effects on different regions of the CNS as shown by lesion profiles (Ch. 12).

These suggestive observations do not of course establish the fundamental importance of the LRS, but they do provide a promising basis for further investigation. Any serious attempt to suggest unifying hypotheses to cover all these findings would result in too many alternatives and require more space than is justified here.

A number of experimental findings can also be considered as discouragements to pursue certain obvious lines of investigation. The complete ineffectiveness of immunosuppression by cyclophosphamide (Worthington and Clark, 1971) and adult thymectomy (McFarlin et al., 1971) to influence i.p. incubation periods tends to exclude the involvement of the B- and T-cell systems as currently understood. Treatments of neonatal mice with normal and scrapie-infected brain does not seem to affect the pathogenesis of scrapie material subsequently injected after weaning (Outram et al., 1973a and unpublished) which goes against the possibility of a 'tolerance' mechanism such as shown by the lymphocytic choriomeningitis virus (LCM). Neonatal treatment of mice with large doses of 6-hydroxydopamine which produces a condition of 'chemical sympathectomy' has so far resulted in significant but only marginal prolongation of i.p. incubation periods of agent given to such mice after weaning (Outram, 1973 and unpublished). This does not support a primary involvement of the sympathetic nervous system of the type suggested for TME by Marsh and Hanson (1975), and considered for scrapie (Outram et al., 1974). Attempts to detect agent activity in circulating lymphocytes and sub-populations of lympho-reticular cells have either been negative or too problematic (because of differential cell fragility) to suggest that anything useful can come from this type of approach to the problems of scrapie circulation and locality (Lavelle et al., 1972; Marsh et al., 1973).

If there is a primary involvement of the LRS in scrapie pathogenesis, our present difficulty in identifying it is due either to the function concerned being beyond the present horizons of knowledge or to the effect being too slight or subtle for current means of detecting it. Meanwhile attempts can be made to identify the cells responsible for the effects of age and steroids on scrapie incubation, using for instance the selective injection of syngeneic sub-populations of cells in an effort to abrogate these effects. Any effective cells discovered by these means may or may not also contain an active *sinc* gene which would be detected by their effect upon the pathogenesis of a variety of agent strains.

14.7 Speculations

The need for operational descriptions of procedures and phenomenological analyses of findings in scrapie research at the present time due to the absence of an acceptable theoretical framework, has been repeatedly emphasised in this chapter, and the frankly speculative ideas which follow must not be allowed to detract from that emphasis. All the same, the speculations of any worker in a field can serve a useful purpose if they stimulate the imagination of others or help to loosen the grip of limiting presuppositions: the following is offered in this spirit.

My guess at the present time is that the scrapie agent is an aberrant form of an informational molecule naturally produced by mammals as part of their long-term control of morphology and development. Such a molecule could become aberrant by mutation or as a result of accidentally crossing a species barrier (or both). It has been suggested that the best understood functions of the LRS, namely immune reaction and surveillance, are in fact specialisations of a phylogenetically ancient system for detecting and rectifying any departure in the phenotype from the genetical programme of development (Fabris et al., 1972). There is a large amount of research at present concerning 'informational molecules' involved in the control of developmental, immunological, inflammatory and neurological functions and responses, and some of these seem to be 'protected' nucleic acids. It has been postulated that mutant forms of such 'regulatory RNA' may give rise to 'viroids' in plants (Diener, 1971) and some kinds of enveloped viruses in animals (Reanney, 1975). The two classes of scrapie agent: ME7- and 22A-groups could represent a kind of functional complementarity amongst informational molecules which exist as part of the mechanisms of cybernetic control of homeostatic and homeodynamic processes of mammals. If scrapie is, as some have suggested (Field, 1967; Field and Peat, 1971), an acceleration of the ageing process (or some particular aspect of it), then one could speculate that long-term, genetically controlled, post-reproductive processes such as may be involved in senility and some diseases of the old, are 'fossil' vestiges of ancestral developmental stages which have been sloughed off, so to speak, in the course of evolutionary steps such as neoteny and paedogenesis. 'Fossil' developmental genes in such cases would tend not to come under the scythe of natural selection, and could therefore be perfect candidates for assuming autonomy by mutation and evolution into successful pathogens.

References

BRUCE, M. and FRASER, H. (1975) Neuropathol. Appl. Neurobiol., 1, 189.
CARP, R. I., MERZ, P. A., LICURSI, P. C. and MERZ, G. S. (1973) J. Infect. Dis., 128, 256.
CHANDLER, R. L. (1959) Vet. Rec., 71, 58.

CLARKE, M. C. (1968) Res. Vet. Sci., 9, 595.

CLARKE, M. C. and HAIG, D. A. (1966) Vet. Rec., 78, 647.

CLARKE, M. C. and HAIG, D. A. (1971a) Res. Vet. Sci., 12, 195.

CLARKE, M. C. and HAIG, D. A. (1971b) Br. Vet. J., 127, xxxii.

DICKINSON, A. G. (1967) Lancet, i, 1166.

DICKINSON, A. G. (1975) Genetics, 79 (suppl.), 387.

DICKINSON, A. G. and FRASER, H. (1969a) J. Comp. Pathol., 79, 363.

DICKINSON, A. G. and FRASER, H. (1969b) Nature, 222, 892.

DICKINSON, A. G. and FRASER, H. (1972) Heredity, 29, 91.

DICKINSON, A. G. and FRASER, H. (1975) In: (ter Meulen and Katz, Eds) Slow Virus Infections of the Central Nervous System. (Springer, Berlin) In press.

DICKINSON, A. G., FRASER, H., MCCONNELL, I., OUTRAM, G. W., SALES, D. I. and TAYLOR, D. M. (1975a) Nature, 253, 556.

DICKINSON, A. G., FRASER, H., MEIKLE, V. M. H. and OUTRAM, G. W. (1972) Nature, New Biol., 237, 244.

DICKINSON, A. G., FRASER, H. and OUTRAM, G. W. (1975b) Nature, 256, 732.

DICKINSON, A. G. and MACKAY, J. M. K. (1964) Heredity, 19, 279.

DICKINSON, A. G., MACKAY, J. M. K. and ZLOTNIK, I. (1964) J. Comp. Pathol. Ther., 74, 250.

DICKINSON, A. G. and MEIKLE, V. M. H. (1971) Molec. Gen. Genet., 112, 73.

DICKINSON, A. G., MEIKLE, V. M. H. and FRASER, H. (1969) J. Comp. Pathol., 79, 15.

DICKINSON, A. G. and OUTRAM, G. W. (1973) J. Comp. Pathol., 83, 13.

DICKINSON, A. G., TAYLOR, D. M. and FRASER, H. (1974) Nature, 248, 510.

DIENER, T. O. (1971) Virology, 45, 411.

EKLUND, C. M., HADLOW, W. J. and KENNEDY, R. C. (1963) Proc. Soc. Exp. Biol. Med., 112, 974.

EKLUND, C. M., KENNEDY, R. C. and HADLOW, W. J. (1964) Scrapie Seminar ARS 91-53, (U.S. Dept. Agriculture) p. 288.

EKLUND, C. M., KENNEDY, R. C. and HADLOW, W. J. (1965) In: (D. C. Gajdusek, C. J. Gibbs. Jr, and M. Alpers, Eds) Natl. Inst. Neurol. Dis. Blindness, Monogr. 2. Slow, Latent and Temperate Virus Infections. (U.S. Government Printing Office, Washington, D.C.) p. 207.

EKLUND, C. M., KENNEDY, R. C. and HADLOW, W. J. (1967) J. Infect. Dis., 117, 15.

FABRIS, N., PIERPAOLI, W. and SORKIN, E. (1972) Nature, 240, 557.

FIELD, E. J. (1967) Dtsch. Z. Nervenheilk., 192, 265.

FIELD, E. J. and JOYCE, G. (1970) Nature, 226, 971.

FIELD, E. J., JOYCE, G. and KEITH, A. (1971) Nature, New Biol., 230, 56.

FIELD, E. J. and PEAT, A. (1971) Gerontologia, 17, 129.

FIELD, E. J. and SHENTON, B. K. (1972) Nature, 240, 104.

FRASER, H., BRUCE, M. and DICKINSON, A. G. (1975) Proc. VII Int. Congr. Neuropathol., in press.

FRASER, H. and BRUCE, M. (1973) Lancet, i, 617.

FRASER, H. and DICKINSON, A. G. (1967) Nature, 216, 1310.

FRASER, H. and DICKINSON, A. G. (1970) Nature, 226, 462.

FRASER, H. and DICKINSON, A. G. (1973) J. Comp. Pathol., 83, 29.

GARDINER, A. C. (1966) Res. Vet. Sci., 7, 190.

GARDINER, A. C. and MARUCCI, A. A. (1969) J. Comp. Pathol., 79, 233.

GIBBS, J. G., GAJDUSEK, D. C. and MORRIS, J. A. (1965) In: (D. C. Gajdusek, C. J. Gibbs, Jr, and M. Alpers, Eds) Natl. Inst. Neurol. Dis. Blindness, Monogr. 2, Slow, Latent and Temperate Virus Infections. (U.S. Government Printing Office, Washington, D.C.).p.195.

GIORGI, P. P., FIELD, E. J. and JOYCE, G. (1972) J. Neurochem., 19, 255.

HAIG, D. A. and CLARKE, M. C. (1965) In: (D. C. Gajdusek, C. J. Gibbs, Jr, and M. Alpers, Eds) Natl. Inst. Neurol. Dis. Blindness, Monogr. 2, Slow, Latent and Temperate Virus Infections. (U.S. Government Printing Office, Washington, D.C.). p. 215.

HEITZMAN, R. J. and CORP, C. R. (1968) Res. Vet. Sci., 9, 600.

HUNTER, G. D. (1972) J. Infect. Dis., 125, 427.

HUNTER, G. D. (1974) Progr. Med. Virol., 18, 289.

HUNTER, G. D., KIMBERLIN, R. H. and MILLSON, G. C. (1972) Nature, New Biol., 235, 31.

HUNTER, G. D. and MILLSON, G. C. (1964) Res. Vet. Sci., 5, 149.

JOSE, D. G. and GOOD, R. A. (1973) J. Exp. Med., 137, 1.

KATZ, M. and KOPROWSKI, H. (1968) Nature, 219, 639.

KIMBERLIN, R. H. (1976) Merrow Monograph, Microbiol. Ser., (Newcastle-upon-Tyne, U.K.) in press.

LAVELLE, G. C., STURMAN, L. and HADLOW, W. (1972) Infect. Immun., 5, 319.

LICURSI, P. C., MERZ, P. A., MERZ, G. S. and CARP, R. I. (1972) Infect. Immun., 6, 370.

LONG, D. A. and MARTIN, A. J. P. (1956) Lancet, i, 464.

MARSH, R. F. and HANSON, R. P. (1975) Science, 187, 656.

MARSH, R. F., MILLER, J. M. and HANSON, R. P. (1973) Infect. Immun., 7, 352.

MCCAY, C. M., CROWELL, M. F. and MAYNARD, L. A. (1935) J. Nutr., 10, 63.

MCFARLIN, D. E., RAFF, M. C., SIMPSON, E. and NEHLSEN, S. H. (1971, Nature, 233, 336.

MIMS, C. A. (1960) Br. J. Exp. Pathol., 41, 52.

MOULD, D. L. and DAWSON, A. MCL. (1968) J. Comp. Pathol., 78, 115.

MOULD, D. L., DAWSON, A. MCL. and RENNIE, J. C. (1970) Nature, 228, 779.

OUTRAM, G. W. (1972) J. Comp. Pathol., 82, 415.

OUTRAM, G. W. (1973) Ph.D. Thesis, University of Edinburgh.

OUTRAM, G. W., DICKINSON, A. G. and FRASER, H. (1973a) Nature, 241, 536.

OUTRAM, G. W., DICKINSON, A. G. and FRASER, H. (1974) Nature, 249, 855.

OUTRAM, G. W., DICKINSON, A. G. and FRASER, H. (1975) Lancet, i, 198.

OUTRAM, G. W., FRASER, H. and WILSON, D. T. (1973b) J. Comp. Pathol., 83, 19.

PATTISON, I. H. (1964) Vet. Rec., 76, 33.

PORTER, D. D., PORTER, H. G. and COX, N. A. (1973) J. Immunol., 111, 1407.

REANNEY, D. C. (1975) J. Theor. Biol., 49, 461.

SAVAGE, R. D. and FIELD, E. J. (1965) J. Anim. Behav., 13, 443.

WALFORD, R. L., LIU, R. K., MATHIES, M., GERBASE-DELIMA, M. and SMITH, G. S. (1973) Mech. Ageing Dev., 2, 447.

WORTHINGTON, M. (1972) Infect. Immun., 6, 643.

WORTHINGTON, M. and CLARKE, R. (1971) J. Gen. Virol., 13, 349.

ZIMMERMAN, F. K. and GUNDELACH, E. (1969) Molec. Gen. Genet., 103, 348.

CHAPTER 15

The subacute spongiform encephalopathies

R. F. MARSH

It will be apparent from the foregoing chapters that scrapie is quite an unusual disease caused by a transmissible agent with properties which are unique to microbiology. Therefore, it should not be too difficult to identify disorders similar to scrapie. There are three such diseases now recognized; two affecting man and one of commercially-reared mink. These diseases, together with scrapie, form a single nosologic entity known as the subacute spongiform encephalopathies (Gibbs and Gajdusek, 1969). This chapter will review these scrapie-like diseases emphasizing their natural histories, experimental development and epidemiologic interrelationships.

15.1 The natural diseases

15.1.1 Kuru

Kuru was first brought to public attention as an endemic degenerative brain disease of New Guinea natives in 1957 (Gajdusek and Zigas, 1957). The disease occurs as a geographically localized illness affecting almost exclusively people of the Fore linguistic group. During the peak incidence of the disease, in the 1950s, 1% of the Fore population would die of kuru each year with the prevalence of active kuru reaching 5–10% of the inhabitants of certain Fore tribes (Klatzo et al., 1959). In the original description of the disease, there was a definite familial pattern of distribution in which children of both sexes were affected, but the highest incidence was in adult females. In recent years, the pattern of disease occurrence has

Slow virus diseases of animals and man, edited by R. H. Kimberlin
© *North-Holland Publishing Company 1976*

Fig. 15.1. This photo was taken in 1960 in the North Fore village of Etesena. These two women are in the intermediate stage of kuru and, because of severe ataxia, can only stand using the support of a stick. Both patients died within six months. (Courtesy of Dr. D. Carleton Gajdusek.)

changed considerably with few children now affected and the incidence becoming more equal in women and men (Gajdusek, 1963). This change in incidence has been accompanied by a sharp decline in the total number of registered cases of kuru, a phenomenon thought to be due primarily to the decrease in ritual cannibalistic practices.

The following clinical synopsis is abstracted from Zigas and Gajdusek (1957) which remains the most descriptive report of the disease. Kuru has a remarkably uniform course beginning with locomotor ataxia becoming insidiously progressive (Fig. 15.1). Ataxia is accompanied by a fine tremor of the trunk, extremities and head which is exaggerated during voluntary activities or when the patient is excited. There is a gradual deterioration of motor skills until the individual can no longer walk and must be balanced in a sitting posture. Convergent strabismus is often present at this time. The patient's sedentary state rapidly progresses to total debilitation with death resulting from starvation or secondary infection. The course of the disease is generally 6–9 months, rarely exceeding one year. It is interesting to note that speech and intelligence are normal during the early months of involvement, with dysarthria appearing and becoming more severe in the terminal stages. During the early months, there is also often exhibited a marked emotionalism usually manifested by excessive laughter. In the later stages, there is withdrawal and masklike faces resembling that of classical Parkinsonism.

There are no reliable records to base speculation on the first appearance of kuru. Glasse and Glasse (1965) have suggested that kuru commenced in relatively recent times and that its arrival coincided with the introduction of European influence around 1910. The earliest association between kuru and scrapie came when Hadlow (1959) recognized pathologic similarities between the two diseases and suggested that kuru might be transmissible to subhuman primates. When Gajdusek et al. (1966) were successful in transmitting kuru to chimpanzees it became an experimental as well as a natural disease.

15.1.2 *Creutzfeldt–Jakob Disease*

In 1920, Hans Creutzfeldt published a description of an unusual case of brain degeneration in a 22-year-old female (Creutzfeldt, 1920). Immediately following this first report, Alfons Jakob (1921) described five similar cases and is given credit for establishing this disease as a distinct entity. Because of Jakob's greater contributions to the delineation of the clinical and neuropathological syndromes, the disorder is often referred to as Jakob–Creutzfeldt disease (Kirschbaum, 1968), disregarding Creutzfeldt's priority.

The disease is rare, the highest estimated incidence being 200 cases a year in the United States (Gajdusek, 1971). Unlike kuru, Creutzfeldt–Jakob disease has a world-wide distribution affecting both men and women almost equally during their 4th to 6th decade of life. The disease has several clinical and pathologic forms making diagnosis more difficult (Roos et al., 1973). Case histories usually include an insidious onset of ataxia followed by a rapidly progressive dementia accompanied by myoclonus, fasciculations and spasticity. The clinical course is less than two years with the debilitated patient often expiring from bronchopneumonia. Clinical diagnosis is based on the presence of a fulmonating presenile dementia associated with characteristic electroencephalographic readings.

Pathologic similarities between Creutzfeldt–Jakob disease and kuru were immediately recognized (Klatzo et al., 1959), as soon as the latter disease was reported. Ten years later, Gibbs et al., (1968) were successful in transmitting the disease from humans to chimpanzees.

15.1.3 Transmissible mink encephalopathy

The first recorded incidence of transmissible mink encephalopathy (TME) was in a Wisconsin mink herd in 1947 (Hartsough and Burger, 1965). The disease was not seen again until 1961 when it was observed on five separate mink ranches sharing a common source of feed. In 1963, TME occurred on three ranches; two in Wisconsin and one in Idaho. The disease has not been reported in the United States since 1963, however, there have been single outbreaks of TME observed on mink ranches in Canada (Hadlow and Karstad, 1968), Finland (Kangas and Marsh, unpublished) and East Germany (Hartung et al., 1970).

The natural disease affects only adult breeder animals, since most commercially raised mink are killed at 6 months of age (see 7.2). Because all mink ranches are structured in the age distribution of their animals, the morbidity in an affected herd will vary depending on the time of year the disease appears. Five of the 12 occurrences of TME have resulted in a near 100% loss of adult animals. Unlike the human diseases, it is possible to calculate a minimum incubation period of 7 months by studying the movement of mink between ranches and the introduction of new animals into the herd (Hartsough and Burger, 1965). More importantly, it is also possible to give a maximum incubation period of 12 months to at least two outbreaks of TME. This is simply because mink one year of age were affected and TME is not congenitally transmitted. This 12-month incubation period becomes highly significant when studying the epizootiology of the disease.

The early clinical signs of TME are mainly behavioral changes indicative of mental confusion. Normal habits of cleanliness disappear, feed is inadvertently walked on, and aimless circling is commonly observed. Changes in appearance are manifested by loss of weight, matting of the fur, and, usually, arching of the tail over the back. Ataxia is most prominent when the mink is at rest as indicated by an inability to maintain the hind quarters in an upright, sagittal plane. As the disease progresses, animals become more and more somnolent often standing for long periods of time with their heads in the corner of the cage. Affected animals become progressively more debilitated, dying of inanition within 6–8 weeks of the onset of the disease.

Epizootiologic studies indicating a long-incubation disease, together with pathologic similarities between TME and scrapie, led to attempts to experimentally transmit the disease to healthy mink. When this was accomplished (Burger and Hartsough, 1965), it was apparent that TME was indeed a scrapie-like disease of carnivores.

15.2 Pathology

15.2.1 Light microscopy

All of these diseases have similar pathologic features which are confined to the CNS and are microscopic in nature. They differ mainly in the topographic distribution of lesions and in their relative intensity (Table 15.1).

TABLE 15.1

Some general pathologic characteristics of natural kuru, Creutzfeldt–Jakob disease, and transmissible mink encephalopathy (TME).

Pathologic feature	Kuru	Creutzfeldt–Jakob*	TME
Diffuse spongiosus of cerebral cortex	rare	common	common
Involvement of cerebellum and spinal cord	common	common	rare
Stellate plaques	common	rare	not reported

* Some patients show particular involvement of the occipitoparietal cortex (Heidenhain's type), some of the pyramidal system (Jakob's pseudosclerosis or amyotrophic type), and others of the cerebellum (transitional type).

There are greater differences observed in the pathologic expression of individual diseases as they are transmitted to other species, than differences between the diseases themselves. The major morphologic changes are found

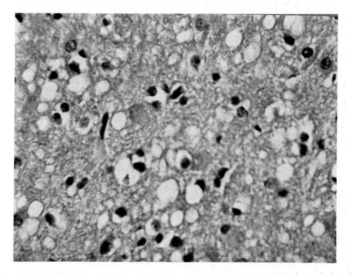

Fig. 15.2. Spongiform degeneration of the parietal cortex from a patient dying of Creutz-feldt–Jakob disease. Hematoxylin–Eosin. × 400.

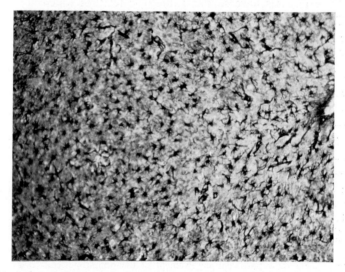

Fig. 15.3. Astrocytic hypertrophy in the frontal cortex of a mink killed in the terminal stages of transmissible mink encephalopathy. Cajal's gold sublimate. × 100.

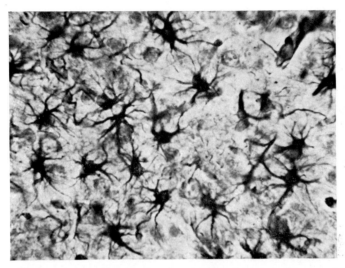

Fig. 15.4. Higher magnification of Fig. 15.3 showing the intense hypertrophy of the stellate-shaped astrocytes. Cajal's gold sublimate. × 400.

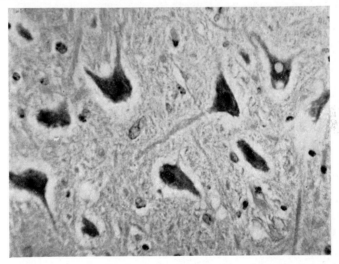

Fig. 15.5. Neuronal degeneration in the brain stem of a mink dying of transmissible mink encephalopathy. Hematoxylin–Eosin. × 400.

in gray matter and consist of spongiform degeneration (Fig. 15.2), astrocytic hypertrophy (Fig. 15.3 and 4), and neuronal degeneration (Fig. 15.5). Lesions are bilaterally symmetrical and there are no consistent cellular infiltrates indicative of inflammation (see Ch. 12).

These diseases are called spongiform encephalopathies, but this is some-what of a misnomer. The spongiform degeneration is a secondary effect not required for their lethal outcome. This is suggested by the fact that spongiform changes are inconsistently present in scrapie (see Ch. 12) and kuru (Klatzo et al., 1959). But the most definitive evidence comes from recent studies on the effect of TME in aged mink of the Chediak–Higashi genotype (Marsh et al., 1976). These animals have a marked reduction in spongiform degeneration with no alteration in length of incubation, clinical signs, astrocytic response, nor in replication of the TME agent as indicated by endpoint titrations. These findings are especially significant when evaluating the pathogenesis of these diseases.

15.2.2 *Electron microscopy*

The basic ultrastructural abnormalities of cerebral cortical tissues are the same in experimental kuru (Lampert et al., 1969), Creutzfeldt–Jakob

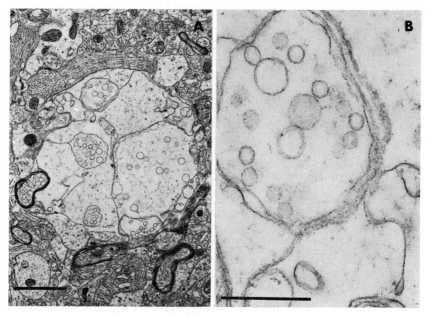

Fig. 15.6. Electron micrographs of cerebral cortex from a mink with experimental scrapie. (A) Focal, vacuolar lesion containing membrane fragments and vesicles. Scale, 2 μm. (B) Higher magnification showing greater detail of the membranous material within the lesion, scale, 0.5 μm (from Hanson, et al., 1971. Copyright 1971 by the American Association for the Advancement of Science).

disease (Lampert et al., 1971a), TME (Zu Rhein et al., 1974), and scrapie (Lampert et al., 1971b). They consist mainly of focal cytoplasmic degradation in nerve cells and their processes. The cytoplasm of neurons, especially their dendritic processes and axon terminals, show focal 'swellings' in which there appears to be an accumulation of membranous material in various forms suggestive of disruption, attempted duplication and repair, or vesiculation (Fig. 15.6). Unfortunately, these focal lesions correspond to vacuoles at the light microscopic level and, therefore, tell us little of the primary cell injury responsible for these apparent secondary changes.

No one has yet positively visualized a virus, although there have been several virus-like structures reported in affected tissues (see 12.10). The most

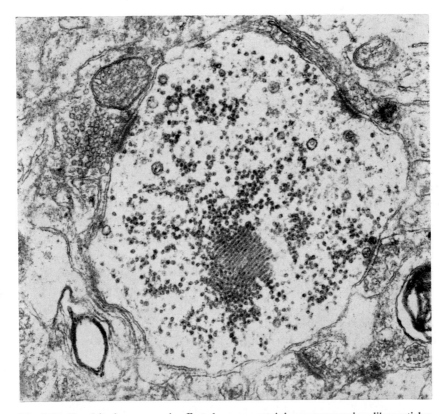

Fig. 15.7. Dendrite from a scrapie-affected mouse containing numerous virus-like particles. Synaptic thickenings and presynaptic processes containing larger synaptic vesicles can be seen in association with the dendrite. × 33,000. (Electron micrograph courtesy of Dr. G. M. Zu Rhein and Dr. John Varakis.)

consistent of these have been particles measuring approximately 35 nm in diameter which have been most commonly reported in dendrites (Fig. 15.7). These particles have been seen in experimental Creutzfeldt–Jakob disease (Lampert et al., 1971a), natural scrapie (Bignami and Parry, 1971), and in mouse scrapie (David-Ferreira et al., 1968; Lampert et al., 1971b; Lamar et al., 1974). The particles are virus-like in that they appear to be circular in at least one dimension, and they are of a uniform size. They are unusual in that some particles are vesicular or 'empty', especially those toward the periphery nearest the plasma membrane. As the particles coalesce toward the center, they become more electron-dense and often produce almost a crystalline arrangement. Because of their appearance, it is possible that their composition may be more lipid than nucleoprotein, as would be expected of a conventional virus. Until these particles can be isolated and characterized, we will not know their aetiologic significance.

15.3 Experimental transmission

15.3.1 Characterization of aetiologic agents

In determining the physicochemical properties of their aetiologic agents, TME and the human encephalopathies must certainly take a back seat to scrapie (see Ch. 11). This is mainly because of the availability of a mouse assay to scrapie researchers for the past 15 years. Studies on TME have been limited to a mink assay until only recently when hamsters were found to provide a more convenient means of measuring the transmissible agent (Kimberlin and Marsh, 1975; Marsh and Kimberlin, 1975). Kuru and Creutzfeldt–Jakob disease have not as yet been transmitted to a laboratory rodent, therefore, investigators have been restricted by the necessity of using non-human primates which are expensive, difficult to handle, and have incubation periods often in excess of one year making titrations almost an endless procedure.

The agents of TME, kuru and Creutzfeldt–Jakob disease are capable of replication as seen by serial passage of endpoint dilutions in experimental animals. The transmissible agents of both human encephalopathies are filterable through 220 nm, but not 100 nm Millipore filters (Gajdusek and Gibbs, 1973). The infectivity of the kuru agent is not significantly reduced after exposure to a temperature of 85 °C for 30 min (Gajdusek and Gibbs, 1973). On the basis of these few data, we can not relate kuru and Creutzfeldt–Jakob disease to scrapie solely by the physico-chemical properties of their

TABLE 15.2

Comparison of some physico-chemical properties of scrapie and transmissible mink encephalopathy (TME) agents.

Treatment	Scrapie	TME
Filtration	< 50 nm (Kimberlin et al., 1971)	< 50 nm (Marsh and Hanson, 1969a)
10% Formalin	Incomplete inactivation (Pattison, 1965)	Incomplete inactivation (Marsh and Hanson, 1969a)
Boiling, 15 min	Incomplete inactivation (Hunter and Millson, 1964)	Incomplete inactivation (Burger and Hartsough, 1965)
Ether	Partial loss of infectivity (Eklund et al., 1963)	Partial loss of infectivity (Marsh and Hanson, 1969a)
UV irradiation (254 nm 1.34 × 10^5 ergs/mm^2)	Substantial resistance (Alper et al., 1967)	Substantial resistance (Marsh and Hanson, 1969a)
90% Phenol	No recovery of infectivity (Hunter et al., 1969; Marsh et al., 1974)	No recovery of infectivity (Marsh and Hanson, 1969a; Marsh et al., 1974)

transmissible agents. Such is not the case with TME. As can be seen in Table 15.2, the transmissible agents of TME and scrapie would be closely related even if nothing else were known about the diseases they produce.

15.3.2 Species susceptibility

It is not possible to recognize any of the agents which can produce spongiform encephalopathies by their effect in in vitro systems. One is limited to a biological assay in an intact animal. Attempts to experimentally transmit

TABLE 15.3

Summary of the experimental host ranges of the spongiform encephalopathies (natural hosts in parentheses).

Species type	Scrapie (sheep, goat)	TME (mink)
Laboratory Rodent:	Mouse (Chandler, 1961) Hamster (Zlotnik, 1963) Rat (Chandler and Fisher, 1963) Gerbil (Chandler and Turfrey, 1972) Vole (Chandler and Turfrey, 1972)	Hamster (Marsh et al., 1969a)
Carnivore:	Mink (Hanson et al., 1971)	Raccoon (Eckroade et al., 1973) Striped Skunk (Eckroade et al., 1973)
Ruminant:		Goat (Zlotnik and Barlow, 1967) Sheep (Hadlow, personal communication)
New World Primate:	Squirrel (Gibbs and Gajdusek, 1973) Capuchin (Gajdusek and Gibbs, 1975) Spider (Gajdusek and Gibbs, 1975)	Squirrel (Eckroade et al., 1970)
Old World Primate:	Cynomolgus (Gibbs and Gajdusek, 1972)	Rhesus (Marsh et al., 1969a) Stump-tailed (Eckroade et al., 1970)

these diseases to species other than their natural hosts serve several functions:

(1) If the natural host is man, transmission to an animal allows for studies which would otherwise not be possible.

(2) Hopefully, this animal model will provide a less expensive, faster, and more accurate means of detecting the transmissible agent while at the same time not losing any relevance to the original disease.

(3) The establishment of experimental host ranges is one means by which to compare and relate the spongiform encephalopathies. This is particularly important because these diseases produce no detectable specific antibody responses.

Kuru (human)	Creutzfeldt–Jakob disease (human)
None	None
None	Cat (Gibbs and Gajdusek, 1973)
None	None
Squirrel (Gajdusek and Gibbs, 1971)	Squirrel (Gajdusek and Gibbs, 1971; Zlotnik et al., 1974)
Spider (Gajdusek et al., 1968)	Spider (Gajdusek and Gibbs, 1971)
Capuchin (Gajdusek and Gibbs, 1971)	Capuchin (Gajdusek and Gibbs, 1971)
Woolly (Gajdusek and Gibbs, 1971)	Woolly (Gajdusek and Gibbs, 1971)
Marmoset (Peterson et al., 1974)	Marmoset (Peterson et al., 1974)
Rhesus (Gajdusek and Gibbs, 1972)	Rhesus (Gibbs and Gajdusek, 1973)
Chimpanzee (Gajdusek et al., 1966)	Chimpanzee (Gibbs et al., 1968)
Mangabey (Gibbs and Gajdusek, 1973)	Cynomologus (Gibbs and Gajdusek, 1973)
Gibbon ape (Gibbs and Gajdusek, 1973)	Stump-tailed (Gibbs and Gajdusek, 1973)
	Mangabey (Gibbs and Gajdusek, 1973)
	Bushbaby (Gibbs and Gajdusek, 1973)
	African green (Gibbs and Gajdusek, 1973)

(4) It enables a comparison of virus interaction with different hosts which may produce different expressions of disease.

(5) In cases where little is known of the natural epidemiologic pattern of disease, as is true of the spongiform encephalopathies, knowledge of the range of species susceptibility may provide an insight into possible sources of infection.

Table 15.3 summarizes the experimental host ranges of the spongiform encephalopathies. These transmitted diseases are characterized by a clinico-pathological syndrome closely resembling that seen in the natural host and which is similar in all four diseases in all experimental animals. There are some differences in pathologic expression which are noticeable. Experimental kuru usually produces more spongiform degeneration of the cerebral cortex in animals and there are no stellate plaques, and experimental TME in the hamster produces lesions in the cerebellum and spinal cord which are not consistent features of the natural disease. However, the nature of these pathologic changes remain constant; astrocytic hypertrophy, neuronal degeneration and microvacuolation of the gray matter. Other reactions to infection, such as inflammation or neoplasia, have not been recognized.

The list of experimental animals which are found to be susceptible to these diseases grows longer each year. It seems as though they are transmissible to almost any animal if enough time and effort are expended. However, there is one major exception to this theme. TME and the human encephalopathies have not as yet been shown to be pathogenic for mice. While there are rare instances when sheep scrapie has not been transmissible to mice (10.8), this difference in experimental host range would seem to represent a major divergence among the spongiform encephalopathies. We do not know whether this variation is due to selection of subpopulations of scrapie or host-induced modification (Hanson and Marsh, 1974; see 10.11). The findings that scrapie agents apparently lose their pathogenicity for mice after passage in mink (Gustafson et al., 1972) or, in at least one instance, sub-human primates (Gibbs, personal communication), would indicate that species susceptibility can be influenced by animal passage and that these diseases may still originate from a common source.

15.4 Pathogenesis

How do these unusual agents produce their lethal effects? Studies on scrapie and TME in the Syrian hamster have shown that there is a steady rise of infectivity in brain tissue during the incubation period reaching 10^9 LD_{50}/g

several weeks before the onset of clinical signs of disease (Marsh and Kimberlin, 1975). Electron microscopic studies on TME in mink have also shown that ultrastructural lesions are present as long as 6 weeks before clinical illness (Zu Rhein and Eckroade, 1970). These results indicate a slow progression of agent replication with associated nerve cell damage. The onset of clinical signs are subtle, often not recognized until there is sufficient nerve cell destruction to cause noticeable changes in behavior and appearance. As the disease continues, more nerves become affected producing the ultimate effect of complete debilitation and death. It is a very slow, insidious process with an invariable course eliciting no apparent host defense mechanisms (Marsh et al., 1970, Gardiner, 1965; see Ch. 14).

We do not as yet understand which cells in the body are susceptible to infection, nor how these agents gain access to the CNS. Experimental studies in hamsters have indicated that after an intracerebral inoculation, the agents of scrapie and TME are disseminated throughout the CNS by cerebrospinal fluid (Marsh and Kimberlin, 1975). However, this is a totally artificial means of infection which may only be significant to natural exposure if there is a viremia. Studies on TME in mink have shown that the transmissible agent can not be detected in serum or blood cell fractions of affected animals (Marsh et al., 1973). There has been recent speculation that the agents of TME (Marsh et al., 1973) and scrapie (Outram et al., 1974) may replicate in nerve endings, progressing to the CNS via connecting nerve fibers (see also 14.6). This is not a new phenomenon to virology since it is known to occur in rabies (Murphy et al., 1973) and herpes simplex (Cook et al., 1973) infections.

After a peripheral inoculation, scrapie agent is first found to replicate in regional lymphatic tissues and in spleen before appearing in the CNS (Eklund et al., 1967). This has not been as accurately determined for the other spongiform encephalopathies; although infectivity can be demonstrated in liver, kidney and spleen in experimental kuru (Gajdusek and Gibbs, 1973), and in these and other tissues in TME (Marsh et al., 1969b). Because of replication in organs other than brain, investigators concluded that these transmissible agents must have both nervous and non-nervous tissue affinities. This need not be true. There are autonomic nerve endings in every visceral organ which have direct or indirect fiber connections to the CNS.

Perhaps the strongest evidence to date that these agents may replicate in nerve endings comes from studies on corneal epithelium of hamsters infected with TME (Marsh and Hanson, 1975). These studies were initiated by a report of the possible person-to-person transmission of Creutzfeldt–Jakob disease by means of a corneal transplant (Duffy et al., 1974). It was found that corneal epithelium from hamsters contained a high level of infectivity which was not maintained after tissue culture. Since corneal epithelium is richly innervated with free, unmyelinated nerve endings (Thomas, 1955; Whitear, 1960), there is some basis for speculating that TME infectivity is mainly associated with a neural component. Mice infected with herpes simplex by corneal scarification have been found to develop encephalitis after progression of the virus in corneal nerve fibers (Knotts et al., 1974).

The inability to infect or produce recognizable cytopathology in a variety of tissue culture systems would suggest that these agents have a narrow range of cell types in which they can replicate. The only cell which appears to be reacting positively to infection is the astrocyte. Astrocytic hypertrophy is the most consistent pathologic change produced by these transmissible agents. We still do not know whether these cells are responding as a direct result of injury, or if they are only reacting to neuronal damage. Lampert has described focal membrane lesions in astrocytes adjacent to affected nerve cell processes, and he has speculated that while nerve cells react to infection by degenerating, astrocytes proliferate (Lampert et al., 1972). Certainly the intimate association between astroglial and nerve cell membranes could allow for direct cell to cell infection. 'Glial' cell cultures from animals affected with either of these four diseases have been found to have a low level of infectivity which can be maintained for an indefinite number of subcultures (Gustafson and Kanitz, 1965; Marsh and Hanson, 1969b; Clarke and Haig, 1970; Gajdusek et al., 1972). Similar cultures from normal animals have not as yet been infected in vitro (Rosenberger, 1972), indicating it is perhaps only the close apposition of glial and nerve cell membranes which allows infection to proceed in vivo.

We do not know the mechanism by which these agents produce neuronal degeneration. Because scrapie infectivity is closely associated with membranes (11.4), we may presume that these agents are likely to elicit their effects at the periphery of the cell. Studies on ultrastructure have shown that the most pronounced lesions are in dendrites and axon terminals. Therefore, if these agents do replicate in nerve fibers, they apparently do not interfere with critical functions until reaching synaptic areas. There have been several reports on reduced ganglioside concentrations in brain tissue from patients (Korey et al., 1961; Suzuki and Chen, 1966; Bass et al., 1974) or chimpanzees (Yu et al., 1974) with Creutzfeldt–Jakob disease, and chimpanzees affected with kuru (Yu et al., 1974). Gangliosides are glycosylceramides which occur as membrane constituents of many tissue cells. The more complex gangliosides are found in the CNS and mainly in neurons, particularly in their dendritic and synaptic regions (Lapetina et al., 1967). There is no general agreement on the physiologic role of gangliosides other than they are probably involved at some level of synaptic transmission.

Since there is a decrease in brain gangliosides in kuru and Creutzfeldt–Jakob disease, it is evident that there is some interference in the normal biosynthesis of these compounds or an increased rate of catabolism. Ganglioside biosynthesis requires a series of reactions in which sugars are transferred by specific glycosyl transferases to a glycolipid receptor. The glycolipid glycosyl transferases in brain are particulate and localized in synaptic membranes (Wolfe, 1972). It is therefore possible that the transmissible agents causing spongiform encephalopathies may interfere with these transferases at the synaptic membrane, thus reducing ganglioside biosynthesis. However, we have an additional experimental observation which must be taken into consideration.

Aged mink of the Chediak–Higashi genotype show a marked reduction in spongiform degeneration with no other apparent alterations in the TME disease process (Marsh et al., 1976). Spongiform degeneration corresponds to the focal 'swellings' and membrane fragmentation found mainly in dendrites and axon terminals, but it is most unlikely that these changes are due to a primary effect of these transmissible agents at the synaptic level (15.2.1). It is possible that the decrease in ganglioside concentration is only a secondary effect due to increased catabolism. Ganglioside catabolism proceeds by a series of hydrolytic cleavages of sialosyl and glycosyl linkages (Wolfe, 1972). These reactions are carried out by various hydrolases which are localized in lysosomes. Since animals of the Chediak–Higashi genotype are known to have lysosomal abnormalities (Windhorst et al., 1968; Root et al., 1972), it is possible that they may be deficient in glycosyl hydrolases or that the disease process is not capable of affecting these lysosomes in the same manner as normal ones. Thus, the decrease in gangliosides reported in kuru and Creutzfeldt–Jakob disease may only be a secondary effect due to lysosomal enzymes causing a breakdown of gangliosides in areas where they are highly concentrated, such as synapses, producing the focal lesions responsible for spongiform degeneration. This hypothesis is supported by studies on mouse scrapie which have shown an increase in certain glycosyl hydrolases before the onset of clinical disease (Millson and Bountiff, 1973; see also 13.3). It can also be tested by measuring brain gangliosides in TME-affected mink of the Chediak–Higashi genotype.

15.5 *Epidemiology*

There are probably few readers of this book on slow virus diseases who have not had some previous knowledge of the spongiform encephalopathies. While you were undoubtedly perplexed as to the physico-chemical nature of these agents and to their mode of action, the epidemiology of the diseases may have appeared to be quite simple. Scrapie is a disease of a meat animal which is consumed by mink and man, both of which are afflicted with a similar disease. This is the most obvious explanation for the origin of TME and Creutzfeldt–Jakob disease. But we should look this gift horse in the mouth very carefully.

It is improbable that kuru originated from scrapie since the Fore natives have had no contact with domestic sheep or goats. It has been speculated that kuru arose from 'a spontaneous case of . . . Creutzfeldt–Jakob disease in a Melanesian native' (Gibbs and Gajdusek, 1974). This is unlikely because it is a virus-induced disease, none of which have been shown to be spontaneously produced. It is more reasonable to suspect that kuru was introduced into the Fore by Europeans and has been perpetuated by cannibalistic practices. If this assumption is valid, kuru represents a form of Creutzfeldt–Jakob disease with a clinicopathologic syndrome which has been 'fixed' by numerous human-to-human passages in individuals with a relatively similar genetic background.

Studies on the epizootiology of mink encephalopathy should offer a better opportunity for determining the source of infection, but they do not. It has not been possible to definitely associate the feeding of sheep tissues to mink with the incidence of TME. There have been suggestive case histories, but nothing substantial. In examining the converse, there are many instances where the commercial feeding of sheep tissues have not resulted in TME. However, this becomes almost meaningless since we know very little of the prevalence of scrapie infection in sheep populations as compared to the incidence of clinical disease.

Mink have been shown to develop TME-like disease 12–14 months after the intracerebral inoculation of one source of scrapie from a Suffolk sheep (Hanson et al., 1971). In the same experiment, scrapie from a Cheviot sheep produced no disease in mink. A later experiment, using two new sources of Suffolk scrapie, failed to produce as short an incubation period and as uniform a response as seen previously (Gustafson, Lamar, Marsh and Hanson, unpublished data). These results indicate that while mink are susceptible to scrapie they do not respond in the same way to different sources of infectivity. It is also important to remember that natural TME has had incubation periods of no longer than 12 months in at least two instances. The natural disease is presumedly induced by oral exposure, a route requiring a longer time to produce disease than intracerebral inoculation (Marsh et al., 1969b; Marsh, 1974). This variation in mink susceptibility to scrapie could be due to the presence of different subpopulations of the agent, some having greater pathogenicity for carnivores and, perhaps, primates. Dickinson (1970) has demonstrated a heterogeneity of scrapie agents in sheep as detected by their distinguishing effects in mice (see 10.6; 12.2.3; 14.4.4d). However, intracerebral inoculation of mink with brain suspensions containing various subpopulations of mouse scrapie (22A, 22C, 87A, 87V, 51C, 138A, ME7, 104A, 79A, or 79V), or their sheep sources, have failed to produce a uniform response of TME-like disease in any experimental group during a 24-month observation period (Marsh, Hanson, and Dickinson, unpublished data). These findings indicate that if TME does arise from the sheep scrapie pool, it is either caused by a subpopulation or combination of agents which we have not yet recognized, or there are genetic factors influencing mink susceptibility to scrapie. Little attention has been given this latter possibility because mink appear to be universally susceptible to TME as indicated by length of incubation period (Marsh et al., 1969b). However, there is a much less uniform response of mink to scrapie and this difference may be genetically programmed.

Creutzfeldt–Jakob disease generally has a sporadic pattern of occurrence but can also present a familial history of susceptibility, the most noted of which is the Backer family which has sustained four cases of the disease over a 20-year period (Kirschbaum, 1968). The familial form of the disease has also been transmitted to a chimpanzee (Gajdusek and Gibbs, 1973), confirming its infectious aetiology. This suggests hereditary predisposition to Creutzfeldt–Jakob disease which may be similar to the well-documented breed predisposition of sheep to scrapie (see 10.7 and 9). It also may indicate a high probability for any one individual coming into contact with the disease agent. Other possibilities include vertical passage, or a high incidence of horizontal transmission through intimate familial contacts (Ferber et al., 1974). The recognition of at least three distinct clinicopathologic forms of Creutzfeldt–Jakob disease suggests a variation in the host–virus interaction which is undoubtedly influenced by genetic background and, perhaps, the existence of subpopulations of the transmissible agent. This has led to speculation that more common degenerative diseases of the brain, such as Alzheimer's disease, Pick's disease, Parkinsonism dementia and Huntington's chorea, may also be caused by similar slow-acting agents (Gajdusek and Gibbs, 1973; 12.9). Certainly the finding of a significant variation in the pathologic response of mink to TME, as effected by the interaction of age and genotype (Marsh et al., 1976), enhances this possibility.

Is Creutzfeldt–Jakob disease caused by the ingestion of scrapie-infected sheep tissue? This is the most compelling question facing investigators today. Epidemiologic studies in Israel have shown that Libyan Jewish immigrants have an incidence of Creutzfeldt–Jakob disease which is approximately 30 times normal (Kahana et al., 1974). It has also been observed that these same immigrants eat sheep brain (Alter, 1974). Have we then discovered that combination of agent exposure and permissive genetic background producing an extraordinary occurrence of this spongiform encephalopathy? Additional studies are needed before any definite associations can be made. Considering the inconclusive results of experiments on mink susceptibility to scrapie, we should be cautious of developing premature conclusions on the origin of this human disease.

References

ALPER, T., CRAMP, W. A., HAIG, D. A. and CLARKE, M. C. (1967) Nature, 214, 764.

ALTER, M. (1974) N. Engl. J. Med., 291, 210.

BASS, N. H., HESS, H. H. and POPE, A. (1974) Arch. Neurol., 31, 174.

BIGNAMI, A. and PARRY, H. B. (1971) Science, 171, 389.

BURGER, D. and HARTSOUGH, G. R. (1965) J. Infect. Dis., 115, 393.

CHANDLER, R. L. (1961) Lancet, 1, 1378.

CHANDLER, R. L. and FISHER, J. (1963) Lancet, 2, 1165.

CHANDLER, R. L. and TURFREY, B. A. (1972) Res. Vet. Sci., 13, 219.

CLARKE, M. C. and HAIG, D. A. (1970) Res. Vet. Sci., 11, 500.

COOK, M. L. and STEVENS, J. G. (1973) Infec. Immun., 7, 272.

CREUTZFELDT, H. G. (1920) Z. Ges. Neurol. Psychiat., 57, 1.

DAVID-FERREIRA, J. R., DAVID-FERREIRA, K. L., GIBBS, C. J., JR and MORRIS, J. A. (1968) Proc. Soc. Exp. Biol., 127, 313.

DICKINSON, A. G. (1970) Proc. Sixth Int. Congr. Neuropathol. (Masson et Cie, Paris) p. 841.

DUFFY, P., WOLF, J., COLLINS, G., DEVOE, A. G., STREETEN, B. and COWEN, D. (1974) N. Engl. J. Med., 299, 692.

ECKROADE, R. J., ZU RHEIN, G. M. and HANSON, R. P. (1973) J. Wildlife Dis., 9, 229.

ECKROADE, R. J., ZU RHEIN, G. M., MARSH, R. F. and HANSON, R. P. (1970) Science, 169, 1088.

EKLUND, C. M., HADLOW, W. J. and KENNEDY, R. C. (1963) Proc. Soc. Exp. Biol. Med., 112, 974.

EKLUND, C. M., KENNEDY, R. C. and HADLOW, W. J. (1967) J. Infect. Dis., 117, 15.

FERBER, R. A., WIENFELD, S. L., ROOS, R. P., BOBOWICK, A. R., GIBBS, C. J., JR. and GAJDUSEK,

D. C. (1974) In: (A. Subirana and J. M. Burrows, Eds) Proceedings of the Xth International Congress of Neurology (Excerpta Medica, Amsterdam) p. 358.

GAJDUSEK, D. C. (1963) Trans. R. Soc. Trop. Med. Hyg., 57, 151.

GAJDUSEK, D. C. (1971) Am. J. Clin. Pathol., 56, 352.

GAJDUSEK, D. C. and GIBBS, C. J., JR. (1971) Nature, 230, 588.

GAJDUSEK, D. C. and GIBBS, C. J., JR. (1972) Nature, 240, 351.

GAJDUSEK, D. C. and GIBBS, C. J. JR. (1973) Perspect. Virol., 8, 279.

GAJDUSEK, D. C. and GIBBS, C. J. (1975) Familial and Sporadic Chronic Neurologic Degenerative Disorders Transmitted From Man to Primates. (Raven Press) in press.

GAJDUSEK, D. C., GIBBS, C. J., JR. and ALPERS, M. (1966) Nature, 209, 794.

GAJDUSEK, D. C., GIBBS, C. J., JR. and ASHER, D. M. (1968) Science, 162, 693.

GAJDUSEK, D. C., GIBBS, C. J., JR., ROGERS, N. G., BASNIGHT, M. and HOOKS, J. (1972) Nature, 235, 104.

GAJDUSEK, D. C. and ZIGAS, V. (1957) N. Engl. J. Med., 257, 974.

GARDINER, A. C. (1965) Res. Vet. Sci., 7, 190.

GIBBS, C. J., JR. and GAJDUSEK, D. C. (1969) Science, 165, 1023.

GIBBS, C. J., JR. and GAJDUSEK, D. C. (1972) Nature, 236, 73.

GIBBS, C. J., JR. and GAJDUSEK, D. C. (1973) Science, 182, 67.

GIBBS, C. J., JR. and GAJDUSEK, D. C. (1974) In: (W. Zeman and E. H. Lennette, Eds) Slow Virus Diseases. (William and Wilkins, Baltimore) p. 39.

GIBBS, C. J., JR., GAJDUSEK, D. C., ASHER, D. M., ALPERS, M. P., BECK, E., DANIEL, P. M. and MATTHEWS, W. B. (1968) Science, 161, 388.

GLASSE, R. M. and GLASSE, S. (1965) Lancet, 1, 1138.

GUSTAFSON, D. P. and KANITZ, C. L. (1965) In: (D. C. Gajdusek, C. J. Gibbs, Jr, and M. Alpers, Eds) Natl. Inst. Neurol. Dis. Blindness, Monogr. 2, Slow, Latent and Temperate Virus. (U.S. Government Printing Office, Washington, D.C.) p. 221.

GUSTAFSON, D. P., MARSH, R. F. and HANSON, R. P. (1972) ASM abstracts (V281).

HADLOW, W. J. (1959) Lancet 2: 289–290.

HADLOW, W. J. and KARSTAD, L. (1968) Can. Vet. J. 9: 193.

HANSON, R. P., ECKROADE, R. J., MARSH, R. F., ZU RHEIN, G. M., KANITZ, C. L. and GISTAFSON, D. P. (1971) Science 172, 859.

HANSON, R. P. and MARSH, R. F. (1974) In: (W. Zeman and E. H. Lennette, Eds) Slow Virus Diseases (Williams and Wilkins, Baltimore) p. 10.

HARTSOUGH, G. R. and BURGER, D. (1965) J. Infect. Dis., 115, 387.

HARTUNG, J., ZIMMERMANN, H. and JOHANNSEN, U. (1970) Monatsh. Vet. Med., 25, 385.

HUNTER, G. D., GIBBONS, R. A., KIMBERLIN, R. H. and MILLSON, G. C. (1969) J. Comp. Pathol., 79, 101.

HUNTER, G. D. and MILLSON, G. C. (1964) J. Gen. Microbiol., 37, 251.

JAKOB, A. (1921) Z. Gesam. Neurol. Psychiat., 64, 147.

KAHANA, E., ALTER, M., BRAHAM, J. and SOFER, D. (1974) Science, 183, 90.

KIMBERLIN, R. H. and MARSH, R. F. (1975) J. Infect. Dis., 131, 97.

KIMBERLIN, R. H., MILLSON, G. C. and HUNTER, G. D. (1971) J. Comp. Pathol., 81, 383.

KIRSCHBAUM, W. R. (1968) Jakob–Creutzfeldt Disease. (American Elsevier, New York).

KLATZO, I., GAJDUSEK, D. C. and ZIGAS, V. (1959) Lab. Invest., 8, 799.

KNOTTS, F. B., COOK, M. I. and STEVENS, J. G. (1974) J. Infect. Dis., 130, 16.

KOREY, S. R., KATZMAN, R. and ORLOFF, J. (1961) J. Neuropathol. Exp. Neurol., 20, 95.

LAMAR, C. H., GUSTAFSON, D. P., KRASOVICH, M. and HINSMAN, E. J. (1974) Vet. Pathol., 11, 13.

LAMPERT, P. W., EARLE, K. M., GIBBS, C. J., JR. and GAJDUSEK, D. C. (1969) J. Neuropathol. Exp. Neurol., 28, 353.

LAMPERT, P. W., GAJDUSEK, D. C. and GIBBS, C. J., JR. (1972) Am. J. Pathol., 68, 626.

LAMPERT, P. W., GAJDUSEK, D. C. and GIBBS, C. J., JR. (1971a) J. Neuropathol. Exp. Neurol., 30, 20.

LAMPERT, P., HOOKS, J., GIBBS, C. J., JR. and GAJDUSEK, D. C. (1971b) Acta Neuropathol., 19, 81.

LAPETINA, E. G., SOTO, E. F. and DE ROBERTIS, E. (1967) Biochim. Biophys. Acta, 135, 33.

MARSH, R. F. (1974) In: (C. A. Brandly and C. E. Cornelius, Eds) Advances in Veterinary Science and Comparative Medicine. Vol. 18 (Academic Press, New York) p. 155.

MARSH, R. F., BURGER, D., ECKROADE, R. J., ZU RHEIN, G. M. and HANSON, R. P. (1969a) J. Infect. Dis., 120, 713.

MARSH, R. F., BURGER, D. and HANSON, R. P. (1969b) Am. J. Vet. Res., 30, 1637.

MARSH, R. F. and HANSON, R. P. (1969a) J. Virol., 3, 176.

MARSH, R. F. and HANSON, R. P. (1969b) Am. J. Vet. Res., 30, 1643.

MARSH, R. F. and HANSON, R. P. (1975) Science, 87, 656.

MARSH, R. F. and KIMBERLIN, R. H. (1975) J. Infect. Dis., 131, 104.

MARSH, R. F., MILLER, J. M. and HANSON, R. P. (1973) Infect. Immun., 7, 352.

MARSH, R. F., PAN, I. C. and HANSON, R. P. (1970) Infect. Immun., 2, 727.

MARSH, R. F., SEMANCIK, J. S., MEDAPPA, K. C., HANSON, R. P. and RUECKERT, R. R. (1974) J. Virol., 13, 993.

MARSH, R. F., SIPE, J. C., MORSE, S. S. and HANSON, R. P. (1976) Lab. Invest., in press.

MILLSON, G. C. and BOUNTIFF, L. (1973) J. Neurochem., 20, 541.

MURPHY, F. A., BAUER, S. P., HARRISON, A. K. and WINN, W. C. (1973) Lab. Invest., 28, 361.

OUTRAM, G. W., DICKINSON, A. G. and FRASER, H. (1974) Nature, 249, 855.

PATTISON, I. H. (1965) J. Comp. Pathol., 75, 159.

PETERSON, D. A., WOLFE, L. G., DEINHARDT, F., GAJDUSEK, D. C. and GIBBS, C. J., JR. (1974) Intervirology, 2, 14.

ROOS, R., GAJDUSEK, D. C. and GIBBS, C. J., JR. (1973) Brain, 96, 1.

ROOT, R. K., ROSENTHAL, A. S. and BALESTRA, D. J. (1972) J. Clin. Invest., 51, 649.

ROSENBERGER, J. (1972) Ph.D. thesis, University of Wisconsin, Madison.

SUZUKI, K. and CHEN, G. (1966) J. Neuropathol. Exp. Neurol., 25, 396.

THOMAS, C. I. (1955) The Cornea (Charles C. Thomas, Springfield) p. 39.

WHITEAR, M. (1960) J. Anat., 94, 387.

WINDHORST, D. B., WHITE, J. G., ZELICKSON, A. S., CLAWSON, C. C., DENT, P. B., POLLARA, B. and GOOD, R. A. (1968) Ann. N.Y. Acad. Sci., 155, 818.

WOLFE, L. S. (1972) In: (N. Marks and R. Rodnight, Eds) Research Methods in Neurochemistry. Vol. 1 (Plenum Press, New York) p. 233.

YU, R. K., LEDEEN, R. W., GAJDUSEK, D. C. and GIBBS, C. J. (1974) Brain Res., 70, 103.

ZIGAS, V. and GAJDUSEK, D. C. (1957) Med. J. Aust., 2, 745.

ZLOTNIK, I. (1963) Lancet, 2, 1072.

ZLOTNIK, I. and BARLOW, R. M. (1967) Vet. Rec., 81, 55.

ZLOTNIK, I., GRANT, D. P., DAYAN, A. D. and EARL, C. J. (1974) Lancet, 2, 435.

ZU RHEIN, G. M. and ECKROADE, R. J. (1970) In: Proc. VIth Int. Congr. Neuropathol. (Masson et Cie., Paris) p. 939.

ZU RHEIN, G. M., ECKROADE, R. J. and GRABOW, J. D. (1974) In: (W. Zeman and E. H. Lennette, Eds) Slow Virus Diseases (Williams and Wilkins, Baltimore) p. 16.

Part V

Conclusions

CHAPTER 16

Some final comments on visna-maedi Aleutian disease and scrapie

Richard H. KIMBERLIN

16.1 Introduction

This book describes only a limited number of slow virus diseases and the reasons for this selectivity have been given in the opening chapter. One of the main aims has been to describe in depth, three diseases which, because they occur naturally in animals and a range of experimental animal models are available, are reasonably well understood at all levels of investigation from epidemiology to molecular biology.

One of the most important points to appreciate about the slow virus diseases is their great diversity – which is why the term 'slow virus diseases' is of such limited use in any taxonomic sense. Hence, we are faced with the problem of understanding not just one but a variety of slowly developing diseases. The diseases described in this book illustrate this diversity extremely well, to the point that each has to be studied independently because an understanding of one contributes little to that of another. At the present state of our knowledge, visna–maedi, AD and scrapie represent extremes but it should not be supposed that all slow virus diseases will differ so dramatically from one another. Indeed, each of these three diseases can be related to some extent to others, and they serve as important models for related but less well understood conditions. Herein lies the real scientific importance of these diseases, because an understanding of their aetiology and pathogenesis is crucial to an understanding of some related human conditions which are less accessible to investigation. This is a classic example of the vital contribution 'veterinary medicine' can make to 'human medicine'.

Slow virus diseases of animals and man, edited by R. H. Kimberlin
© *North-Holland Publishing Company 1976*

Because of the great differences between visna–maedi, AD and scrapie, a simple summary of the forgoing chapters is not possible – rather three long summaries would be needed. However, it will be apparent from the style and content of many chapters that as much emphasis has been placed on describing problems and posing questions as in reporting data and giving answers. It might therefore be useful to conclude this book by summarising and discussing some of the general points to emerge from the preceeding text which may help to put the subject into perspective.

The format of *Disease, Virus* and *Virus–Host Interactions* was deliberately chosen because this is the sequence through which studies of disease problems initially progress. It is convenient to use the same format for the remaining discussion.

16.2 The diseases

It is at this level that visna–maedi, AD and scrapie have most in common. Apart from the fact that they are all slow virus diseases, they are also diseases which have been, and still are, important veterinary problems. Furthermore control of these diseases is still rather crude and inefficient.

16.2.1 Diagnosis

With any disease the initial problem to be resolved is that of diagnosis. Clinical signs alone are seldom adequate and the field observer is only too well aware how rarely 'typical signs' occur in a really clear-cut fashion. Histopathological examination of affected tissues helps considerably in diagnosis although there is often a spectrum of damage, such that some lesions may resemble those found in other diseases caused by different agents because there are only a limited number of ways an affected tissue can manifest cellular damage. Virus isolations from affected animals by tissue culture techniques are often invaluable, although there can be problems here with concurrent virus infections which may have little to do with the disease in question. Tests for circulating antibodies are also extremely useful but they may be subject to similar limitations. There are two general points to be emphasised. Firstly, accurate diagnosis is often only possible when several independent criteria are available. Secondly, it is important to recognise the natural range of variation of each of these criteria and if possible to define this range in terms of host factors (physiological and genetic) and also of virus strains.

Both visna–maedi and AD can be diagnosed adequately using a number of independent methods. However, with scrapie there are still serious problems of diagnosis and for practical purposes the disease can be defined only in clinical and histopathological terms. In some cases the clinical period can be very brief, often with minimal signs of pruritus. The wide range in the character and severity of histological lesions is well recognised in scrapie. Since cases of experimental scrapie with negligible brain lesions (of diagnostic value) have been described, it is almost certain that sheep may die of natural scrapie which goes unrecognised.

16.2.2 Virus strains

It is interesting that although serological and tissue culture techniques are not available in scrapie research, strain typing of natural isolates is reasonably well developed using biological and histopathological criteria in inbred mice. These techniques are important to studies of the epidemiology of scrapie and in future could be vital in the investigation of the range of scrapie-like agents which are associated with transmissible mink encephalopathy (TME) and the two diseases of man, kuru and Creutzfeldt–Jakob disease. The problem to date has been that none of the agents associated with these scrapie-like diseases has been transmitted to mice and strain typing techniques have not yet been developed in other host species.

Very little virus typing work has been carried out with visna–maedi and AD, although the existence of biologically different strains is suspected in both diseases. The recent success in growing AD virus in cell culture should provide a new impetus to virus-typing studies in this disease and, in view of the wide variety of interactions between different scrapie agents and the *sinc* gene in mice, it would be interesting to investigate the effect of different strains of AD virus on the development of disease in mink of different genotypes. With visna–maedi, more work needs to be done on virus typing in relation to the large number of maedi-like diseases which have been described throughout the world under various names. It would also be interesting to study the possible relationship between individual virus strains and the occurrence of visna in flocks affected with maedi.

16.2.3 Genetic variation in the host

With all three diseases, genetic variation in the host is either known or suspected to be important in influencing the development of clinical disease.

This is best seen in AD and in scrapie. The situation is less clear with visna–maedi although the genetic constitution of the host is likely to be important in visna–maedi as well – the dramatic appearance of maedi in Icelandic sheep after the importation of clinically normal Karakul sheep provides some suggestive evidence for this possibility.

However, in none of the naturally occurring diseases has it been possible to achieve a satisfactory analysis of the genetic factors controlling disease development. There are many problems in trying to obtain such information, such as measuring the degree of exposure of animals to the infectious virus and dissociating the effect of vertical transmission of virus from the overall picture. The latter point is particularly important since vertical transmission can easily give the illusion of 'genetic control' because the disease appears in a familial pattern.

Analysis of the genetic control of disease development following experimental infection is a sounder proposition because one can control many of the important variables which would otherwise confuse the picture. There is some information on the genetic control of experimental and natural AD (e.g., Aleutian mink have a shorter incubation period than non-Aleutian mink) and in scrapie, two long-term selection experiments with sheep have indicated that the length of incubation in animals infected subcutaneously with the SSBP/1 source of agent is largely controlled by a single gene. The data are consistent with the existence of 2 alleles and the allele for a shorter incubation period is dominant. However, this type of analysis is an extremely daunting proposition partly because of the need for a large number of animals and partly because the long incubation period of these diseases makes the initial selection and testing of different lines a very long-term task indeed.

16.2.4 Natural transmission

All three diseases (visna–maedi, AD and scrapie) are infectious and horizontal spread from one animal to another can account for a high proportion of cases. The main route of spread varies with the disease. With visna–maedi the respiratory route is almost certainly the most important one, whereas with AD and scrapie the oral route is probably the predominant route of spread. It is interesting to note that the relatively high physico-chemical stability of AD virus and of many scrapie agents may be an important factor in permitting horizontal spread of infection.

There is evidence that all three diseases can be transmitted vertically from mother to offspring. With scrapie, epidemiological and other evidence suggests that this is the predominant route of transmission of the natural disease, but it is not known to what extent this operates by transplacental infection of the foetus or by oral infection of the lamb at the time of parturition or during the subsequent period of intimate association between mother and offspring. With AD, the vertical route of transmission is probably at least as important as the horizontal route, and in this case there is firm evidence of intrauterine infection which can lead to foetal death. The vertical route is less important in transmitting visna–maedi than the horizontal route. There is little evidence of foetal infection in visna–maedi but the virus is known to be present in milk.

With diseases of such long incubation period it is not difficult to see how vertical transmission of infection can provide an efficient means of maintaining the virus in an animal population, since an infected female can pass the infection to many offspring. Nor is it difficult to explain how an epidemic can arise quite rapidly from perhaps a single infected animal. Vertical transmission (supplemented by the horizontal route) provides an excellent means of amplifying the infection until a whole flock or herd becomes infected. This is especially true for the sheep diseases since the 'commercial life span' may be nearly as long as the 'natural life span'.

16.2.5 Methods of disease control

Control of these diseases presents many problems. There are no immunological or chemotherapeutic ways of preventing or curing them: nor indeed is there any immediate prospect of so doing. The pathogenesis of AD is known to be immunologically mediated and the attempts at vaccination carried out so far either had no effect or else enhanced the severity of lesions. Similar attempts to vaccinate against visna–maedi infection in sheep have failed and there is no sound basis on which to found further attempts; there is evidence that this disease may also be immunologically mediated. The situation is equally bleak for scrapie since no specific immunological response to infection is known to exist and attempts to suppress infection in mice by either repeated doses of interferon or by stimulating interferon production in vivo have failed.

The evidence that host genetic factors can influence the course of these diseases raises the possibility of breeding resistant animals. There are, however, many difficulties associated with this approach. First there is question of what is meant by resistance. Resistance to clinical disease is one criterion but how can this be defined in practical terms, i.e., how long does one keep animals after experimental infection before saying they are resistant? Studies with all three diseases have clearly shown that the infection can persist for extremely long periods of time. Hence, it is probably quite meaningless to talk of 'resistance to clinical disease' in any absolute sense. All that can be said is that, in a particular set of circumstances, the incubation period of the disease is so long that animals are unlikely to succumb within a specified time. However, nothing in nature is static and even if one accepted such restricted criteria of resistance there would always be the possibility that 'resistance' might break down. An obvious way this could happen is if the persistent virus mutated to a variant which had a much shorter incubation period. Extensive studies of mouse scrapie and also of experimental sheep scrapie have shown that resistance to clinical disease is a function of the agent strain; for example the 22A strain of agent has a short incubation period in VM mice ($sinc^{p7}$) and a long incubation period in C57BL mice ($sinc^{s7}$) but the 22C strain behaves in the opposite manner.

Genetic selection for individuals which are non-permissive for virus multiplication would provide a better approach to resistance if there were any evidence that such a genetic block existed for these diseases in their natural hosts. There is evidence that some non-Aleutian mink can eventually clear the viraemia and consequently do not develop lesions and clinical disease but this property of the host does not appear to be genetically determined in any simple manner.

Another problem concerns the way in which animals are experimentally exposed to a virus in order to select those which are operationally resistant. If different routes of infection involve different steps in early pathogenesis, then it is quite possible that genetic control operating on one route of infection may be irrelevant for another. This kind of situation is known to occur in experimental sheep scrapie, where it has been shown that some animals resistant to disease after subcutaneous infection may be susceptible if infected intracerebrally. Ideally the selection should be carried out using natural routes of infection.

Disease control by breeding is clearly difficult and unlikely to provide a permanent solution to the problem. However, provided the limitations of this method are appreciated, it may have a place in controlling some animal diseases where no easier method is available.

Fortunately, with visna–maedi and AD there are alternatives; the main one being to dispose of all animals that are infected. The success of this method depends on being able to detect infection early by methods that are suitable for screening large numbers of animals. This approach can be effective in controlling AD because the rapid development of very high antibody levels in infected mink makes early detection relatively easy. Initial screening can be carried out with the very simple iodine test and non-reactors can be retested by specific and more sensitive antibody tests such as counter-immunoelectrophoresis.

A similar approach can be used to control visna–maedi, although in this case the levels of circulating antibodies are not as high as commonly found in AD. Unless very frequent testing is employed, some infected animals may remain undetected. Total eradication still demands the slaughter of all animals in an affected flock. The eradication of visna and maedi in Iceland is a classic example of this drastic but effective measure of control.

A slaughter policy is the only sure way of eliminating scrapie from flocks. A reasonable control measure is to cull infected animals but the problem with scrapie is the absence of any practicable method for identifying infected animals which are not showing clinical signs; as mentioned earlier, the disease can only be diagnosed clinically and histopathologically. However, the predominant route of transmission is vertical and the culling of affected maternal lines is a simple effective procedure. When combined with improved husbandry to minimise horizontal spread of infection when animals are kept together (e.g., at lambing time), it can certainly reduce the incidence of natural scrapie in the field. The basic requirements are a simple and efficient means of identifying animals and keeping accurate mating records, and also time, since it may take a few years for selective culling to have a marked impact on disease incidence in flocks where such recording procedures are not already routine.

There remains an urgent need for methods of detecting scrapie infection pre-clinically. Direct testing of infectivity in blood is impracticable because the levels of agent are almost non-existent, the incubation periods of scrapie in laboratory animals are much too long and cell culture methods are not available. Hopefully an indirect test may become available, but so far all attempts based on a conventional immunological response to infection have been negative and biochemical markers for the presence of the agent remain to be discovered.

16.3 The viruses

From here on, the similarities between visna–maedi, AD and scrapie vanish, since these diseases are caused by quite different agents. Visna–maedi virus is the best characterised of the three and it closely resembles the oncogenic RNA viruses (although it is not serologically related to them and the nucleic acid genome appears to be different). Recent developments in the purification of AD virus suggest it is a picorna virus and one can expect a full characterisation of this virus in the very near future. The scrapie-like agents are still far from being adequately characterised and progress in this area has been painfully slow. The main reasons for this are the lack of success in purifying the agent and the total dependence on a single, very time-consuming assay for scrapie agents. Despite these limitations there is little doubt that scrapie infectivity is closely associated with cell membranes. The existence of several biologically stable strains of scrapie indicates that genetic information must be encoded in scrapie-specific molecules and the target size of scrapie agent to ionising radiation suggests that these molecules may have a size of about 10^5 daltons. Furthermore there is no reason why these molecules should not be nucleic acid. Therefore, the close association of a putative scrapie-specific nucleic acid with certain components of membranes (some of which could be under the control of the *sinc* gene) would appear to be the best available working hypothesis for the nature of the scrapie agents. The problem is to identify the nucleic acid and define its association with membrane components.

It is useful to remember that the ultimate aim of the work described in this book is to understand and control certain diseases, and it is salutory to reflect how little knowledge of a virus *per se* tells us about the disease it causes. A good example of this is visna–maedi virus which structurally resembles the oncogenic RNA viruses but produces visna and maedi in sheep, which are far from being neoplastic diseases. It is interesting that this virus can produce two clinically distinct diseases depending on whether it is the lung or the CNS which is the more severely damaged.

It is generally recognised that individual viruses are capable of a wide variety of interactions with different cell types in different host species. At the same time it is apparent that there may only be a limited number of fundamentally different ways that cell and organ systems can respond to infectious or other traumatic agents. Consequently one may find instances of a virus which can cause more than one type of disease (depending on the target cell and the nature of the interaction with it), and also instances of a similar type of disease (seen in clinical or pathological terms) caused by a number of different agents. An example of the last point can be seen in some of the difficulties encountered in making a histopathological diagnosis of maedi, because some similar lesions can also be found in lungs associated with bacterial infections and parasite infestations.

Against this background it is clear that characterisation of viruses that cause disease is of great practical importance, particularly in answering questions about virus multiplication and in providing specific antigens for the production of more efficient vaccines. However in many cases, and especially with the slow virus diseases under discussion where conventional vaccination procedures seem to be so inappropriate at the moment, a precise understanding of virus–host interactions will be crucial to understanding and controlling these diseases.

16.4 Virus–host interactions

16.4.1 Visna

Much work has been done on the properties of the visna–maedi viruses in cell culture because of the ease with which they can be grown in vitro. This work has been invaluable in studying virus multiplication and in describing the range of virus–cell interactions that can occur, but it has contributed little direct information on the pathogenesis of visna in sheep. For example the characteristic cytopathic effect of visna virus in vitro is cell fusion and the formation of multinucleate stellate cells but nothing equivalent to this is seen in infected animals. However, visna virus can infect other cells to produce carrier cultures which can synthesise virus but show no cytopathic effect. Systems of this kind could be usefully studied to suggest possible ways in which visna virus can persist in vivo.

Direct studies of the pathogenesis of visna in animal hosts have been restricted to sheep (the only susceptible species) and have been inhibited by the long and variable incubation period after experimental infection. Despite these limitations there have been a number of single-minded attempts to study the disease in sheep. Low levels of virus can be detected in blood and lymphoid tissues soon after intracerebral infection. This is followed by the slow appearance of serum antibody but this host response does not appear to suppress infection nor to contain the histological lesions. Indeed, there is suggestive evidence that the early lesions in brain and spinal cord may be immunologically mediated (possibly by T cells), although more studies on the effects of immunosuppression are needed to confirm and extend this possibility.

The precise relationship between the immune response, the development

of lesions and the occurrence of clinical disease are far from being understood. The reasons for the slowness of disease development are also unknown, although some possibilities have been largely eliminated, for example,
immunological damping of infection and defectiveness of the virus. The
limited permissiveness of sheep cells to visna–maedi virus is an attractive
possibility which could be explored by studying cell cultures derived from
tissues of infected animals. Clearly a more uniform experimental animal
model of visna would be a great help in unravelling the subtleties of virus–
host interaction but there seems to be no immediate prospect of this.

16.4.2 Aleutian disease

The pathogenesis of AD is reasonably well understood in general terms,
partly because of the availability of a good experimental host (i.e., Aleutian
mink) and partly because of the dramatic nature of the host response to
infection.

   The disease is characterised by an early and rapid proliferation of plasma
cells and hypergammaglobulinaemia in which extraordinarily high levels
of antibody are produced. At the same time virus is rapidly synthesised
and can persist in the circulation for indefinite periods. The very high
antibody levels may arise because continued virus synthesis provides a
chronic stimulation of the immune system. Alternatively, the virus may
cause a defect in the normal homeostatic mechanism that controls antibody
production. This kind of interference is suggested by the immuno-depressed
state of infected mink when tested against other antigens.

   The important point is that the vigorous immune response to Aleutian
disease virus does not clear the infection but leads to the formation of virus–
antibody complexes. It is the accumulation of these complexes at various
sites which is the direct cause of the arteritis and glomerulonephritis seen
in AD. The deposition of immune complexes in the kidneys causes stimulation and proliferation of the mesangial cells leading eventually to renal
failure and death.

   In AD there is a good correlation between the immune response, kidney
lesions and death of infected mink. This is convincingly demonstrated by
the effects of immunosuppression with cyclophosphamide which will prevent
the development of lesions even though virus is present. Cessation of cyclophosphamide therapy leads to a resumption of lesion development. Observations on the disease process in infected ferrets also support this relationship.
Ferrets develop similar lesions but rarely die of the disease. The significant

difference from Aleutian mink is that glomerular alterations are minimal or absent in ferrets.

The slowness of disease development is not understood, but it seems possible that it may be related in part to some properties of the host. For example, there is no difference between Aleutian and non-Aleutian mink in the nature of lesions in animals with clinical signs but there is a difference in the rate of development which is invariably faster in the Aleutian genotype. The association between the Chediak–Higashi syndrome in mink and the Aleutian gene suggests that this difference could be related to an impaired ability to catabolise the immune complexes that are deposited in the kidneys.

The absence of arteritis and the rare occurrence of glomerulonephritis in infected ferrets suggests that the relative resistance of these animals to AD may be due to their ability to remove immune complexes. This is obviously an important area to investigate in the future. Equally important will be an investigation of why a proportion of non-Aleutian mink do not develop the disease. There is evidence that such animals are susceptible to infection but are able to eliminate the virus. An understanding of this phenomenon could suggest a new approach to the prevention or control of Aleutian disease of mink.

16.4.3 Scrapie

Compared to AD the pathogenesis of scrapie is poorly understood, despite the availability of a wide range of experimental mouse models of the disease. One of the main difficulties is the limited number of appropriate techniques for investigating the problem. Studies of some of the short-incubation models have shown that agent replication initially takes place in tissues such as spleen, lymph nodes and thymus, following a peripheral route of infection but titres of infectivity only reach modest levels. Later, the agent reaches the CNS (by an unknown route; there is no significant 'viraemia') and the titre in brain steadily increases throughout the remaining incubation period until the clinical disease develops. There is no evidence that a specific immune response is stimulated at any time during the incubation period; moreover long-term immunosuppressive treatments have had no effect on pathogenesis.

However, there is evidence that some kind of non-specific defence mechanism may operate at the time of infection and it has been shown that the injection of a number of anti-inflammatory drugs can delay the incubation period of peripherally inoculated scrapie when given close to the time of infection. These results provide important clues for the role of the lympho-reticular system in scrapie pathogenesis but this role needs to be defined much more fully.

A remarkable biological feature of mouse scrapie (with all combinations of agent strain and mouse strain that have been tested) is the uniformity of incubation period under standard conditions of infection. This uniformity suggests a clockwork-like precision in disease development. The *sinc* gene plays an important role in controlling the pathogenetic mechanism and it is known to have a major effect on agent multiplication in both spleen and brain. Moreover, there are clear indications of a close relationship between agent multiplication in brain and onset of clinical disease but (in contrast to AD) the relationship between brain lesions and disease is vague. There is in fact enormous variation in the histological lesions in clinically affected mice; variation in terms of the type of lesion (grey and white matter vacuola-tion, amyloid plaques, gliosis etc), the distribution and the severity.

This variation in lesions is also true of the scrapie-like diseases such as TME which, in non-Aleutian mink, produces extensive vacuolation whereas scarcely any vacuolar lesions are produced in aged Aleutian mink. In view of these observations the use of the phrase 'subacute spongiform encephalo-pathies' to describe scrapie, TME, kuru and Creutzfeldt–Jakob disease may be unwise, since it attempts to classify these diseases in terms which are perhaps too restricted and imprecise. 'Scrapie-like diseases' is a useful alternative which does not seek to limit the histological characteristics which link these four diseases.

Despite the paucity of knowledge of the details of scrapie pathogenesis it is not too early to consider why the disease is so slow. There is accumulating evidence from biological experiments in mice that the long incubation period is mainly due to the slow rate of agent multiplication which in turn is restricted by the small number of *sinc*-controlled replication sites. At present, the best single piece of evidence for this suggestion is the well-documented phenomenon of agent competition. It has been shown that the inoculation of mice of a particular genotype with an agent that is operation-ally 'slow' in that genotype, can completely block an operationally 'faster' agent inoculated later. This kind of interaction between agents immediately suggests a novel way of preventing scrapie (at least in mice). However, a great many practical problems must be solved before this can be seriously considered as a possible way of controlling scrapie in sheep.

16.5 The last word

Most reviews of 'slow virus diseases' appear in specialised volumes devoted to microbiology, pathology or immunology. This is as it should be, but at the same time it is useful to present these diseases against a broader background because their investigation requires a multidisciplinary approach. Hopefully this volume will help to foster this approach particularly in those who were not familiar with this fascinating but difficult area of research. A great deal has been achieved since Sigurdsson gave his series of lectures in London over 20 years ago, but those who have read the preceeding chapters will appreciate that there is still a long way to go before we have all the answers.

Subject index

Viral encephalitides, 274, 316
Viral enteritis of mink, 137, 145
Viral hepatitis, 202
Visna (*see also* Visna–maedi virus)
  antibodies, 35–6, 38, 55, 108, 112, 117,
    119–21, 125, 391
  clinical course, 5, 22, 25
  clinical signs, 21–2, 25, 81
  comparison with other diseases, 6–7, 10,
    129–30, 389
  concurrent infections, 19, 25, 32
    maedi, 19, 32
  control and eradication, 33–8, 41–62,
    387–9
  CSF pleocytosis, 22, 25, 63–4, 81, 117,
    120–1
  diagnosis, 25, 36, 385, 388
  eradication, (*see* Control and Eradica-
    tion)
  geographical occurrence, 19, 32–3, 62
    Iceland, 19, 32–3, 62
  histopathology of CNS, 24, 25, 62–85,
    110–1, 119–120, 124–8, 391
    astrocytic changes, 70–1, 74, 76, 78
    clinical correlations, 81–2
    comparative, 82–5, 124–5
    demyelination, 11, 68, 74–5, 77–9,
      84–5, 110–1, 120, 126–8
    evolution of lesions, 70–80, 119, 125–6
    glial reaction, 24, 71–2, 74
    inflammation of choroid plexus, 67,
      77, 80–1
    inflammation of grey matter, 67, 70, 72
    inflammation of white matter, 67–79
    lymphocytic infiltration, 24, 77, 80, 83,
      111
    meningitis, 63–8, 80–1
    necrosis, 68–9, 72, 74–5, 77, 85, 119
    perivascular infiltration, 66–7, 71–3,
      75–6
    subependymal infiltration, 24, 66–71,
      77, 81, 84
  histopathology of extraneural lesions,
    62–3, 83
  histopathology of peripheral nervous
    system, 80–1
  immune response, 38, 95, 110, 112, 116,

120–2, 125–9, 391
  incubation period, 8, 111–2, 115, 121–3
    391
  pathogenesis, 62, 95, 110–2, 115–29, 391
  serology (*see* Visna–maedi virus)
  sheep breed susceptibility, 37, 386
  sheep farming practices, 37
  virus (*see* Visna–maedi virus)
Visna-like diseases, 32–3, 38, 41, 82–5
  comparative histopathology, 82–5
  geographical occurrence, 32–3, 38, 41
Visna–maedi virus (*see also* Visna *and* Maedi)
  antibodies, 30, 35–6, 38, 55, 108, 112,
    117, 119–21, 125, 391
  antigens, 55, 57, 100, 107, 110
  assay, 46–7, 57, 107
  biochemical properties, 102
  cell and tissue culture, 11, 25, 34–5,
    45–6, 48–9, 91, 97–112, 116, 391
    cytopathic effect, 91, 97–101, 104,
      107–112, 116, 391
    persistent infection, 104–6, 109
    relation to disease, 110–12
  chemical properties, 34, 47–8, 109
  components, 51–5
    nucleic acids, 51–4, 102–4
    proteins, 54–5
  defectiveness, 122
  DNA polymerase activity, 53–6, 102–3,
    109
  genome, 52–3, 56, 58, 103
  growth cycle, 100–9
  maturation of virions, 48–9, 51–2, 57,
    99, 100, 102, 109–10
  oncogenic potential, 11, 83, 111
  physical properties, 34, 47–8, 109
  purification, 47, 50, 116–20, 124
  replication, 55–8, 117, 122
  serology, 25, 28, 32, 35–6, 44, 55, 57–8,
    83, 108–9, 117
  strains, 46, 62, 107–8, 116
  transmission, 33–7, 46, 61–2, 81, 108–9,
    112, 117–8, 121, 386
    experimental, 33–5, 46, 61, 81, 108–9,
      112, 117–8, 121
    natural, 33–7, 61–2, 386
  ultrastructure, 48–53, 56–7

Index prepared by Karen J. Beaton.
A. R. C. Institute for Research on Animal Diseases, Compton, Newbury,
Berkshire, U.K.